D0081543

——— ᢞ PIERRE-SIMON LAPLACE, 1749–1827 ᢞ ———

NATIONAL UNIVERSITY
LIBRARY   SAN DIEGO

PIERRE-SIMON LAPLACE

1749–1827

A Life in Exact Science

CHARLES COULSTON GILLISPIE

With the Collaboration of

ROBERT FOX *and* IVOR GRATTAN-GUINNESS

PRINCETON UNIVERSITY PRESS · PRINCETON, NEW JERSEY

Copyright © 1997 by Princeton University Press
Published by Princeton University Press, 41 William Street,
Princeton, New Jersey 08540
In the United Kingdom: Princeton University Press,
Chichester, West Sussex
All Rights Reserved

Library of Congress Cataloging-in-Publication Data
Gillispie, Charles Coulston.
Pierre-Simon Laplace, 1749–1827: a life in exact science/
Charles Coulston Gillispie; with the collaboration of Robert Fox
and Ivor Grattan-Guinness.
p. cm.
Includes bibliographical references and index.
ISBN 0-691-01185-0 (alk. paper)
1. Laplace, Pierre Simon, marquis de, 1749–1827. 2. Physics—
History. 3. Astronomy—History. 4. Scientists—France—Biography.
5. Physicists—France—Biography. I. Fox, Robert, 1938–
II. Grattan-Guinness, I. III. Title.
Q143.L36G55 1997
509.2—dc21
[B]   97-8331 CIP

Frontispiece: Pierre-Simon Laplace.
Posthumous portrait (1842) by Madame Feytaud.
(Courtesy of the Académie des Sciences, Paris.)

This book has been composed in Times Roman

Princeton University Press books are printed
on acid-free paper and meet the guidelines
for permanence and durability of the Committee
on Production Guidelines for Book Longevity
of the Council on Library Resources

Printed in the United States of America

1 3 5 7 9 10 8 6 4 2

Contents

Preface   vii

Acknowledgments   xi

PART I: EARLY CAREER, 1768–1778

Chapter 1. Youth, Education, and Election to the Academy   3

Chapter 2. Finite Differences, Recurrent Series, and Theory
of Chance   7

Chapter 3. Probability of Events and of Their Causes:
The Origin of Statistical Inference   13

Chapter 4. Universal Gravitation   29

Chapter 5. Distribution of Comets   38

Chapter 6. Partial Differential Equations, Determinants,
and Variation of Constants   44

Chapter 7. The Figure of the Earth and the Motion of the Seas   51

PART II: LAPLACE IN HIS PRIME, 1778–1789

Chapter 8. Influence and Reputation   67

Chapter 9. Variation of Constants, Differential Operators   70

Chapter 10. Probability Matured   72

Chapter 11. Generating Functions and Definite Integrals   86

Chapter 12. Population   93

Chapter 13. Determination of the Orbits of Comets   96

Chapter 14. Lavoisier and Laplace: Chemical Physics of Heat   101

Chapter 15. Attraction of Spheroids   109

Chapter 16. Planetary Astronomy   124

PART III: SYNTHESIS AND SCIENTIFIC STATESMANSHIP

Chapter 17. The Revolution and the Metric System   149

Chapter 18. Scientific Work in the Early Revolution   156

Chapter 19. *Exposition du système du monde*   166

Chapter 20. A Scientific Eminence   176

Chapter 21. *Traité de mécanique céleste*   184

PART IV: LAPLACIAN PHYSICS AND PROBABILITY

Chapter 22. The Velocity of Sound
(Robert Fox)   199

Chapter 23. Short-Range Forces
(Robert Fox)   203

Chapter 24. The Laplacian School
(Robert Fox)   209

Chapter 25. Theory of Error   216

Chapter 26. Probability: *Théorie analytique* and
Essai philosophique   224

Chapter 27. Loss of Influence
(Robert Fox)   243

Chapter 28. The Last Analysis   250

PART V: THE LAPLACE TRANSFORM

Chapter 29. Laplace's Integral Solutions to Partial
Differential Equations
Ivor Grattan-Guinness   259

Conclusion   271

Abbreviations   280

Bibliography   281

Index   319

PIERRE-SIMON LAPLACE (1749–1827) was among the most influential scientists in all history. His career was important for his technical contributions to exact science, for the philosophical point of view he developed in the presentation of his work, and for the part he took in forming the modern scientific disciplines. The main institutions in which he participated were the Académie Royale des Sciences, until its suppression in the Revolution, and then its replacement, the scientific division of the Institut de France, together with two other Republican foundations, the École Polytechnique and the Bureau des Longitudes. It will be convenient to consider the scientific life that he led therein as having transpired in four stages, the first two in the context of the Old Regime and the latter two in that of the French Revolution, the Napoleonic period, and the Restoration.

The boundaries must not be taken more categorically than biography allows, but in the first stage, 1768–1778, we may see Laplace rising on the horizon, composing memoirs on problems of the differential and integral calculus, mathematical astronomy, cosmology, theory of games of chance, and causality, pretty much in that order. During this formative period, he established his style, reputation, philosophical position, certain mathematical techniques, and a program of research in two areas, probability and celestial mechanics, in which he worked mathematically for the rest of his life.

In the second stage, 1778–1789, he moved into the ascendant, reaching in both those areas many of the major results for which he is famous and that he later incorporated into the great treatises *Traité de mécanique céleste* (1799–1825) and *Théorie analytique des probabilités* (1812 and later editions). They were informed in large part by the mathematical techniques that he introduced and developed, then or earlier, most notably generating functions, the transform since called by his name, the expansion also named for him in the theory of determinants, the variation of constants to achieve approximate solutions in the integration of astronomical expressions, and the generalized gravitational function that, through the intermediary of Poisson, later became the potential function of nineteenth-century electricity and magnetism. It was also during this period that Laplace entered on the third area of his mature interests, physics, in his collaboration with Lavoisier on the theory of heat, and that he became, partly in consequence of this association, one of the inner circle of influential members of the scientific community. In the 1780s he began serving on commissions important to the government and affecting the lives of others.

In the third stage, 1789–1805, the Revolutionary period and especially that of the Directory brought him to his zenith. The early 1790s saw the completion of the great series of memoirs on planetary astronomy and involved him centrally in the preparation of the metric system. More important, in the decade from 1795 to 1805 his influence was paramount for the exact sciences in the newly founded Institut de France; and his was a powerful position in the counsels of the École Polytechnique, where the first generation of mathematical physicists had their training. The educational mission attributed to all science in that period of intense civic consciousness changed the mode of scientific publication from academic memoir to general treatise. The first four volumes of *Mécanique céleste* (Laplace himself coined the term), generalizing the laws of mechanics for their application to the motions and figures of the heavenly bodies, appeared from 1799 through 1805. The last parts of the fourth volume and the fifth volume, really a separate work that appeared in installments from 1823 to 1825, contain important material (on physics) not already included in the sequence of Laplace's original memoirs published previously by the old Academy.

Laplace accompanied both *Mécanique céleste* and *Théorie analytique des probabilités* by verbal paraphrases addressed to the intelligent public in the French tradition of *haute vulgarisation*. The *Exposition du système du monde* preceded *Mécanique céleste* and initially appeared in 1796. The *Essai philosophique sur les probabilités*, first published in 1814 as an introduction to the second edition of *Théorie analytique* and printed separately earlier in the same year, originated in a course of lectures at the École Normale in 1795.

The work of the fourth stage, occupying the period from 1805 until 1827, exhibits elements of culmination and of decline. It was then that the mature—perhaps the aging—Laplace, in company with the chemist Claude-Louis Berthollet, surrounded himself with disciples in the informal Société d'Arcueil. But the science that he set out to shape was not astronomy. The center of their interest, following Volume IV of *Mécanique céleste*, was in physics—capillary action, the theory of heat, corpuscular optics, and the speed of sound. The Laplacian school of physics, as it has come to be called, has had a mixed scholarly press. But whatever else may be said about it, there can be no doubt about the encouragement that it gave to the mathematization of the science.

Beginning in 1810, Laplace turned his attention to probability again, moving back by way of error theory into the subject as a whole. Mathematically speaking, *Théorie analytique des probabilités* (1812) may be said to belong to the previous phase of drawing together and generalizing the researches on special topics of his younger years. There were important novelties in the application, however, notably in the treatment of least squares, in the extension of probability in later editions to analysis of the credibility of witnesses and the procedures of judicial panels and electoral bodies, and in the increasing sophistication of the statistical treatment of geodetic and meteorological data.

Before proceeding further, readers may wish to turn to the Bibliography and familiarize themselves with its several categories. Its organization reflects the complexity of Laplace's life work, and inevitably so. The central portion (Section I, pp. 284–306) constitutes a dual chronology of his writings, by order of composition and by order of publication. References to the former sequence (Section I, Part 1) are given in parentheses thus: (23). References to the latter sequence (Section I, Part 2) are given by bracketing the date, and the several memoirs published in any given year are distinguished by letters of the alphabet, thus: [1777a]. The great majority of Laplace's publications were reprinted in his *Oeuvres complètes* (see Section H, pp. 283–84). That edition is more accessible than the original papers and treatises. Where the footnotes indicate *OC*, the reference is to the location in those volumes of the passage cited. Other abbreviations are listed facing the Bibliography, p. 280. Secondary works are cited in the footnotes by the name of the author and the date of publication. The full reference will be found in the alphabetical section (O) of the Bibliography, pp. 309–17.

Acknowledgments

THE ORIGINAL edition of this book appeared in 1978 in Volume 15 of the *Dictionary of Scientific Biography*, published in sixteen volumes by Charles Scribner's Sons from 1970 to 1980. I had the primary editorial responsibility for that collection and also undertook the entry on Laplace. The purpose is to give an account of the sequence, range, and results of Laplace's scientific work, explained in terms of his own time and in his own notation. This book does not aspire to be a critical analysis of his mathematical achievements and limitations per se. For that, and for the entire mathematical movement of his generation and the succeeding one, the reader may and should turn to the magisterial study by my colleague and collaborator, Ivor Grattan-Guinness, *Convolutions in French Mathematics, 1800–1840* (1990).

Throughout I have benefited even more than normally from the advice and assistance of others. The preface to the *DSB* article specifies the help received from colleagues and students in preparing the earlier version. Dr. Grattan-Guinness composed chapter 29 on the history of the Laplace transform, while Robert Fox contributed chapters 22, 23, 24, and 27 on the Laplacian school of physics. Both collaborators have generously accepted the request to revise their contributions for the present edition, and both are assisting with suggestions for revision of the entire work. So also is Stephen M. Stigler, who gave invaluable guidance on Laplace's earliest memoirs, and on statistical aspects in general.

In undertaking this revision, I am incorporating material from a number of studies of various aspects of Laplace's life and work that have appeared since the initial publication in 1978, all of which are included in the Bibliography. Two of the authors, Bernard Bru and Curtis Wilson, have very kindly reviewed and suggested improvements in the sections that concern their specialties, those on probability and on the inequality of Jupiter and Saturn. A current graduate student, David Attis, is serving as a very capable research assistant. As always Emily Gillispie has read drafts and proof. Her eye for style and language has improved matters throughout. Finally, and also as always, I resolutely claim responsibility for errors and infelicities that have escaped the scrutiny of all these guardian angels. Unless otherwise indicated, all translations are my own.

A special word of thanks is due to Jack Repcheck, editor at Princeton University Press, whose idea it was to republish this monographic article as a scientific biography in its own right. The American Council of Learned Societies, which holds the copyright to the *Dictionary of Scientific Biography*, has granted the requisite permission.

Madame Christiane Demeulenaere-Douyère, Conservateur of the Archives de l'Académie des Sciences, Institut de France, unfailingly affords scholars and scientists the most cordial and expert guidance and assistance in consulting the invaluable materials under her care, and like all who work with these documents, I am correspondingly grateful to her and to her staff. On behalf of the Academy she has graciously authorized reproduction of the portrait that serves as frontispiece. Its date is 1842, and it seems highly probable that the painting would have been commissioned by Madame Laplace, who, with the very active assistance of Arago, was even then arranging for publication of the first collection of Laplace's *Oeuvres* (see Section G, p. 283).

Princeton
September 1996

Part I

EARLY CAREER, 1768–1778

CHAPTER 1

Youth, Education, and Election
to the Academy

GENEALOGICAL RECORDS of the Laplace family in lower Normandy go back to the middle of the seventeenth century.[1] Laplace's father, Pierre, was a syndic of the parish, probably in the cider business and certainly in comfortable circumstances. The family of his mother, Marie-Anne Sochon, were well-to-do farmers of Tourgéville. He had one elder sister, also called Marie-Anne, born in 1745. There is no record of intellectual distinction in the family beyond what was to be expected of the cultivated provincial bourgeoisie and the minor gentry. One paternal uncle, Louis, an abbé although not ordained, is said to have been a mathematician and was probably a teacher at the college (a secondary school) kept at Beaumont-en-Auge by the Benedictines. He died in 1759, when his nephew was ten. Laplace was enrolled there as a day student from the age of seven to sixteen. Pupils usually proceeded to the church or the army; Laplace's father intended him for an ecclesiastical vocation.

In 1766 he went up to the University of Caen and matriculated in the Faculty of Arts, still formally a cleric. During his two years there he must have discovered his mathematical gifts, for instead of continuing in the Faculty of Theology, he departed for Paris in 1768. Apparently, he never took a degree, although he may briefly have been a tutor in the family of the marquis d'Héricy and may also have taught at his former college. The members of the faculty at Caen who opened his eyes to mathematics, and their own to his talent, were Christophe Gadbled and Pierre Le Canu. All that is known about them is that they were points of light in the philosophic and scientific microcosm of Caen, professors with the sense to recognize and encourage a gifted pupil.

On Laplace's departure for Paris at the age of nineteen, Le Canu gave him a letter of recommendation to d'Alembert, who immediately set him a problem and told him to come back in a week. Tradition has it that Laplace solved it overnight. Thereupon d'Alembert proposed another, knottier puzzle, which Laplace resolved just as quickly.[2] The

---

[1] See Boncompagni (1883) and G. A. Simon, "Les origines de Laplace: sa généalogie, ses études," in Pearson (1929), pp. 202–16.
[2] Bigourdan (1931).

story may be apocryphal, but there is no doubt that d'Alembert was somehow impressed and took Laplace up, as he had other young men in the evening of his own career, although none of comparable merit mathematically. The next question was a livelihood, and d'Alembert himself answered to that necessity, securing his new protégé the appointment of professor of mathematics at the École Militaire. Imparting geometry, trigonometry, elementary analysis, and statics to adolescent cadets of good family, average attainment, and no commitment to the subjects afforded little stimulus, but the post did permit Laplace to stay in Paris. He taught there from 1769 to 1776.

It was expected of Laplace that he should concentrate his energies on making a mathematical reputation in order to win election to the Academy of Science. If its records are complete, he presented thirteen papers in just under three years, beginning in March 1770. The topics were extreme-value problems; adaptation of the integral calculus to the solution of difference equations; expansion of difference equations in a single variable in recurrent series and in more than one variable in recurro-recurrent series; application of these techniques to theory of games of chance; singular solutions for differential equations; and problems of mathematical astronomy, notably variation of the inclinations of the ecliptic and of planetary orbits, the lunar orbit, perturbations produced in the motion of the planets by the action of their satellites, and "the Newtonian theory of the motion of the planets" (9). Of these papers, four were published (1, 5, 8, 13). Laplace translated the first two into Latin and placed them in the *Nova acta eruditorum*, where the second was printed before the first.[3]

Laplace read the earlier paper, "Recherches sur les maxima et minima des lignes courbes," before the Academy on 28 March 1770, five days after his twenty-first birthday. After a review of extreme-value problems, he proposed several improvements in the development that Lagrange had given to Euler's *Methodus inveniendi lineas curvas maximi minimive proprietate gaudentes* (1744). One modification concerned Lagrange's finding in a 1761 paper that there was no need to follow Euler in assuming a constant difference.[4] If the assumption was justified, the number of equations might be reduced by at least one, and otherwise the problem was unsolvable. Laplace found the same result by a method that his commissioners called "less direct, less rigorous in appearance, but simpler and fairly elegant" (1). In cases where a

---

[3] [1774a], [1771a].

[4] "Essai d'une nouvelle méthode pour déterminer les maxima et les minima. des formules intégrales indéfinies," *OL*, **1**, 334–62.

difference is not constant, the difficulty was shown to arise from a faulty statement of the problem. A variable was concealed that should have appeared in the function representing the curve to be optimized. When it was identified, the equations became determinate. If the solutions yielded maximum values, the equations involved double curvature. Laplace further gave a general analytic criterion for distinguishing a true maximum or minimum from instances in which two successive values happen to be equal, and he appeared to have regarded this as his chief contribution.

It was, however, with the other Leipzig paper, "Disquisitiones de calculo integrale," that Laplace made his debut in print [1771a]. The subject is a particular solution for one class of ordinary differential equations. The method he developed subsequently led to enunciation of a theorem, the statement of which he later annexed without proof to his first memoir on probability although it has nothing to do with that subject.[5] Reworking this material two years later, in 1773, Laplace repudiated this earliest publication, or very nearly so, apologizing for grave faults that he blamed on the printer.[6] That was the only reference he ever made to either of these youthful ventures into Latin.

Laplace won election to the Academy of Science on 31 March 1773, after five years in Paris. Six years his senior, Condorcet had become acting permanent secretary earlier that month. In that capacity he composed the preface to the volume containing the first of Laplace's memoirs to be published in Paris, one on recurro-recurrent series, the other on probability of causes.[7] The Academy had never, observed Condorcet, received from so young a candidate in such a short time so many important papers on varied and difficult topics as the sequence submitted by Laplace.[8] On two previous occasions his candidacy had been passed over: in 1771, in favor of Alexandre Vandermonde, fourteen years his elder, and the following year in favor of Jacques-Antoine-Joseph Cousin, ten years older and a professor at the Collège Royal de France.

Evidently, Laplace felt slighted, despite his youth. On 1 January 1773 d'Alembert wrote to Lagrange asking whether there was a possibility of obtaining a place in the Prussian Academy and a post at Berlin, since

---

[5] [1774c]. The equation was Equation (23), in chapter 6 below. For Laplace's statement of the theorem, see *OC*, **8**, pp. 62–63, and for the demonstration, see [1775a], Section VI, *OC*, **8**, pp. 335–46.

[6] [1774c], *OC*, **8**, p. 63, note, where Laplace refers the reader to [1777a], discussed in chapter 6.

[7] [1774b, 1774c].

[8] *SE*, **6**, (1774), "Histoire," p. 19.

the Paris Academy had just preferred a person of markedly inferior ability.[9] The approach lapsed three months later when Laplace was chosen an adjunct member in Paris.[10]

[9] *OL*, **13**, pp. 254–56.

[10] Bigourdan's statement that he was admitted directly to the second rank of *associé* is incorrect. (Bigourdan [1931] p. 384).

CHAPTER 2

Finite Differences, Recurrent Series,
and Theory of Chance

AMONG LAPLACE'S early interests, it turned out to be a memoir on the solution of difference equations that marks the beginning of one of the main sequences of his lifework [1771b]. An unpublished sequel of the following year is entitled "Sur les suites récurrentes appliquées à la théorie des hasards" (10). We may reasonably surmise that the applicability of such series to problems of games of chance, and not any prior penchant for that subject matter, was mainly responsible for the appearance given by the record of his publications that probability attracted him more strongly than did celestial mechanics in this opening phase of his career.

The appearance is misleading, however, or at least ironic. For at the outset of the earlier paper, "Recherches sur le calcul intégral aux différences infiniment petites, et aux différences finies," Laplace observed that the equations he was studying turn up more frequently than any other type in applications of the calculus to nature. A general method for integrating them would be correspondingly advantageous to mechanics, and especially to "physical astronomy."[1] That science looms largest in his unpublished, as distinct from his published, record (1–13). Judging from a report in the archives of the Academy of Sciences, a paper of 27 November 1771, "Une théorie générale du mouvement des planètes," may even have been the germ of *Mécanique céleste* (9). The draft does not survive, but the referees, d'Alembert, Bezout, and Bossut, say that the subject was the Newtonian theory of the planets. Their report is full enough to permit the conclusion that Laplace expanded this memoir into his first general treatise on celestial mechanics, *Théorie du mouvement et de la figure elliptique des planètes* (1784).[2]

Since Laplace was still only knocking at the door of the Academy in 1771, he submitted "Recherches sur le calcul intégral aux différences infiniment petites, et aux différences finies" to the Royal Society of Turin for publication in its *Mélanges* [1771b]. It was almost surely the expansion of a paper on difference equations alone that he had read on

[1] [1771b], p. 273.
[2] [1784a], Chapter 15.

18 July 1770 in his second appearance before the Academy in Paris (2). Laplace now reserved difference equations for the second part of the memoir and proposed adapting to their solution—or as he said in the looser terminology of the time, to their "integration"—the method he had developed for infinitesimal expressions, presumably in a paper "Sur le calcul intégral des suites récurrentes," dated 13 February 1771 (6). He began by confirming in his own manner a theorem that Lagrange had recently proved concerning integration of equations of the following form:

$$ X = y + H\frac{dy}{dx} + H'\frac{d^2y}{dx^2} + H''\frac{d^3y}{dx^3} + \cdots + H^{n-1}\frac{d(d^n y)}{dx^n}, \quad (1) $$

where X, H, H', H'', ..., are any functions of x. Lagrange had shown that such equations can always be integrated if integration is possible in the homogeneous case when $X = 0$. His proof was of a type classical in eighteenth-century analysis. It involved introducing a new independent variable, z; multiplying both sides of the equation by $z\,dt$; supposing integration of the resulting adjoint equation accomplished; and examining the steps needed to reduce the order one degree at a time until a solvable form should be reached. The approach worked for differential equations. The procedure presupposed the validity of infinitesimal methods in analysis, however, and therefore the operations were inapplicable to the solution of difference equations.

Laplace's approach appeared to be more cumbersome and turned out to be more general. Instead of introducing a multiplying factor and supposing the subsequent integration accomplished, Laplace employed integrating factors directly. The problem is to integrate equations of the form (1). To that end, he let

$$ \omega \frac{dy}{dx} + y = T, \quad (2) $$

where T and ω are functions of x. Differentiating Equation (2) successively n times, he then multiplied the first of the resulting equations by ω', the second by ω'', the third by ω''', and so on. He next added all these equations to Equation (2), grouped the terms by orders of y, and compared the enormous resulting expression to Equation (1). Lengthy manipulation allowed him to determine the multipliers ω', ω'', ω''', ..., in terms of ω and H', H'', H''', ... Not only was Laplace able to prove Lagrange's theorem by this method, but he could also write equations equivalent to Equation (1) and evaluate them generally. In the second part of the memoir, and this was its motivation, Laplace adapted his approach to equations in finite differences by showing that all steps could be modified to conform to the rules of algebra.

Although the memoir had opened with mention of mechanics and astronomy, Laplace introduced its raison d'être, the solution of difference equations, with a reminder that their calculus was the foundation of the entire theory of series.[3] Coherently enough, therefore, he continued the discussion with a determination of the general term of series of the important class[4]

$$y^x = A\phi^x y^{x-1} + {}'A\phi^x \phi^{x-1} y^{x-2} + {}''A\phi^x \phi^{x-1}\phi^{x-2} y^{x-3} + \cdots , \quad (3)$$

where A, $'A$, $''A$ are constants, and ϕ is a function of x. In the simplest case, in which $\phi = 1$, Equation (3) reduces to the recurrent form,

$$y^x = Ay^{x-1} + {}'Ay^{x-2} + {}''Ay^{x-3} + \cdots + {}^{n-1}Ay^{x-n} \quad (4)$$

The memoir ends with an application of the calculus of finite differences to a solution of this equation, and Laplace gives a method for determining the constants.

The episode is largely typical of the relation of Laplace's point of departure to the work of elders and near contemporaries. There is the not quite ritual obeisance to a principle or practice, in this case the formulation of problems in terms of differential equations in general, attributed to d'Alembert, the patron. There is the tactful nod to a result found quite differently by Condorcet, the well-placed official. There is the pioneering analytical breakthrough achieved by Euler, although in restricted form. There is the formal mathematical theorem stated by Lagrange, emphasizing analyticity. There is, finally, the adaptation imagined and executed by Laplace, his motivation being the widest applicability to problems in the real world.

His finding that equations of the form (3) are always integrable became the starting point of the next memoir, which deals with what Laplace called "recurro-recurrent" series and their application to the theory of chance [1774b]. Still not a member of the Academy, he submitted it for their judgment on 5 February 1772, and they placed it in the *Savants étrangers* collection. Recurrent series of the familiar form (3) were restricted to a single variable index, the definition being that "every term is equal to any number of preceding terms, each multiplied by a function of x taken at will.[5] It had been while investigating certain

---

[3] [1771b], p. 299.
[4] Ibid., p. 330.
[5] [1774b], *OC*, **8**, p. 5.

problems in the theory of chance, so Laplace said a little later, that he had come upon equations in finite differences of another, novel type.[6] They are the analogues in finite analysis of partial differential equations and give rise to a complex set of series, the general term of which has two or more variable indices. In such series as Equation (4), if ϕ is a function of x and n rather than of x alone, and if the integers 1, 2, 3,..., are substituted for x and n, then for each value of n, a series results in which $^ny^x$ designates the term corresponding to the numbers x and n. The definition of a recurro-recurrent series is that $^ny^x$ is equal to any number of preceding terms, taken in rank or in order, in any number of such series, each multiplied by some function of x. Here is the example that Laplace displayed:

$$^1y^x + A'\,y^{x-1} + B'\,y^{x-2} + \cdots + N = 0,$$

$$^2y^x + A''\,^2y^{x-1} + B''\,^2y^{x-2} + \cdots + N'' = H''^1\,y^x + M''^1y^{x-1}$$

$$+ P''^1y^{x-2} + \cdots,$$

$$\dots\dots\dots\dots\dots\dots\dots\dots\dots\dots,$$

$$^ny^x + A^{n\,n}y^{x-1} + B^{b\,n}y^{x-2} + \cdots + N^n = H^{n\,n-1}y^x$$

$$+ M^{n\,n-1}y^{x-1} + \cdots. \quad (5)$$

The value of $^ny^x$ must be determined when $A^n, B^n, \ldots, N^n, H$ are any functions of n; when also $A'', B'', \ldots, A''', B''', \ldots$ are what those functions become on substituting 1, 2, 3,... for n; and when, finally, $A, B, \ldots, N$ are any constants.

The solution consists in showing that Equation (5) can always be transformed as follows:

$$^ny^x = a^{n\,n}y^{x-1} + b^{n\,n}y^{x-1} + c^{n\,n}y^{x-3} + \cdots + u^n, \quad (6)$$

where $a^n, b^n, \ldots, u^n$ are functions of n and of constants that can be determined.[7] This Equation (6), in turn, is an expression for a recurrent series precisely of the type (3) that he had shown to be integrable in his Turin memoir.[8] A further problem of a third-degree equation in finite differences is then shown to be reducible successively to the forms (5) and (6), and Laplace went on to propose a general procedure for reducing an equation of any degree r to a lower degree, the requirement being that by the assumption made for the value of n, the equation of degree r become one of degree $r - 1$.

[6] [1776a, 1°], *OC*, **8**, p. 71.
[7] *OC*, **8**, p. 7.
[8] [1771b].

A passage included many years later in the *Essai philosophique sur les probabilités* makes clearer to the layman how he was visualizing these series.[9] A recurrent series is the solution of a difference equation with a single variable index. Its degree is the difference in rank between its two extreme terms. The terms may be determined by means of the equation, provided the number of known terms equals its degree. These terms are in effect the arbitrary constants of the expression for the general term or (which comes to the same thing) of the solution of the difference equation. The reader is next to imagine a second series of terms arranged horizontally above the terms of the first series, a third series above the second, and so on to make a display infinite upward and to the right. It is supposed that there is a general equation between the terms that are consecutive both horizontally and vertically and the numbers that indicate their rank in both directions. This will be an equation in finite partial differences, or recurro-recurrent.

The reader is finally to imagine that on top of the plane containing this pattern of series there is another containing a similar pattern, and so on to infinity, and that a general equation relates the terms that are consecutive in the three dimensions with the numbers indicating their rank. That would be an equation in finite partial differences with three indices. Generally, and independently of the spatial model, such equations may govern a system of magnitudes with any number of indices. Some eight years later, Laplace replaced the use of recurro-recurrent series with the more efficient tool of generating functions for solving problems in finite differences, which he encountered mainly in the calculus of probability.[10]

It is important biographically to notice his passing remark that investigations in the theory of chance had led him to the formulation of recurro-recurrent series. The latter part of the memoir illustrates how they might be applied to the solution of several problems concerning games of chance. In the first such example, two contestants, A and B, play a game in which the loser at each turn forfeits a crown to his opponent. Their relative skills are as a to b. At the outset A has m crowns, and B has n. What is the probability that the game will not end with x or fewer turns? Laplace found the answer by substituting values given by the conditions of the problem in a series of equations of the form (5), first for the case in which $a = b$, $m = n$, and n is even, and then for all possible suppositions about the parameters.

This problem, like many others that Laplace adduced, appears in the writings of De Moivre, who seems to have furnished his first reading in

[9] *OC*, **8**, pp. xxvi–xxvii.
[10] [1782a], Chapter 11.

the subject, but whose solutions were less direct.[11] A further example was suggested to Laplace by a bet made on a lottery at the École Militaire. What is the probability that all the numbers $1, 2, 3, \ldots, n$ will be taken after x draws? That, too, he found by formulating the problem in a recurro-recurrent series in two variable indices, observing that the approach could clearly have wide applicability in the theory of chance, where the most difficult problems often concern the duration of events.

Laplace chose precisely this juncture for defining probability. The statement occurs immediately after his development of the method of recurro-recurrent series and just before its application to the foregoing examples:

> The probability of an event is equal to the sum of each favorable case multiplied by its probability, divided by the sum of the products of each possible case multiplied by its probability, and if each case is equally probable, the probability of the event is equal to the number of favorable cases divided by the number of all possible cases.[12]

The definition is noteworthy, not for its content, which was standard, but for its location in the development of his work and for its phrasing. The wording should serve to temper the criticism often made of Laplace, particularly in respect to inverse probability (which, to be sure, he had not yet started), that he gratuitously assumed equal a priori probabilities or possibilities. It is true that he often did—although not, as will appear, when he had any data. It is also interesting that in the passage immediately preceding the definition, Laplace should have written "duration of events" rather than "duration of play." This was the first remark he ever printed about the subject as a whole, and already he was thinking about its applicability to the world, and not merely to games of chance.

[11] De Moivre (1756). The problem in question is LVIII, p. 191.
[12] [1774b], *OC*, **8**, pp. 10–11.

CHAPTER 3

Probability of Events and of Their Causes: The Origin of Statistical Inference

HISTORICALLY and biographically, the instinctive choice of words is often more indicative than the deliberate. There is much anachronism in the literature, and it may be well, therefore, to take this, the juncture at which Laplace was entering upon one of the central preoccupations of his life, as the occasion to venture an observation about the early history of probability. More substance has sometimes been attributed to it than the actual content warrants. It is true that adepts of the subject were much given to celebrating its applicability, but prior to Laplace what they were praising was more prospect than actual accomplishment. Jakob Bernoulli's *Ars conjectandi* (1713) is justly famous mathematically, although it is seldom mentioned that part IV, headed "Usum et applicationem praecedentis doctrinae in civilibus, moralibus, & oeconomicis," contained simply the law of large numbers and otherwise remained uncompleted.

In the Dutch and English insurance industry in the eighteenth century, risks were estimated empirically, and the tables on which actuarial transactions depended were numerically insecure. De Moivre and others did employ probability in calculation of annuities, and Daniel Bernoulli undertook a theoretical analysis of the risks attending inoculation for smallpox.[1] These were particular problems, however, and the only extended field of application open to probabilistic analysis was the theory of games of chance. Even there, most of the experiments were thought experiments.

The statement that games of chance provided the principal subject matter in which theorems could be demonstrated and problems solved mathematically will be confirmed by close attention to contemporary usage, which refers to *théorie des hasards*, or "theory of chance." The word "probability" was not used to designate the subject. That word appears in two ways, one more restricted and the other vaguer than in the post-Laplacian science. In the mathematical "theory of chance,"

---

[1] Daniel Bernoulli, "Essai d'une nouvelle analyse de la mortalité causée par la petite vérole et des avantages de l'inoculation pour la prévenir," *MARS* (1760/1766), part 2, pp. 1–79.

probability was a quantity, its basic quantity, that which Laplace defines in the passage quoted in the preceding chapter. *"Calcul des probabilités"* refers to calculation of its amount for certain outcomes in given situations. The phrase *"théorie des probabilités"* rarely, if ever, occurs.

The word "probability" had its second and larger sense, one that would have befitted a theory if any had existed, rather in the philosophic tradition started by Pascal's wager on the existence of God.[2] The changes rung on that idea belong to theology, epistemology, and moral philosophy, some pertaining to what is now called decision making and others to political economy. Such is the discourse in the article "Probabilité" in the *Encyclopédie* (1765).[3] It was largely skepticism about the mathematical prospects for that sort of thing that inspired d'Alembert's overly deprecated hostility to the subject, and optimism about it that inspired Condorcet's overly celebrated enthusiasm.[4] Laplace himself was clear about the difference, although in other terms. Indeed, he insisted upon it in the distinction between mathematical and moral expectation in one of the pair of papers to be discussed next, which between them did begin to join mathematical theory of games with philosophic probability and scientific methodology.

The bibliographical circumstances need to be discussed before the significance of these two papers can be fully appreciated. Both were composed before Laplace became a member of the Academy. They were thus printed in the *Savants étrangers* series, in successive volumes. The more famous was entitled "Mémoire sur la probabilité des causes par les événements" [1774c]. The second, and lengthier, was delayed for two years and then combined with an astronomical memoir under the title "Recherches, 1°, sur l'intégration des équations différentielles aux différences finies, et sur leur usage dans la théorie des hasards. 2°, sur le principe de la gravitation universelle, et sur les inégalités séculaires des planètes qui en dépendent" [1776a, 1° and 2°].

As we shall see, this early coupling of probability with astronomy was no mere marriage of convenience. Laplace spent his entire professional life faithful to the pattern that it started. Before discussing that, however, we shall need to consider the way in which the first part of the dual memoir completed his earlier application of recurro-recurrent series to solving problems in the theory of chance and at the same time complemented his new departure into the determination of cause. He

---

[2] For the historiography of probability, see Bibliography, pp. 308–9.

[3] Long thought to have been written by Diderot himself, the article is now known to have been composed for the most part by the Swiss contributor, Gabriel Cramer. See Bru and Crépel (1994), p. 225.

[4] For d'Alembert's reservations, see Daston (1988), chapter 2, section 3, and for Condorcet's way around them, Brian (1994), part I, chapters 4–5.

submitted this resumption of his work on difference equations to the Academy on 10 March 1773.[5]

These were the writings in which Laplace began broadening probability from the mathematics of actual games and hypothetical urns into the basis for statistical inference, philosophic causality, estimation of scientific error, and quantification of the credibility of evidence, to use terms not then coined. The preamble of the paper on cause declares, and cross-references in both essays confirm, that they were conceived as companion pieces on the subject that Laplace was now beginning to call probability. The one breaks new ground in what was later called its inverse aspect. The other extends and systematizes a direct approach. In preferring the word "probability" to suggest the wider scope that he was giving the subject itself as a branch of mathematics, Laplace may well have been following the precept of Condorcet, newly the acting permanent secretary of the Academy, who, in prefatory remarks to the volume in which the memoir on causes appeared, praised it for its approach to predicting the probability of future events. "It is obvious," wrote Condorcet, "that this question comprises all the applications that can be made of the doctrine of chance to the uses of ordinary life, and of that whole science; it is the only useful part, the only one worthy of the serious attention of philosophers."[6]

The memoir on causes opens with a preamble, most of which might more appropriately have belonged to the concurrent piece on the solution of difference equations. Laplace referred readers to the latter memoir after reviewing what De Moivre and Lagrange had contributed to these problems and the way in which they had then involved him in the theory of chance. The present memoir had a different object, the determination of the probability of causes, given knowledge of events. Uncertainty, the reader is told, concerns both events and their causes (notice that at the outset Laplace took probability to be an instrument for repairing defects in knowledge). When it is given that an urn contains a set number of black and white slips in some definite ratio, and the probability is required of drawing a white one, then we know the cause and are uncertain about the event. But if the ratio is not given, and after a white slip is drawn the probability is required that it be as p is to q, then we know the event and are uncertain about the cause. All problems of theory of chance could be reduced to one or the other of these classes.

[5] The date appears in the procès-verbaux of the Academy (13). A marginal note in the printed text dating it 10 February must be erroneous.

[6] SE, 6 (1774), "Histoire," p. 18. On the interplay between Condorcet and Laplace here, see Gillispie, (1972) and Brian (1994), part I, chapter 5.

Laplace here proposed to investigate problems of the second type. He began on the basis of a theorem that, like the definition of probability in the previous memoir, he enunciated verbally:

> If an event can be produced by a number *n* of different causes, the probabilities of the existence of these causes, given the event (*prises de l'événement*), are to each other as the probabilities of the event, given the causes: and the probability of each cause is equal to the probability of the event, given that cause, divided by the sum of all the probabilities of the event, given each of the causes.[7]

In substance, this theorem is the same as that published posthumously by Thomas Bayes in 1763, eleven years previously. Not only is it now named "Bayes's theorem" or "Bayes's rule," but in the twentieth century the entire approach to probability and statistics depending on it has come to be called Bayesian. That usage derives ultimately from the early-nineteenth-century vindication by Augustus De Morgan and George Boole of their obscure countryman's priority, which was also recognized by Poisson in his *Recherches sur la probabilité des jugements* (1837).[8] Laplace nowhere mentioned Bayes in the memoir on probability of causes. Later he did refer to Bayes in one sentence in the *Essai philosophique sur les probabilités.*[9] In 1774 he had almost certainly not read Bayes's paper in the *Philosophical Transactions*. In this period continental mathematicians seldom read or referred to their British contemporaries, and the statement and approach in Bayes's piece are very different.

On the other hand, Laplace may well have heard of Bayes. His paper was submitted to the Royal Society after his death by his friend and fellow minister, the theologian and liberal moral philosopher Richard Price, who accompanied it with a covering note and an appendix.[10] Price was known on the Continent, especially among political theorists, including Condorcet, and Condorcet did mention Bayes in a comment of 1781.[11] Moreover, in introducing the analysis of cause, Laplace did not claim that it was an altogether new subject. He said it was novel "in many respects." More important, he echoed Condorcet's prefatory

---

[7] [1774c], *OC*, **8**, p. 29.

[8] Bayes's paper, "An essay towards solving a problem in the doctrine of chances," is reprinted in Pearson and Kendall (1970), pp. 134–53. For its importance to Laplace, see Stigler (1978) and (1982).

[9] *OC*, **7**, p. cxlviii.

[10] For a discussion of the motivation and significance of Bayes's paper, and of the relations between Bayes and Price, see Gillies (1987).

[11] In the prefatory comment to Laplace's later "Mémoire sur les probabilités," [1781a], *HARS*, (1778–1781), pp. 43–46.

remarks to the effect that the approach "the more merits being developed in that it is mainly from that point of view that the science of chance can be useful in civil life." It should also be said, finally, that in the sequel the analysis of inverse probability derived from Laplace's memoir and from his further work. Bayes remains one of those pioneers remembered only after the subject they intrinsically might have started had long been flourishing thanks to work of others that did have consequence.

However that may be, Laplace proceeded from the statement of the theorem to an example. From an urn containing an infinite number of white and black slips in unknown ratio, $p + q$ slips are drawn, of which p are white and q black. What is the probability that the next slip will be white? The above theorem gives the following formula for the probability that x is the true ratio[12]

$$\frac{x^p(1 - x)^q \, dx}{\int x^p(1 - x)^q \, dx}, \tag{7}$$

where the integral is taken from $x = 0$ to $x = 1$. Laplace calculated that the required probability of drawing a white slip on the next try is

$$\frac{p + 1}{p + q + 2}. \tag{8}$$

A second example from the same urn leads to the application that was the point of this analysis. After pulling p white and q black slips in $p + q$ draws, what is the probability of taking m white and n black slips in the next $m + n$ draws? For that probability, Laplace obtained the expression

$$\frac{\int x^{p-m}(1 - x)^{q-n} \, dx}{\int x^p(1 - x)^q \, dx}, \tag{9}$$

where the limits are again 0 to 1, and went on to ask a more significant question. How large would $(p + q)$ need to be, and how small would $(m + n)$ need to remain, in order to permit calculation of the probability of the ratio of m to n given the ratio of p to q? Clearly, the solution of that problem would permit calculating the probability of a future event from past experience, which is to say for statistical inference (although Laplace never used that phrase). He did not try to solve the problem in general, pleading the lengthiness of the calculation. Instead,

[12] [1774c] *OC*, **8**, p. 30.

he proceeded to the demonstration of a limit theorem. In effect, he showed, in what he called an interesting (*curieux*) proof, that the numbers $p + q$ can be supposed large enough to bring as close to certainty as one pleases the probability that the ratio of white slips to the total lies between $p/(p + q) + \omega$ and $p/(p + q) - \omega$, where ω is less than any given magnitude.

Laplace had two changes to ring on one of the classic problems, the division of stakes after a game is interrupted. First, he referred the reader to the companion memoir for the method of employing recurro-recurrent series in deducing the canonical solution for the standard case, in which the relative skills of the two players are given. He further promised a general solution in the case of three or more players. So far as he knew, that problem had never been solved. (Never reticent about claiming credit, Laplace was mistaken in this instance, for De Moivre had already solved the problem.)[13] Second, Laplace addressed the case in which the relative skill of the players is unknown, pointing out that it pertained to the probability of future events. Laplace solved it for a two-person game in this inverse example, without attempting the generality of three or more players.

Altogether more significant is the next topic: first, for its subject, which was the determination of the mean value among a series of observations; second, for the area from which Laplace took it, which was astronomy; and third, for the application, which was to theory of error. In view of the whole development of Laplace's later career, this article may be considered highly indicative, not least in his manner of introducing it. Two years previously, he wrote, he had worked out a solution for taking the mean value among several observations of the same phenomenon. Thinking it would be of little use, he had deleted it from the memoir on recurro-recurrent series, where he had originally intended it as a postscript. He had since learned from an astronomical journal that Daniel Bernoulli and Lagrange had both investigated the same problem.[14] Their memoirs remained unpublished, and he had never seen them.

This announcement, together with what he now calls the utility of the matter, had led him to set out his own ideas. It would appear that he must have given them some further development, for the approach turned on treating the true value as the unknown cause of three observed values, taken for effects. Afterward, the mean value giving the minimum probability of error would be determined. It is evident from

---

[13] Todhunter (1865) p. 468.

[14] [1774c], *OC*, **8**, pp. 41–42. He says *Journal astronomique*, but the name in fact was *Recueil des astronomes*. On this matter, see Stigler (1978), 247–48.

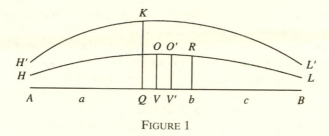

FIGURE 1

the outset that the observations Laplace had in mind were astronomical, for the point to be fixed indicated the time at which the event occurred. Laplace now constructed a graph and a posterior distribution curve, *HOL* in Figure 1. The line *AB* represents time. Three observations of the phenomenon occur at points *a*, *b*, and *c*. The interval between *a* and *b* is *p* seconds, and between *b* and *c* it is *q* seconds. The problem is to find the point *V* at which to fix the mean between the three readings *a*, *b*, and *c*.

Laplace begins by considering the error distribution in general. It is more likely that an observation deviates from the true value by two seconds than by three, by three seconds than by four, and so on, but the law relating probability of error inversely to its degree is unknown. The probability that an observation differs from the true value *V* by the amounts *Vp* and *Vp'* may be represented along the curve *RMM'* in Figure 2. If *x* stands for the abscissa *Vp* and *y* for the corresponding ordinate *pM*, the equation of the curve may be written $y = \phi(x)$. The curve is symmetrical around *VR* since it is equally probable that the deviation from the true value is to the left as to the right.

Next Laplace returns to Figure 1, where the point *a* is at distance *x* from the true value *V*. The probability that the three observations deviate by the distances *Va*, *Vb*, and *Vc* will be $\phi(x) \cdot \phi(p - x) \cdot \phi(p + q - x)$. If it be supposed that the true instant is at *V'* and that $aV' = x'$, this probability will be equal to $\phi(x') \cdot \phi(p - x') \cdot \phi(p + q - x')$. Then by the fundamental principle of inverse probability stated at the outset (i.e., Bayes's rule), the probabilities that the true instant of

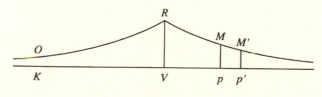

FIGURE 2

the phenomenon is at V or V' will be to each other in the proportion of these two expressions. The equation of the curve HOL is $y = \phi(x) \cdot \phi(p - x) \cdot \phi(p + q - x)$, and its ordinates represent the probabilities of the corresponding points on the abscissa.

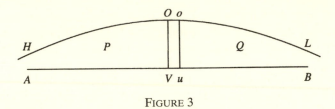

FIGURE 3

In speaking of the mean to be taken among several such observations, either of two things may be intended. One is the instant at which it is equally probable that the true time of the phenomenon occurs before or after it. That may be called the "probabilistic mean." The other is the instant at which the sum of errors to be incurred ("feared" in Laplace's terminology) multiplied by their probabilities is a minimum. Laplace named it the "mean of error" or "astronomical mean," the latter because it is the one that astronomers should prefer.

Laplace now shows by a nice piece of geometrical reasoning that the probabilistic and astronomical means come to the same value, which may be found by bisecting the area under the curve HOL in Figure 3. The probabilistic mean may be found by determining the ordinate OV, which divides the area under the curve in two equal parts, since it is then as probable that the true instant falls to the left as to the right of the midpoint V. For the probabilistic mean, the point V has to be located where the sum of the ordinates of the curve HOL, multiplied by their distance from the point V, is a minimum. To prove that there is no difference between the two, Laplace introduces the ordinate ou infinitely close to OV and lets $VU = dx$ and $OV = y$. He also lets Q be the center of gravity of the part uol of the curve, M be the area under this part, and z be the distance from the point Q to the ordinate OV. If the point V is taken as the mean, the sum of the ordinates multiplied by their distances from V will be

$$ Mz + Nz' + \frac{1}{2} y \, dx^2, $$

whereas if u is taken as the mean, then the sum of the ordinates multiplied by their distance from u will be

$$M(z - dx) + N(z' + dx) + \frac{1}{2}y\,dx^2.$$

The difference between these two quantities is $Ndx - Mdx$, and that must be equal to zero in the case of a minimum. In that case, $M = N$, so that the ordinate OV bisects the area under the curve HOL. Thus, the astronomical and probabilistic means are one and the same.

In an illuminating commentary accompanying a translation of the essay, Stephen Stigler considers that this argument characterizes, in the terminology of modern statistics, a posterior median (the probabilistic mean) as optimal for a certain loss function (the astronomical mean), and he takes it to be one of the earliest results that pertains to mathematical statistics as distinct from theory of probability.[15]

In order to choose a mean value in practice, Laplace needed to specify the form of the probability curve $\phi(x)$. It is reasonable to suppose, he argued, that the ratio of two consecutive infinitesimal differences is the same as that of the two corresponding ordinates, so that $d\phi/\phi$ is constant. If so, the equation relating the ordinates to their infinitesimal differences is

$$\frac{d\phi(x + dx)}{d\phi(x)} = \frac{\phi(x + dx)}{\phi(x)}, \qquad (10)$$

whence

$$\frac{d\phi(x)}{dx} = -m\phi(x); \quad \text{thus } \phi(x) = \beta e^{-mx}. \qquad (11)$$

The parameter m is constant, and since the area under the curve ORM is supposed to equal unity, and the curve is symmetrical, $\beta = m/2$, and

$$\phi(x) = \frac{m}{2}e^{-mx}. \qquad (12)$$

Thus Laplace did not here arrive at the curve that would lead to the famous least-squares rule, first published by Legendre in 1805 (chapter 21). Laplace's purpose was to convince astronomers that their normal practice of taking an arithmetical average was erroneous, although he had to acknowledge that his method was difficult to use. A further application of the principle of inverse probability calculates the probabilities that the different values of m are to each other as the

---

[15] Stigler (1986b), p. 360.

probabilities that the values when obtained will be in the proportion of their respective distances. That calculation (for which the curve $H'KL'$ of Figure 1 was introduced) was very intricate and yielded a fifteenth degree equation.[16] Resorting to iteration, therefore, Laplace gave a table for the correction that he advised applying to whichever of the two extreme observations was farther from the middle one.

The final topic in the probability of causes may be thought equally significant in view of the importance that the finding assumed in its later application and, indeed, in the philosophy of probability. It concerns the effect of inequalities in the prior probabilities that are unknown to the gamblers who are tossing coins or throwing dice. In the normal theory, the assumption is that either heads or tails, or any of the six sides of a die, are equally probable. In physical fact, of course, everyone recognizes that there are no such things as perfectly balanced coins or ideally symmetrical dice; but it was further assumed that the game is nevertheless fair, provided that both players are ignorant of the actual inequalities.

On the contrary, Laplace found—and claimed (in this instance correctly) to be the first to have noticed and demonstrated the fallacy—that the latter assumption is valid only for the first throw, that is to say, for situations involving simple probabilities. For example, if B should agree to give two crowns to A if A tosses heads on the first flip of a coin, then before the game begins, a fair division of the stakes would be 50–50 even if the coin is weighted, since neither player knows which side is heavier. The assumption ceases to be valid, however, as soon as the game continues under the same rules, according to which B will then give A four crowns if a head turns up only at the second turn, six crowns at the third turn, and so on to x turns. On the supposition that the coin has a probability of $(1 + \pi)/2$ to fall on one side, and $(1 - \pi)/2$ on the other (we don't know which is which), Laplace's calculation reaches one of those surprising results that, in defying common sense, lend piquancy to probability. In x tosses, A's expectation is less than x when x is between 1 and 5, equal to x when x is 5, and greater than x when x is over 5, increasing with x to infinity. In other words, the outcome becomes more inequitable as probabilities are compounded.

Laplace pursued the matter in the case of throwing dice. He had never, he observed, found any dice that were perfectly cubical, and he mentions particularly the imperfection of English dice, though whether because of faulty manufacture or deliberate weighting he does not say. At all events, A and B play, the rule being that if A throws a given side in n tosses of one die, B pays him the sum a. What, then, should A

---

[16] Stigler (1986b), p. 361, finds that Laplace made a subtle mistake in the calculation that prevented him from reaching a simpler solution.

forfeit to B if he fails? By the classical theory of chance, the amount equals the expectation of A, which is $a(1 - 5^n/6^n)$. In fact, however, given any asymmetry at all, A's expectation will be less than that, for B has the advantage as the game continues. Determining the correct value gave Laplace a very lengthy calculation, issuing in a complicated function of the degree of asymmetry.

Laplace introduced this discussion with a warning that "the science of chances must be used with care and must be modified when we pass from the mathematical case to the physical." The concluding sentence repeated the admonition with respect to "the different objects of civil life." D'Alembert was his great patron, and it would have been out of character for the young Laplace to take explicit issue with the great man's view that the calculus of probability was in principle inapplicable to a world where physics is approximate and politics contingent. Here at the outset he says only that the theory had to be modified.

Throughout his career Laplace proceeded to modify it. This analysis presages not only the general thrust of Laplace's later work but also two of the most characteristic techniques he would employ. Both pertain to inverse probability and the study of causes. First, the occurrence of seemingly aberrant patterns would invite investigation of the source of departures from the results that equipossibility would entail. Second, the multiplication of observations would provide the basis for calculating from experience the value of the prior probabilities when they are unequal. By the law of large numbers, it could then be determined how large the sample needed to be in order to reduce the probability of error within given limits.

There are textual grounds for thinking that the implications of asymmetries were an important factor in opening Laplace's eyes to the wider prospects for probability. He enlarged on them in a philosophical, or methodological, article strategically situated in the midst of the companion memoir on difference equations and theory of chance, to which we now turn (deferring for a moment the question of its combination with the first gravitational memoir).[17] The analysis of chance, says Laplace in this crucial article, has objects of two kinds: first, the probability of happenings about which we are uncertain, whether in their occurrence or in their cause (it is worth noting that probability is still subsidiary to theory of chance in the way he puts it here); and second, the hopes that attach to their eventuality. Laplace acknowledged that questions could legitimately be raised about the very enterprise of applying mathematical analysis to situations of both sorts. In concerns of the former type, he attributed the difficulty to the mode of application, not to the definition. There is no ambiguity about the

---

[17] [1776a, 1°] Article XXV, *OC*, **8**, pp. 144–53.

definition of probability itself, and no legitimate objections could be lodged against its calculus unless equal prior probabilities were to be assigned to cases that are not in fact equally probable. He had to admit that, unfortunately, all applications yet attempted to problems of civil life entailed precisely that fallacy. Laplace proceeded to illustrate it by a simplified repetition of the demonstration that the false supposition of symmetry in coins or dice is unfair to one of two players in any game involving composite probabilities.

The question of equipossibility was also involved in exposing the error committed by commentators who argued that a run of heads or tails is less likely than any alternation whatever in a sequence of the same number of tosses of a coin. Proceeding from a mistaken notion of common sense, they supposed that each time a head turns up, the odds increase against another. In effect, they were saying that past events influence future ones. Laplace characterized that idea as "inadmissible" and proceeded to refute it in a general statement, his earliest, on the relation between regularity, chance, and causality. Why does so-called common sense give us to suppose that a sequence of twenty heads is not due to chance, whereas we would think nothing of any equally possible mixture of heads and tails in a total of twenty? The reason is that wherever we encounter regularity, we intuitively take it for the effect of a cause acting in an orderly manner. If we were to come upon letters from a printer's font lying in the order INFINITESIMAL on some composing table, we should be disinclined to think the arrangement random, although if that were not a word in some known language, we should pay it no heed. The example may in all probability be adapted from the chapter in the Port-Royal *La Logique* (1662) on judgment of future events, where Arnauld points out that it would be stupid to bet twenty sous against ten thousand livres that a child playing with printer's type would arrange the letters to compose the first twenty lines of Virgil's *Aeneid*.[18]

Laplace even made a little calculation to show how our intuition that order bespeaks causality is itself conformable to probability. A symmetrical event has to be the result either of cause or of chance. Let $1/m$ be its probability in case it is due to chance, and $1/n$ its probability in case it is due to a regular cause. By his basic principle the probability of the cause will then be

$$\frac{\dfrac{1}{n}}{\dfrac{1}{m} + \dfrac{1}{n}} \quad \text{or} \quad \frac{1}{1 + \dfrac{n}{m}}.$$

[18] Antoine Arnauld and Pierre Nicole (1664), ed. Louis Marin, 1970, p. 429.

The greater *m* in relation to *n*, the greater the probability that a symmetrical event is due to a regular cause.

In retrospect, it seems clear that an element of wanting it both ways was always lurking at the bottom of Laplace's outlook. On the one hand, the regularity of the universe as a whole bespeaks the rule of natural law. Order governs amid the infinity of its combinations. On the other hand, where there do appear to be disturbing factors, for example, in the results of real games with imperfect dice, or (to anticipate for a moment) in anomalies of planetary motion, indeterminacies in the shape of the earth, widely varying inclinations among the planes of cometary orbits, and inequality in the partition of births between boys and girls, such out-of-line data call for identification of a particular cause. We do not seek out one that will be an exception to the larger realm of causality but, rather, one that, if properly calculated, will vindicate it. Thus, both apparent regularity and exceptions to it bespeak causality, general or particular as the case may be, and the goal of analysis is to make the two cases one. Everything that a modern student wants to read into Laplace's outlook says that he should have attributed the larger regularity to randomness, and many things that he actually said point toward that anachronism. He did not, however. The notion of an order of chance would have been a contradiction in terms to Laplace:

> Before going further, it is important to pin down the sense of the words *chance* and *probability*. We look upon a thing as the effect of chance when we see nothing regular in it, nothing that manifests design, and when furthermore we are ignorant of the causes that brought it about. Thus, chance has no reality in itself. It is nothing but a term for expressing our ignorance of the way in which the various aspects of a phenomenon are interconnected and related to the rest of nature.[19]

In regard to the second set of objections to probability as a branch of analysis, namely that hopes, fears, and states of mind cannot be quantified, Laplace considered that reservations of this sort arose from a fundamental misunderstanding and not from a mere fallacy in the procedures. They could, therefore, be obviated by making clear distinctions in the definition of terms. Such, in Laplace's view, had been the root of d'Alembert's resistance to the calculus of probability. Now that Laplace did mention his patron, it was most respectfully. In the future, practitioners of the science would feel obligated to d'Alembert for having forced upon it clarification of its principles and recognition of its proper limitations. For his objections could be obviated by making clear distinctions in the definition of terms.

---

[19] [1776a, 1°], *OC*, **8**, p. 14.

Laplace took the limits of probability to be identical with those of all the "physico-mathematical sciences. In all our research, it is the physical cause of our sensations that is the object of analysis, and not the sensations themselves."[20] It is obvious that out of estimates of the likelihood of future happenings come hopes and fears, and that the prospect for calculating probability was the reason that the science of chance had long been heralded for its potential utility in civil life. Taking advantage of that opportunity in the measure possible depended, however, upon seizing the distinction between *espérance morale* and *espérance mathématique*, between aspiration and expectation. (The French makes the contrast rhetorically more effective, but perhaps the English makes the difference more inescapable.) In the theory of chance, expectation is simply the product of the amount to be gained by the probability of winning it. It is a number, like probability itself— nothing more. No doubt something similar might be said about aspiration in the ordinary concerns of life, but only in a qualitative way; for that always depended on such indefinable factors that it was illusory ever to think of calculating it. Even Daniel Bernoulli's suggestion for measuring personal gain by the quotient of the value of the winnings (or other profit) divided by the total worth of the winner, although an ingenious idea, was one incapable of generalized mathematical application. (Laplace was too dismissive in this remark, for Bernoulli's approach has proved to be the root of modern utility theory. He recurred to it himself in *Théorie analytique*.)

The article that makes Laplace's apology for probability in general opens with language famous from its reemployment almost verbatim nearly forty years later in the *Essai philosophique sur les probabilités*.[21] We give these thoughts of his youth as he set them down when he was twenty-six:

> The present state of the system of nature is evidently a consequence of what it was in the preceding moment, and if we conceive of an intelligence that at a given instant comprehends all the relations of the entities of this universe, it could state the respective position, motions, and general affects of all these entities at any time in the past or future.
>
> Physical astronomy, the branch of knowledge that does the greatest honor to the human mind, gives us an idea, albeit imperfect, of what such an intelligence would be. The simplicity of the law by which the celestial bodies move, and the relations of their masses and distances, permit analysis to follow their motions up to a certain point; and in order to determine the state of the system of these great bodies in past or future

[20] [1776a, 1°], *OC*, **8**, p. 147.
[21] *OC*, **7**, p. vi.

centuries, it suffices for the mathematician that their position and their velocity be given by observation for any moment in time. Man owes that advantage to the power of the instrument he employs, and to the small number of relations that it embraces in its calculations. But ignorance of the different causes involved in the production of events, as well as their complexity, taken together with the imperfection of analysis, prevents our reaching the same certainty about the vast majority of phenomena. Thus there are things that are uncertain for us, things more or less probable, and we seek to compensate for the impossibility of knowing them by determining their different degrees of likelihood. So it is that we owe to the weakness of the human mind one of the most delicate and ingenious of mathematical theories, the science of chance or probability.[22]

Long afterward, Laplace observed, also in the *Essai philosophique*, that his early investigations of probability were what had led him to the solution of problems of celestial mechanics in the first place.[23] The historian is bound to temper respect for creative people with skepticism about their reminiscences of how they came to do their work. Nevertheless, it is at least interesting that Laplace's first general statement about nature and knowledge should have been consistent with the entire configuration of his oeuvre. That configuration first appears in the relation of the two parts of the dual memoir under discussion. The structure of the first part is clear. It consists of thirty-five articles. Articles I–XXIV develop the integration of equations in finite differences by means of recurro-recurrent series. They constitute a comprehensive resume of what Laplace and others had already done, incorporating many improvements. Article XXV then treats probability in general, opening with the sentences just quoted and concluding with a repetition of the exordium about the two classes of problems in probability from the companion "Mémoire sur la probabilité des causes par les événements" [1774c]. The reader is directed there for inverse probability, while Laplace goes on in Articles XXVI–XXXV to apply the methods developed in the first twenty-four articles to the solution of problems of direct probability in the theory of games of chance.

Article XXXVI, although it continues the numbering, is subheaded "Sur le principe de la gravitation universelle, et sur les inégalités séculaires des plantes qui en dépendent." Matters of mathematical astronomy are treated throughout this second half of the dual memoir in a sequence of twenty-nine further articles, which we shall discuss in the next section. We do not know precisely when Laplace wrote them, though it is quite possible that he developed them from the paper

[22] [1776a, 1°], *OC*, **8**, pp. 144–45.
[23] *OC*, **7**, p. lxv.

entitled "Une théorie générale du mouvement des planètes," which he presented to the Academy in November 1771 (9). They must at all events have been completed by early 1774, for on 27 April Laplace submitted another investigation of secular inequalities in planetary motion, which is described as a sequel although it was printed earlier, in 1775, because of accidents in the academic publishing schedule.[24] Almost certainly, therefore, he must have turned to the second part of the dual memoir immediately after completing the paper on the probability of causes and the concurrent revision of the first part, which had occupied him between March and December 1773 (13). It also seems likely that the methodological Article XXV, with its philosophic propositions about the relation between probability and astronomy, would have been interpolated in the midst of the aleatory Part 1 during the time when Laplace was composing the astronomical Part 2. As we shall see, passages introducing the latter are as prophetic, or rather programmatic, of his celestial mechanics as the paragraphs quoted above are of his epistemology. It appears, therefore, that Laplace framed the main questions that his enormous lifework refined, extended, and largely answered during a crucial period of about a year following his election to the Academy in March 1773.

[24] (15), [1775a].

CHAPTER 4

Universal Gravitation

THE QUESTIONS that occupied Laplace were slightly but significantly different from what they are sometimes said to have been. A close reading of the astronomical part of the dual memoir, Laplace's first comprehensive piece on the mechanics of the solar system, serves to temper the conventional image of a vindicator of Newton's law of gravity against the evidence for decay of motion in the planets.[1] Nothing is said about apparent anomalies gathering toward a cosmic catastrophe; on the contrary, the state of the universe is assumed to be steady. The problem is not whether the phenomena can be deduced from the law of universal gravity, but how to do it. Since that appeared to be impossible on a strict Newtonian construction of the evidence, Laplace proposed modifying the law of gravity slightly. He proceeded to try out the notion that gravity is a force propagated in time instead of instantaneously. Its quantity at a given point would then depend on the velocity of bodies as well as on their mass and distance. Even more interesting, the reasoning in this argument was not that of normal mathematical astronomy but was of the type that Laplace brought to physics in other, much later writings. Lastly, in a problem that he did handle in the tradition of theoretical astronomy, namely the secular variations in the mean motions of Jupiter and Saturn, the conclusion is that the mutual attraction of the planets cannot account for them, contrary to what we expect from *Mécanique céleste*. Let us, therefore, examine these matters more fully.

As in other early papers, Laplace's point of departure was an analysis by Lagrange. In a memoir that had won the prize set by the Paris Academy for 1774, Lagrange had argued that it was impossible to derive from the theory of gravity an equation for the acceleration of the mean motion of the moon giving values large enough to agree with observation.[2] Lagrange then wondered whether resistance of the ether might be slowing the rotation of the earth enough to resolve the apparent discrepancy. Laplace for his part took the question to be one involving the sufficiency of the law of gravity with implications for all of cosmol-

---

[1] [1776a, 1° and 2°]; see chapter 3.
[2] "Sur l'équation séculaire de la lune," *OL*, **6**, pp. 355–99.

ogy. (In the same manner, it is worth noting, he expanded the significance of problems in the theory of chance into a discussion of probability in relation to knowledge in general.)

Announcing at the outset his intention of doing a series of memoirs on physical astronomy, he began by deriving general equations of motion for extended bodies referred to polar coordinates in a form especially adapted to analyzing problems of secular inequalities, which is to say the discrepancies increasing over time between the predicted and observed positions of planetary bodies. Thereupon, he turned to the "principle of universal gravitation" in general, which he called the most incontestable truth in all of physical science. It rested, in his view, upon four distinct assumptions generally accepted among *géomètres*, or persons doing exact science. Given their importance in marking out the main lines along which Laplace developed his celestial mechanics, we shall state them in the form of a close paraphrase:[3]

1. The force of attraction is directly proportional to mass and inversely proportional to the square of the distance

2. The attractive force of a body is the resultant of the attraction of each of the parts that compose it

3. The force of gravity is propagated instantaneously

4. It acts in the same manner on bodies at rest and in motion.

The plan was to examine the respective physical consequences of these assumptions. Since Laplace continued that examination throughout his life, the tactics of this memoir in effect became the strategy of much of his subsequent research. We need to note, therefore, what he believed it was that followed physically from each of those four propositions.

The inverse-square law of attraction came first. Laplace asserted roundly that it was no longer permissible to doubt its applicability to the solar system. It is evident to the modern reader that this law was the one ultimately vindicated by his eventual demonstration that apparent irregularities in planetary motion, far from requiring qualification of the law of gravity, may be deduced from it. Only later, however, in a series of memoirs composed from 1785 through 1787 did he succeed in bringing off those investigations.[4] Here at the outset, deductions from the inverse-square law engaged his interest in a different order of effects. In a speculative vein, he took issue with philosophers who doubted whether it holds for forces acting at very short ranges. Even

---

[3] [1776a, 2°], Article XLII, *OC*, **8**, pp. 212–13.
[4] See Chapter 16.

though the radius of the earth was the smallest distance over which its validity had been confirmed by observation, analogy and canons of simplicity still gave reasonable grounds for supposing that the gravitational-force law obtains universally. If it be asked why gravity should diminish with the square of the distance rather than in some other ratio, Laplace—after duly objecting that such questions make mathematicians uncomfortable—would consent to say only that it is pleasing to think that the laws of nature are such that the system of the world would be the same whatever its size, provided the dimensions be increased or decreased proportionally.

It is perhaps somewhat surprising to find that in the first stage of Laplace's astronomical work, his analysis of the consequences following from the second of these assumptions, the principle that the attractive force exerted by a body is the resultant of the attractions exerted by all of its parts, took precedence over the theory of planetary motion. Although a less famous topic, it may have been an even more fruitful preoccupation, if not necessarily for astronomy itself, then certainly for mathematical science in general, since the problem of the attraction of a spheroid was the main source of the potential function.[5] It may also be surprising to learn that in the later eighteenth century, skepticism persisted about whether the force of attraction really operates between all the particles of matter individually or only between centers of mass of macrocosmic bodies of which the shape and internal structure are governed by other, unknown laws. Again, Laplace saw no reason to suppose that there exists some least measure of distance below which analogy becomes an unconvincing mode of argument. For the present purpose, that did not matter, however, since this second assumption, unlike the inverse-square law of the intensity of short-range forces, could be subjected to analysis and the results submitted to the test of observation on the scale of terrestrial physics. D'Alembert had already derived the precession of the equinoxes and the nutation of the axis of the earth from the principle of universal attraction among all the particles of the globe, and he found the prediction confirmed by data. The tides were too complicated a phenomenon to tackle yet—although Laplace attacked them soon afterward in a major calculation.[6]

There remained the shape of the earth.[7] By Newtonian theory, it should be an ellipsoid with the polar and equatorial axes in a ratio of 229/230. Two sets of independently measured values existed. The first derived from geodetic surveys of the length of meridional arcs in

---

[5] See Chapter 15.
[6] See Chapter 7.
[7] Essential for the background of this important topic is Greenberg (1995).

Lapland by the Maupertuis expedition in 1735–1736, at the equator in Peru by the Bouguer–La Condamine expedition in 1736–1737, in France itself on the meridian of Paris by Lacaille in 1739–1741, and at the Cape of Good Hope again by Lacaille in 1751–1752. The other set of values derived from determinations of the length of the pendulum beating seconds at various latitudes during those expeditions and on other occasions.

The geodetic data gave a flattening greater than an axial ratio of 229/230 would produce, and the pendulum yielded a smaller departure from the spherical than Newton required. Laplace considered it likely that simplifications introduced into the calculation were at fault rather than the theory itself. The most serious were the assumption of uniform density and the neglect of irregularities of the surface. However that might be, the failure of the earth's longitudinal profile to fit an elliptical curve would not disprove the mutual attraction of all its particles, unless it were shown that an ellipsoid was the only solid that could satisfy the equilibrium conditions for such a force, or else that every theoretically possible figure had been tried without satisfying the observations. Since neither of those propositions had ever been demonstrated, the principle of universal interparticulate attraction held its ground. Thus, because of the relevance to that principle of the shape of celestial bodies considered as solids of revolution, Laplace entered upon this, another of his continuing preoccupations.

So much for the long-range importance of the first two gravitational assumptions in Laplace's work; the latter two, on the other hand, are interesting mainly for the discussion of them that he gave in the ensuing articles of the present memoir, and it was there that he least resembled the celestial dogmatist for which he has sometimes been taken. The third and fourth assumptions have a different standing from the inverse-square law and the principle of attraction. Anyone doing planetary astronomy would have thought it necessary to write down those first two principles in the axiomatic structure of a treatise. To most of Laplace's colleagues, on the other hand, the instantaneous propagation of gravity and its indifference to motion would have seemed prior, self-evident truths, like the rectilinear transmission of light or the three-dimensionality of space, something not usually needing statement. Not so the young Laplace, who stated these assumptions for the sake of taking issue with them. It is unreasonable, he observed immediately, to suppose that the power of attraction or any other force acting at a distance should be propagated instantaneously. Our sense is rather that it should correspond in its passage to all the intervening points of space successively. Even if communication should appear instantaneous, what really happens in nature may well be different, "for it is infinitely far from an unobservable time of propagation to one that is absolutely nil."[8] (It later became a distinctive characteristic of Laplace's physics that the phenomena he analyzed should occur in the realm of the unobservable.) Thus, he would try what followed from the supposition that gravitation does take place in time. On the amount of time, however, he would disagree with Daniel Bernoulli, who in a piece on tidal motion had advanced the proposition that the action of the moon takes a day or two to reach the earth.

Broaching the matter in a more abstract manner, Laplace posited a corpuscle to be the bearer of gravitational force. In his analysis, the effect of weight in a particle of matter is produced by the impulse of such a gravitational corpuscle, infinitely smaller than the particle, moving toward the earth at some undetermined velocity. Given this model, the received hypothesis, according to which gravity has an identical effect whether bodies are at rest or in motion, is equivalent to supposing that velocity to be infinite. Laplace supposed it indefinite and determinable by observations. His reasoning was similar to that employed to account for the aberration of light. Let us see how he set up the problem.

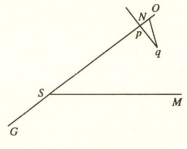

FIGURE 4

The calculation analyzes the motion of an infinitely small body p describing any orbit around S in the plane pSM. $Sp = r$; the angle $pSM = \phi$; and N is the corpuscle causing p to gravitate toward S (see Figure 4). Resolving the forces and motions along the radius vector Sp and the perpendicular direction pq, Laplace introduced the expression θ/α for the measure of the distance described by the corpuscle N in time T. T and α are constants, α being an extremely small numerical coefficient and αe the ratio of the original eccentricity to the mean

[8] [1776a, 2°], OC, **8**, p. 220.

distance; t is time, and θ is a distance varying according to some function of the distance Sp. Calculation yields as equations of the orbit:

$$r = a\left(1 + \alpha e \cos nt - 2\frac{a\alpha}{\theta}Tn^2t\right) \qquad (13)$$

and

$$\phi = nt - 2\alpha e \sin nt + \frac{3}{2}\frac{a\alpha}{\theta}Tn^3t^2, \qquad (14)$$

where a and n are terms introduced in the change of variables

$$r = a(1 + \alpha y) \quad \text{and} \quad \phi = nt + \alpha x, \qquad (15)$$

which refers the equations to rectangular coordinates for manipulation.[9]

From the above equations, it appears that the mean motion of p is governed by a secular equation proportional to the square of the time. In the normal assumption of instantaneous gravitation, $a\alpha/\theta$ is infinitely small, and the secular equation vanishes. If that term were not zero, however, its effect would appear in the mean motion of planets and satellites. At this point, therefore, Laplace turned to the observations and particularly to the lunar tables, knowing that ancient and modern records of eclipses showed that the earth's satellite has been increasing its mean speed of revolution.

Laplace placed more confidence in the German Tobias Mayer than in any other practical astronomer. In Mayer's view, the acceleration of the mean lunar motion had amounted to one degree in two thousand years. D'Alembert had shown that this acceleration could not be explained within the ordinary theory by any calculation involving the sun, earth, and moon alone. Lagrange's prize paper of 1774 on the secular equation of the moon had then demonstrated that the oblate shape of the earth and of the moon, taken together with the influence of other planets, failed to account for the acceleration.[10]

It might be worthwhile, therefore, to try out the notion that gravitation takes time, which is to say that the term $\alpha t/\theta$ is not nil. To recall the conditions of the problem, θ/α is the space traversed by the gravitational corpuscle that is impelling the moon toward the earth during one revolution of the moon in its orbit. The velocity then is $\theta/\alpha T$. Substituting Mayer's values for the acceleration of the mean motion in these expressions, Laplace found that the velocity of the gravitational corpuscle is 7,680,000 times as great as the velocity of light.

[9] [1776a 2°], *OC*, **8**, p. 224.
[10] Memoir cited in n. 2.

Since Mayer's values were thought to be off by 12′ in two thousand years, the correct velocity was 6,400,000 times that of light.

Discussing this, which he admitted to be a conjecture, Laplace acknowledged that the abbé Bossut's postulate of a very subtle fluid in space also permitted a calculation agreeing closely with the secular equation of the mean motion of the moon. What could that fluid be, however, unless it were light itself emanating from the sun? If so, the consequences could be computed, for the orbit of the earth would have expanded and its mean motion would have been retarded. Unfortunately for Bossut's scheme, that has not happened. Its advocates might still say that sunlight has dilated the terrestrial atmosphere so as to produce the trade winds and has thus retarded the rotation of the earth in another manner. Such a mechanism would indeed explain the apparent acceleration of the moon, but Laplace had also found, by a method that he promised to give elsewhere, that the rotation of the earth cannot be retarded by the friction of those winds in any detectable degree [1779b]. Thus, we are left with the force of gravity, "astonishing" in its activity but finite in velocity.

Following this analytic flight of fancy, Laplace "returned" to the normal assumptions about gravity and addressed himself to traditional problems of mathematical astronomy concerning inequalities or variations in the elements of planetary motion. He thought of them in four main classes: positions of nodes and apsidal lines, eccentricities, inclinations of the orbital planes, and mean motions. The last were the most significant, and even they were less well determined than Laplace could have wished. On so signal a matter as the irregularities of the mean motions of Jupiter and Saturn, Euler and Lagrange disagreed. Euler had found them to be about equal and Lagrange very different.[11] Laplace was worried about the applicability even of Lagrange's differential equations, for terms involving sines and cosines of very small angles were dropped since the coefficients were extremely small. But the coefficients became large on integration, which moved them to the denominator of expressions, and Laplace feared lest the resulting formulas for determining the true movement of the planets would hold good for limited periods only. In his own calculation, Laplace took account of those terms and obtained formulas agreeing with Lagrange's in the values for apsides, eccentricities, and inclinations but differing drastically for the mean motions. Although his equations did contain terms proportional to the time and the square of the time, he would not claim that they succeeded in representing the true motions rigorously.[12]

---

[11] "Solution de différents problèmes de calcul intégral," *OL*, **1**, pp. 471–668, see p. 467.
[12] [1776a, 2°], *OC*, **8**, p. 241.

That question was interesting only mathematically and of no practical importance for astronomy throughout the historic span of recorded observations. More important were the results he obtained for Jupiter and Saturn.

Over the centuries most astronomers had come to consider that the observations show an acceleration for Jupiter and a much larger deceleration for Saturn. But when Laplace substituted values from Halley's tables in the expression that he had just derived for the secular equation of the mean motion of a planet, it reduced approximately to zero, leaving him to conclude that if an alteration existed in the mean motion of Saturn, it could not be caused by the influence of Jupiter.[13]

Thereupon Laplace applied a different method, d'Arcy's principle of conservation of areas, to the same problem. The chevalier Patrick d'Arcy, an Irish emigré, soldier, and mathematician, was something of a maverick in rational mechanics in that he took angular momentum to be fundamental. In a system of bodies revolving about a fixed center, he asserted, the sum of the products of the mass of each by the area that its radius vector describes is always proportional to the time.[14] Taking this approach to the possible effect of the mutual influence of Jupiter and Saturn on variations in their mean motions, Laplace obtained the same null result as in the previous analysis.[15] It seemed unreasonable to suppose that a virtually complete canceling out of positive and negative terms could be due to particular circumstances, and Laplace concluded that mutual gravitation between any two planets and the sun failed mathematically to account for inequalities in their mean movements and that some other cause must be responsible for the observed anomalies. As a candidate for the disturbing factor, he suggested—again by way of conjecture—the action of the comets.

He recognized that the conclusion was "contrary to what all mathematicians who have worked at the subject have hitherto supposed."[16] This statement will be equally surprising to students of *Mécanique céleste*, still more so to those who know only the *Exposition du système du monde*, and most of all to those who are told in many textbooks that Laplace rescued the Newtonian planetary system from increasing instability by proving that precisely such mutual interactions do resolve the apparently cumulative inequalities that are actually periodic. That this finding, which he emphasized for its importance, should seem contrary

---

[13] Ibid., p. 252.

[14] "Principe général de dynamique, qui donne la relation entre les espaces parcourus et les temps, quel que soit le système de corps que l'on considère," *MARS*, (1747/1752). On d'Arcy's principle and its place in history of mechanics, see Truesdell (1964).

[15] [1776a, 2°], Article LVIII, *OC*, **8**, pp. 254–58.

[16] Ibid., p. 258.

to what he is most famous for is explicable largely in consequence of the mathematics. He had not yet developed perturbation functions, and there was no provision in his expressions for long-term periodicity. Moreover, he nowhere worried about instability, unless that was what he had in mind in a passing remark that his conclusion would be less convincing if the secular inequalities increased proportionally to the square of the time, for that would mean that a continuously acting force was at work—and none other than universal gravitation was known.

The remaining articles of this, the first memoir on gravitation, will be somewhat anticlimactic to the modern reader, although there is no indication that Laplace thought them so. He considered that Euler in treating the secular inequalities in the motion of the earth had neglected significant terms and thus had given an incomplete analysis, particularly with respect to the eccentricity and apogee of the sun. The latter value, determined with precision from Halley's tables, could serve to establish the mass of Venus with sufficient accuracy for Laplace to calculate the share in the decrease of the obliquity of the ecliptic that resulted from the action of the planets. In the brief Articles LXII and LXIII, he gave a method for determining that variation in general by means of analytic geometry.[17] They may possibly represent the memoir on that problem which Laplace had read before the Academy in November 1770, one of his earliest (3). In principle the method could have been applied to calculating the position of the equinoxes and hence the degree of that inclination at any time past or future, but the ancient observations were too imprecise to render the exercise worthwhile. Another conjecture about cometary influence speculated that it may be a factor offsetting the attraction of the planets in their influence on the decreasing inclination of the ecliptic. Finally, a postscript reviewed the calculations of Lagrange that had left their author uncertain about the acceleration of the moon.[18]

Laplace found, on the contrary, that they strengthen the case for it.

---

[17] Ibid., pp. 268–72.
[18] "Sur l'équation séculaire de la lune," *OL*, **6**, pp. 335–99.

CHAPTER 5

Distribution of Comets

IN THE SAME volume with the dual memoir on probability and gravitation discussed in the preceding two sections, Laplace published the tripartite "Mémoire sur l'inclinaison moyenne des orbites des comètes, sur la figure de la terre, et sur les fonctions," the last paper that he published in the *Savants étrangers* series [1776b]. The three parts indicated in the title are altogether independent of each other. We shall discuss the first here and defer the second and third for consideration in connection with the further development that he gave to the respective topics (see Chapters 7 and 9). Occupying over two-thirds of the entire memoir, Laplace's treatment of comets consists in an application of his work in probability to analysis of the distribution of their orbital planes in space.

Let us begin by recalling Laplace's remark about having been led into astronomy through probability (Chapter 3). In construing what he meant, it will be useful to distinguish between weak and strong interactions in the evolution of his interests. To the weak, or philosophical, sort belong the regulative remarks about order, causality, and knowledge; to the strong, or technical, belong the probabilistic analysis of the phenomena themselves and also of distributions and errors in the data, and more rarely the application of mathematical techniques conceived for one set of problems to another. The memoir on comets illustrates both aspects. Its motivation pertained to the weaker, philosophical or cosmological, aspect of probability and its execution, to the stronger mathematical aspect.

Uncertainty still bedeviled the status of comets in the solar system. Did they fully belong to it or not? Like many others, Laplace was undecided. We have just seen, in Chapter 4, how at this early stage he invoked their action as a possible deus ex machina to explain the secular inequalities of the mean motions of Jupiter and Saturn, which appeared inexplicable on the basis of forces operating among the planets. He opened the present memoir by pointing to the anomaly that the motions of comets exhibit if they do belong to the planetary system. The known planets and their satellites, some sixteen bodies in all, revolve about the sun in the same direction and almost in the same plane. The probability that a common cause lies behind the arrangement could be calculated and was approximately equal to certainty (although to the further question of what that cause might be, Laplace confessed that he had never found a convincing answer).[1]

Comets are quite another matter. They move in any direction in very eccentric orbits inclined at all angles to the plane of the ecliptic. Laplace's senior colleague, Dionis du Séjour, one of the few aristocratic amateurs who actually contributed to astronomy, had analyzed the problem of whether the cause of planetary motion—whatever it might be—had also produced the phenomena of comets. He calculated the mean inclination of the sixty-three known orbits and found it to be 46°16'. The departure from 45° was so small as to be insignificant. Moreover, the ratio of forward to retrograde motions was five to four, or nearly one to one. Dionis du Séjour had concluded, therefore, that the comets serve a principle of indifference characteristic of the universe at large, within which the causal system of the planets constitutes a distinctive ordering.

The conclusion was reasonable enough in Laplace's view, but a further calculation was needed to establish the degree of certainty that it held. Suppose the comets randomly projected into space. What, then, are the probabilities that the mean inclination of their orbits, and also the ratio of clockwise to counterclockwise revolutions, will be contained within given limits? For if the value of the mean inclinations were $45° + \alpha$, and if there were very large odds (say, a million to one) to bet that it should be less, it could plausibly be concluded that some particular cause does account for their moving in one plane rather than another. The same argument could be applied to the ratio of forward to retrograde motions. There the calculation is easy, involving only two possibilities. Not so the problem of the orbits. Perhaps the mention of a bet hints that Laplace was thinking in terms antedating the analysis of causes from events that he might have been expected to apply. At any rate, he went on to qualify the problem as one of the most complicated in the entire theory of *hasards*, especially so since the goal was a general formula applicable to any number of comets.

The problem was as follows: Given an indefinite number of bodies randomly projected into space and revolving around the sun, what is the probability that the mean inclination of the orbits with respect to a given plane, say, that of the ecliptic, lies between two limits? Laplace

---

[1] Laplace does not cite, though he most probably knew, a 1734 paper by Daniel Bernoulli on the application of probability to the inclination of planetary orbits, and also d'Alembert's comment on the analysis. See Todhunter (1865), pp. 222–24, 273. I am indebted to Stephen Stigler for this reference. It is less likely that Laplace knew a 1757 calculation of random errors by Thomas Simpson, though it was essentially the same as the one he here develops for the distribution of cometary orbits. See Stigler (1986a), pp. 95–97.

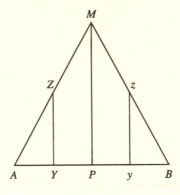

FIGURE 5

approached it by constructing probability curves for the simpler cases, much as he had done in discussing the mean value of a series of astronomical observations in the memoir on probability of causes. Here, however, the mean is the arithmetical average, and the probability sought is direct, not inverse. The first case is that of two bodies only, M and N. The probabilities are represented in Figure 5, and perhaps the reasoning is worth following since it exhibits in the most elementary case how Laplace thought about the probability that a certain quantity would be comprised within given limits.

The line $AB = a$ represents 90°, which is the maximum possible mean inclination of the two orbits. On it Laplace constructs a "curve of probability," $AZMB$, along which every ordinate is proportional to the probability that the mean inclination is equal to the corresponding abscissa. Let the mean inclination of the two orbits equal x, where x is less than $1/2a$. Then since $AY = x$ and $YZ = y$, y will be proportional to $2x$ from A to the midpoint P of AB. Beginning at zero, the value of the inclination of the orbit of comet M may increase by infinitesimal increments dx to a maximum of $2x$. Thus $YZ = 2AY$, and AZM has the form of a straight line and is the hypotenuse of a right triangle APM in which $PM = 2AP = a$.

Since it is as probable that the mean inclination approaches the limit A as the limit B, the ordinates equidistant from PM on both sides must be equal, and thus the line BM is entirely equal to AM. In the case of two bodies, therefore, the "curve of probabilities" $AZMB$ consists of two right lines AM and AB so joined that $PM = a$. Finally, in order to calculate the probability that the mean inclination is comprised within the limits Y and y, the area $YZMzy$ must be divided by the entire area AMB, and the quotient (i.e., field or number of favorable cases divided by field or number of possible cases) gives the probability sought.

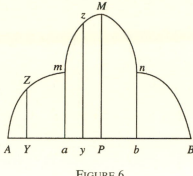

FIGURE 6

Laplace then turned to the case of three bodies (or comets) M, N, and P, where $AB = a$ (Fig. 6) is divided into three equal parts, Aa, ab, bB. In order to find the probability that the mean inclination is equal to any abscissa AY, he draws the probability curve $AmMnB$. The reasoning is similar to that in the previous case of two bodies. $AY = x$, less than Aa or $1/3a$. Any of the three comets, say, M, has an inclination f. The mean inclination of the other two will then be $(3x - f)/2$ since the mean inclination of all three is x. In order to have the total number of cases in which the mean inclination may be x, the integral of $(3x - f)\,df$ is to be taken from $f = 0$ to $f = 3x$. That comes to $9x^2/2$. The ordinate YZ is then equal to $9/2 \cdot x^2/a$.[2]

It remained to determine the nature of the curve mMn, composed of two equal parts, mM and Mn. Laplace lets $ay = z$, and by a similar train of reasoning arrives at the expression $1/a(1/2\,a^2 + 3az - 9z^2)$ for the value of the ordinate yz. The equation of the curve AZM and mMn are then respectively

$$ay = \frac{9x^2}{2} \quad \text{and} \quad ay = \frac{1}{2}a^2 + 3az - 9z^2. \tag{16}$$

Again the required probability that the mean inclination falls between certain limits is given by the quotient of the area under the curve between those limits divided by the whole area AMB.

[2] Bernard Bru points out (private communication) that the calculations are correct but that the curves here, and also in Figure 6, are erroneous since AZm would have a curvature the reverse of that shown, so that the overall figure should be bell-shaped. Much later [1810b] Laplace noted that the error curve would approximate a normal distribution.

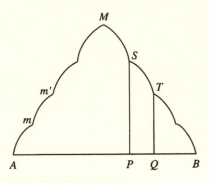

FIGURE 7

In a figure corresponding to a system of four comets, the equations for the curves corresponding to *am* and *mM* respectively are

$$a^2 y = \frac{32}{3} x^3 \quad \text{and} \quad a^2 y = \frac{1}{6} a^3 + 2a^2 z + 8az^2 - 32z^3. \quad (17)$$

It would have been possible to continue inductively, dividing AB into five or more equal parts and obtaining the equations of the curves corresponding to n bodies from the results for $(n - 1)$. To that end Laplace supposed the line AB of Figure 5 divided into n equal parts and sought an expression for the curve relative to the rth part, as shown in Figure 7.

In finding it, he let $a/n(r - 1)$ be the distance of one of its ordinates from A (z being less than a/n), and represented by $_r y_{n,z}$ the number of cases in which the mean inclination of n bodies is $a/n(r - 1) + z$. That general expression could be given a value in the form (with Laplace's notations):

$$_r y_{n,z} = {}_r A_n z^{n-1} + {}_r B_n z^{n-2} + {}_r C_n z^{n-3} + \cdots + {}_r G_n z + {}_r H_n, \quad (18)$$

where $_r A_n, \,_r B_n, \ldots$, etc., are functions of r and n to be determined. Now it becomes evident what Laplace had been up to all the time. He could determine the value and solve the problem by a method that he had published in Article XII of the double memoir on difference equations and gravitation.[3] Essentially, the problem was to find the general expression for quantities subject to a law governing their formation. In this instance, when $AB = a = 90°$, and AB is divided into

[3] [1776a, 1°], *OC*, **8**, pp. 97–102.

n equal parts, the method gives the equation for the curve corresponding to the rth part:[4]

$$a^{n-2}y = {_r}y_{n,z}.\tag{19}$$

The probability that the mean inclination of n orbits lies between any
two points P and Q is then given by the quotient of the area $STQP$
divided by the entire area $Amm'MSTB$.

In a complete application of the theory, the value for n would be 63,
that being the number of comets for which astronomers had calculated
the orbits. Leaving that computation to whoever wished to undertake it,
Laplace worked out the case for twelve comets. He supposes the line
AB divided into twelve equal parts of $7°30'$, and computes the probability that the mean inclination of the comets lies between $45°$ and $37°30'$
or between $45°$ and $52°30'$. The resulting value is 0.339. That gives 839
to bet against 161 (to use his language) that the mean inclination is
above $37°30'$ and the same that it is below $52°30'$, and the odds that it
falls within those limits are 678 to 322. Referring to data, Laplace found
that the mean inclination of the twelve most recently observed comets
was $42°31'$. There could be no reason to assign a cause tending to make
these comets move in the plane of the ecliptic unless the contrary
assumption of random projection resulted in enormous odds that the
mean inclination should be above $42°31'$. In fact, the odds are less than
six to one that it would be above $37°30'$ and are very much smaller that
it would exceed $42°31'$. Hence, no such cause can be supposed to exist.

The important element here is not the support that this heaviest of
probabilistic artillery brings to the position that Dionis du Séjour had
already occupied in a comparatively straightforward and lightweight
analysis. What is interesting is Laplace's virtuosity in applying to the
distribution of the orbits of comets a method for formulating in a
general manner a problem encountered in the theory of chance.

[4] [1776b], *OC*, **8**, p. 299.

CHAPTER 6

Partial Differential Equations, Determinants, and Variation of Constants

THE REMAINING mathematical investigations in Laplace's early period involved techniques of analysis developed in a group of three memoirs, two of them concerned also with astronomical problems. It will be convenient to begin with the piece that had no special astronomical relevance, "Recherches sur le calcul intégral aux différences partielles," which, although published later than the other two, was started earlier. Laplace read a version in 1773 but did not submit it for publication until December 1776.[1] It consists of a reworking of his youthful *Acta eruditorum* paper, for which he had already apologized in the paper on probability of causes.[2] He had there appended a preliminary statement of the main theorem that he was now demonstrating. The long delay suggests that he may have had trouble with it. In any case, he had arrived at a general method (since called the cascade method) for solving linear partial differential equations by an approach that yielded the complete integral when it could be obtained and further served to identify equations that are insoluble. As originally printed, the memoir employs Euler's bracket notation for partial differential coefficients, in which a general second-order linear partial differential equation is

$$0 = \left(\frac{d\,dz}{dx^2} \right) + \alpha \left(\frac{d\,dz}{dx\,dy} \right) + \beta \left(\frac{d\,dz}{dy^2} \right) + \gamma \left(\frac{dz}{dx} \right) + \delta \left(\frac{dz}{dy} \right) + \lambda z + T,$$

(20)

α, β, γ, δ, λ, and T being functions of x and y.[3] Laplace then simplified the form by changing the variables x and y into others, ω and θ, which are functions of x and y, and by regarding z as a function

---

[1] (19), [1777a].
[2] See chapter 1, notes 8, 10.
[3] For discussion, see Grattan-Guinness (1990a), pp. 324–28.

of these new variables. He could thereby obtain the equation in a new form:[4]

$$0 = M\left(\frac{\partial\,\partial z}{\partial\omega \cdot \partial\theta}\right) + N\left(\frac{\partial z}{\partial\omega}\right) + L\left(\frac{\partial z}{\partial\theta}\right) + Rz + T'. \quad (21)$$

It followed that any second-order linear partial differential equation is reducible to the simple form

$$0 = \left(\frac{\partial\,\partial z}{\partial\omega \cdot \partial\theta}\right) + m\left(\frac{\partial z}{\partial\omega}\right) + n\left(\frac{\partial z}{\partial\theta}\right) + lz + T. \quad (22)$$

Since m, n, and l are functions of ω and θ, ω and θ can be determined and sustituted in Equation (21), thus transforming it into Equation (22), which in turn has all the generality of Equation (20) but is simpler and easier to investigate.

Laplace was now in a position to restate the theorem that is the heart of the matter. In cases where the complete integration of Equation (22) is possible in finite terms (and those cases could always be identified), one of the two integral signs affecting the arbitrary functions $\phi(\omega)$ or $\psi(\theta)$ can be eliminated and the remaining integration performed. Moreover, the cases wherein such a solution is possible include almost all the problems encountered in mechanics. For that was the field in which Laplace expected the method to be applicable. Among the many uses of linear partial differential equations, he singled out for special mention the determination, whatever the state of the system, of infinitesimal oscillations among an infinite number of corpuscles interacting in any manner whatsoever.[5]

A paper started at about the same time, "Mémoire sur les solutions particulières des équations différentielles et sur les inégalités séculaires des planètes," begins with the demonstration of a second theorem that Laplace had stated without proof in an appendix to the paper on probability of cause.[6] It concerns the theory of differential equations and, in particular, Euler's discovery of singular solutions, of which the properties were not well understood. Certain functions can satisfy a differential equation and still not be contained in its general integral, and the consequence is that integration of an expression is not necessarily tantamount to the complete solution of a problem. All the singular

[4] [1777a], *OC*, **9**, pp. 21–22.
[5] Ibid., p. 7.
[6] [1775a]. For the initial statement of the theorem (Chapter 3), see [1774c], *OC*, **8**, p. 64.

solutions also have to be determined, and the method that Euler gave was restricted to first-degree equations. Laplace went on to find further methods of determining special solutions. For example, given the differential equation $dy = p\,dx$, if $\mu = 0$ is a particular solution, then μ is a factor common to the expressions

$$p + \frac{\dfrac{\partial^2 p}{\partial x\,\partial y}}{\dfrac{\partial^2 p}{\partial y^2}} \quad \text{and} \quad \frac{1}{\dfrac{\partial p}{\partial y}}, \qquad (23)$$

and, reciprocally, any factor common to both expressions is a particular solution to the equation when it can be set equal to zero.[7] Laplace also obtained results for equations of higher degree.

In a hastily printed "Addition" [1775b] to the paper, Laplace introduced a powerful method of solving a collection of simultaneous differential equations that became known as "successive approximations." It consisted of a means whereby under certain conditions the solution to each equation could be built up by solving its reduced form and then successively reintroducing each omitted term. Laplace gave only a summary idea of the method in this postscript, reserving its full development for the next volume of the *Mémoires*.[8] It was specially suited to planetary perturbation theory, and probably his inspiration came from there.

This remark may mitigate the otherwise fair judgment of Condorcet, in the preface of the volume in which this memoir appeared, to the effect that the two topics in Laplace's title, particular solutions of differential equations and secular inequalities of the planets, were entirely distinct and had totally different objects.[9] Formally, that was no doubt true; and indeed the mathematical part was submitted in July 1773 and the astronomical in December 1774 (14, 17). More significant, however, is the evidence that throughout these years Laplace's mind was moving back and forth between problems of analysis and astronomy, and that the relation of emulation and competition with Lagrange embraced both domains. The astronomical section resumes Laplace's discussion of the long-term invariance of the mean motions and hence the mean distances of the planets in respect to any gravitational interaction.[10] Receipt of a manuscript memoir from Lagrange dealing

---

[7] [1775a], *OC*, **8**, p. 339.

[8] [1775b], *OC*, **8**, pp. 361–66. See Grattan-Guinness (1990a), p. 154.

[9] *HARS* (1772, part 1/1775), p. 70.

[10] [1776a, 2°], See Chapter 4. Although written earlier, this piece was printed later than [1775a].

with secular inequalities in the motions of the nodes and also in variations of orbital inclinations had stimulated him to return to the subject after laying it aside in some frustration. By means of a transformation of variables, now named the Lagrange variables, Lagrange succeeded in reducing his problem to the integration of as many linear differential equations as there were unknowns and then to the determination of the constants, whatever the number of planets.

Laplace hastened to apply the transformation, which he greatly admired, to his own formulas for secular inequalities of mean motion and succeeded in deriving the same equations. With considerable enthusiasm, he also tried whether he could determine the secular inequalities of the eccentricities and motions of aphelion by an analogous method. Happily he could. He apologized for rushing into print before Lagrange himself could publish.[11] Between them, however, they had the makings of a complete and rigorous theory of all the secular inequalities, and Laplace for his part now proposed to draw together just such a general theory in a further study.

This intimation is the earliest clear hint of the famous astronomical program eventually realized in the late 1780s. The astronomical part of Laplace's dual probability-gravitation memoir, published in 1776 but composed prior to April 1774, argued only that gravitational interactions of Saturn, Jupiter, and the sun failed to account for the secular inequalities of the major planets. Laplace did not then raise the question of the stability of the planetary system, which he appeared to take for granted.[12] Clearly, therefore, it was in the course of these exchanges that Lagrange and Laplace set out to supersede Newton's conclusion, and Euler's after Newton, that the planetary system would become unstable over time, and also the recourse to divine intervention as the agency that would put it back in order. The job was to demonstrate mathematically that, on the assumption that all planetary bodies revolve in the same direction around the sun, the stability of the system follows from Newton's laws. In order to accomplish that, Lagrange and Laplace analyzed both the eccentricities and inclinations of orbits, the former to prove that no planet could escape into space, the latter that none could tip continuously away from the plane of the ecliptic. If Lagrange initiated the program, and it seems clear that he did, it was Laplace who eventually brought it off.[13]

The essay "Recherches sur le calcul intégral et sur le système du monde" constitutes the third item in this mainly mathematical series

[11] [1775a], *OC*, **8**, pp. 355–56n.
[12] [1776a, 2°], See Chapters 4 and 16.
[13] See Grattan-Guinness (1990a), pp. 324–29.

[1776c]. Perhaps it was the most important, for the approach became a mainstay of Laplace's later work in planetary theory and thereby of positional astronomy in general. For readers unfamiliar with that calculus, Laplace explained how the complexity of planetary motions rules out all hope of achieving rigorous solutions to the problems they present. Called for instead were simple and convergent methods for integrating the differential equations by approximation. The most widely practiced technique, which Laplace attributed to d'Alembert, consisted in an adaptation of Newton's method for approximating to the roots of polynomials. Since the values were known with only a small degree of uncertainty, accurate quantities could be substituted in their place and a very small indeterminate term added. Then the squares and higher powers of this new indeterminate might be neglected and the problem reduced to the integration of as many linear differential equations as there were variables.

All available methods for that, however, entailed one or both of two drawbacks. They were inapplicable to planets with more than one satellite and to the motion of two or more planets in the same direction around the sun. Moreover, in some cases, notably that of the moon, the second approximation in the process of integration introduced secular terms with no physical meaning—then called "arcs of circles"—that spoiled the convergence. The method that he had just imagined avoided these inconveniences. It was specifically applicable to expressions in which the variables are functions of periodic quantities and also of other quantities that increase very slowly. That is precisely the situation in physical astronomy. The trick consisted in varying the arbitrary constants in the approximate integrals and then determining their values for a given time by integration. In the preface Condorcet despaired of conveying an idea of it verbally.[14] Agreeing that examples would convey a better notion than any generalities, Laplace produced a large number in a memoir of some one hundred pages followed by several additions.

Andoyer gives a summary account, which may be of interest.[15] Given a differential equation

$$\frac{d^2x}{dt^2} = P, \tag{24}$$

in which P, a function of x and of dx/dt, is periodic with respect to appropriate linear arguments in t, it is supposed that its general integral

[14] *HARS* (1772, part 2/1776), pp. 87–89.
[15] Andoyer (1922), pp. 54–55.

is obtained and that it takes the form

$$x = X + tY + t^2 Z + \cdots .\qquad(25)$$

The functions X, Y, Z,..., depend on two arbitrary constants C_1 and C_2 and are periodic in t, as well as in any other arguments introduced by integration. Here Laplace restricted the conditions so that a function can be developed consistently with the above equation for x in only one way. The expression

$$x' = X' + (t - \theta)Y' + (t - \theta)^2 Z' + \cdots .\qquad(26)$$

will then satisfy the equation, where θ is any arbitrary quantity and X', Y', Z',... are the forms taken by X, Y, Z,... when the constants C_1 and C_2 are replaced by similar variants, C_1' and C_2'.

If C_1' and C_2' are now properly determined as functions of C_1, C_2, and θ, the expressions for x and x' become identical. Since x' will then be independent of θ, any value can be assigned to θ at will; and if θ is set equal to t, the expression for x reduces to

$$x = X'\qquad(27)$$

when C_1' and C_2' are assigned corresponding values. If those are periodic, the manipulation will have eliminated the secular terms from the initial expression for x. In order to determine those values, it can be asserted that the value of x' becomes independent of θ in such a way that

$$\frac{dx'}{dC_1'} \frac{dC_1'}{d\theta} + \frac{dx'}{dC_2'} \frac{dC_2'}{d\theta} + \frac{dx'}{d\theta} = 0.\qquad(28)$$

That equation permits the determination of C_1' and C_2', given the further condition that for $\theta = 0$, these quantities reduce to C_1 and C_2.

Among his examples, Laplace applied the method to the determination of inequalities in eccentricities and the inclinations of the orbit; he derived anew his theorem from the probabilistic-astronomical memoir concerning the long-term invariance of the mean motions with respect to gravitational interactions;[16] and he tried out an analysis of a new sort of problem, the secular inequalities that a planet would exhibit if it moved in a resisting medium. His examples amount to the elements of just such a theory as he had promised in the previous memoir, and it was certainly the astronomical possibilities opened to him by the variation of constants that motivated the present paper. The reader who

---

[16] [1776a, 2°], Article LIX, *OC*, **8**, pp. 258–63.

wishes an abbreviated recapitulation by Laplace himself may turn to Articles XVI–XIX of the great memoir on the "Théorie de Jupiter et de Saturne."[17] Yet it is characteristic of Laplace that along the way he should have developed a method for solving problems capable of much more general applicability. Early in the course of the analysis Laplace had found himself confronting a set of linear equations from which n quantities had to be eliminated. The technique he there introduced involved an expansion since called by his name in the theory of determinants.[18]

[17] [1788a], *OC*, **11**, pp. 131–40. See Chapter 16.

[18] [1776c], Article IV, *OC*, **8**, pp. 395–406. Muir (1906–23) gives a summary of the importance of the Laplace expansion in the development of determinants (**1**, pp. 25–33).

CHAPTER 7

The Figure of the Earth and the
Motion of the Seas

GEOPHYSICAL TOPICS occupied Laplace in the final investigation to be assigned to his early period. In the program of the probability-gravitation memoir, measurements of the figure of the earth constituted the one set of data by which the principle of universal inter-particulate attraction might be tested.[1] He did not then develop the matter, however, and instead reserved the subject for preliminary treatment in a single article of the tripartite memoir that opened with his analysis of the distribution of cometary orbits.[2] He there conceived the problem to be one of determining the equilibrium conditions governing the form of planets considered as spheroids of revolution. The table of contents of the volume in which the paper appeared classifies the discussion under "Hydrostatics" rather than geodesy. The attribution was appropriate, however, for Laplace analyzed the same model as Newton had, a homogeneous fluid mass spinning on its axis with all the particles serving the inverse-square law of attraction among themselves.[3] He considered Newton's assertion that such a body would be an ellipsoid to be very shaky, however. Although it was easy to show that an ellipsoid, or indeed some other solid of revolution, satisfied the conditions a posteriori, it was enormously difficult to determine a priori what those conditions must be.

Laplace undertook that task in a further brief analysis tacked on to the memoir on variation of constants in the integration of differential equations pertaining to astronomy.[4] Although he failed to determine the equation for the curve described by a meridian, he did there discover the law of gravity in a generalized homogenous spheroid: On the surface of any such body in a state of internal gravitational equilibrium, the variation of gravity from the equator to the poles follows the same law as in an ellipsoid. Its force at any point on the surface is

[1] [1776a, 2°]; see Chapter 4.
[2] [1776b], Article X, *OC*, **8**, pp. 302–13.
[3] *SE* (1773/1776), p. xv.
[4] [1776d], attached to [1776c], *OC*, **8**, pp. 481–501.

given by

$$P = P'\left(1 + \frac{5}{4}\alpha m \cos^2 \theta\right), \qquad (29)$$

where P' is the value at the equator, αm is the ratio of centrifugal to gravitational force at the equator (α being very small), and θ is the complement of the latitude.[5]

Those two small pieces were forerunners of "Recherches sur plusieurs points du système du monde," which Laplace submitted to the Academy in three installments in 1777 and 1778 and which, taken together, comprise a treatise of over two hundred pages in its *Mémoires*.[6] The investigation had three objects:

   1. The law of gravitational force on the surface of homogeneous spheroids in equilibrium
   2. The motion of the tides, together with the precession of the equinoxes and the nutation of the earth's axis
   3. The oscillations of the atmosphere caused by the gravitational action of the sun and the moon.

The first section is entirely formal and quite in the d'Alembert tradition, except that Laplace dispensed with the stipulation that the solid be generated by the revolution of a curve. He set up the problem by means of a diagram in which an element of surface fluid in the form of an infinitesimal parallelogram exerts a force upon the center of mass, and after eight pages of analysis he obtained the equation[7]

$$P = P' + \frac{5}{4}\alpha f \cos^2 \theta - \frac{1}{2}\alpha V + \alpha R \qquad (30)$$

where again θ is the complement of latitude, P' is a constant of integration, R is a centrally directed force at a point on the surface and hence a function of latitude and longitude, V is a term expressing the departure (assumed to be small) of the spheroid from the spherical, and αf is the centrifugal force at the equator. None of this would be very interesting, perhaps, except for what emerges as the motivation. Only at the end does it become evident why Laplace had reversed directions and determined the attraction of a surface element for the center. For he there adduced a special case in which the forces are produced by the attraction of any number of bodies near to or far from the spheroid and

[5] [1776d], *OC*, **8**, p. 493.
[6] [1778a], [1779a], [1779b].
[7] [1778a], *OC*, **9**, p. 85.

subject to the conditions that they change its form negligibly, and that they either participate in its rotation or are infinite in number and disposed in a uniform ring.

Laplace frequently indulged in the practice of specifying some peculiarity of the world in highly abstract terms in order to make it appear to follow from a general analysis. This example is an instance, for the case is that of Saturn's rings. The conditions give a series expression for the last two terms in Equation (30):

$$\Sigma\left(\frac{S}{2rs} - \frac{3}{2}Sb - \frac{S}{r^2s}\frac{\partial r}{\partial s} \right), \qquad (31)$$

where S is the mass of a discrete body in an assembly constituting a ring, r is its distance from the point M on the surface of the planet, s is the radius of the point M, and b is a function of θ and ω, respectively the complement of the latitude and the longitude of M. If the shape and density of the ring, and its position relative to the axis of rotation of Saturn, were known, and also if the planet were supposed to be uniform in density, the law of gravity at the surface could be determined. For S would then be infinitesimal, and summing the series would become a simple exercise in integration. Laplace attempted no solution, however, and the general significance of the analysis is its applicability to systems of particles external to the surface of an attracting spheroid.

It is the second topic in this series of memoirs that is the most interesting, as much for its subject matter as for the example it affords of Laplace's mode of thinking. The phenomenon analyzed is the tidal ebb and flow of the seas. In the program of the dual probability-gravitation memoir, he had put that problem aside as too complicated for calculation.[8] Now that he felt ready to take it on, the question (as always) was not whether the motion of the tides is subject to gravitational forces, which fact no one doubted, but how the law could be derived. He started with the assumption classically brought to the problem in the study it had received from Newton through Daniel Bernoulli and Maclaurin to d'Alembert, namely that the spheroid covered by a fluid should differ very little from a sphere; that the axis of rotation should be invariant; that the centrifugal force of the fluid particles and other forces, such as attraction of the stars, should be small compared to self-gravity; and that the sidereal motion should be slow compared to the speed of rotation of the sphere. To these a priori conditions, Laplace added two others that emerged in the course of his analysis and that offer interesting illustrations of his sense of the relation between the operations of analysis and physical facts.

[8] [1776a, 1° and 2°]; see Chapter 4.

The new conditions emerging from the analysis were that the seas be approximately uniform in depth and that the depth be about four leagues (twelve miles). If one were to read no further in Laplace than this lengthy memoir, one would be bound to suppose that he took these assertions to be statements about the way the world is made emerging from a mathematical analysis of tidal action under gravitational force. He acknowledged that if the case were different, his expressions would have been unmanageable. He repeatedly referred to it as "the case in nature," however, and nowhere stipulated that it was a simplification introduced in order to make the analysis possible at all.

On the contrary, he took his predecessors to task for having simplified matters in a manner that was unfaithful to the facts. Beginning with Newton, the form they had investigated was that which would be assumed on a stationary planet by a thin covering of fluid in equilibrium under the gravitational pull of a stationary star. They then assumed that if the star were given a real or apparent motion around the planet, it would simply drag the bulge around without changing the form. Laplace, for his part, would now give the problem a realistic, dynamical analysis. The most obvious discrepancy with fact in the classical picture arose from the consequence it entailed that the two full tides occurring in a single day should differ markedly in height and most notably when the declination between the sun and moon was greatest. Actually, however, it was common knowledge at any seaside that consecutive high tides differ very little in level. A second difficulty was formal. The traditional analysis overlooked the variation of "the angular motion of rotation" of molecules of seawater with latitude, which would produce longitudinal displacements of the same order as those induced by the direct gravitational action of the sun and moon.[9]

From the frequency with which Laplace recurred to the former anomaly, it would appear to have been that which had fixed his attention on the problem at just this juncture. A letter of 25 February 1778 to Lagrange confirms that the lack of conformity between theory and observation over the inversion of the tides had struck him very forcibly.[10] In his own analysis Laplace formed expressions containing terms for the actual momentum of fluid elements at all latitudes on a rotating globe. When he developed them, he was pleased to be able to argue that the theory yielded a convincing explanation for the near equality of consecutive high tides. Indeed, he said, it exhibited any number of reasons for it. What those expressions were, we shall show in a moment.

[9] [1778a], *OC*, **9**, p. 90. Aiton (1953) is valuable for tidal theory.
[10] Laplace to Lagrange, 25 February 1778, *OL*, **14**, pp. 78–81.

Meanwhile, his analysis had led him to notice another deficiency in existing theory, this one affecting the precession of the equinoxes and also the motion of nutation discovered by James Bradley in 1748. Heretofore, calculations of these variations in the direction of the earth's axis had presupposed the action of the sun and moon on a solid globe. To be fully accurate, however, account would also need to be taken of the gravitational interaction of both bodies with the seas. If the sun and the moon were in the plane of the equator, the pull on the waters in the northern and southern hemispheres would be symmetrical and no problem would arise. In actuality, the inclination of the plane of the ecliptic is bound to result in uneven tidal distributions, and there can therefore be no warrant a priori for supposing that the unequal reaction of the waters in the two hemispheres will fail to produce disturbances in the direction of the polar axis. Quite the contrary.

As it happened, Laplace's calculations showed (he did not say to his surprise) that introducing these considerations produced no changes in the overall laws of precession and nutation. It turned out that only the ratio of the quantity of the nutation to that of precession is influenced by including terms for tidal asymmetry between the hemispheres. That composite effect is very small, moreover, when calculated on the basis of reasonable hypotheses concerning the depth and density of the seas. Nor was Laplace disappointed, for its amount came out to be proportional to the very small difference between consecutive high tides. Thus, the validity of the four-league uniform depth of the sea, which had been introduced to facilitate calculation of the tides, was confirmed by its applicability to the interrelation between the tides and the variations in direction of the axis of the earth.

This piece of research makes a very nice example of the juncture at which Laplace came into the history of astronomical and geodetic problems, and equally of the approach that he brought to them. The first stage, the discovery of the laws themselves, had pertained to Newton for the most part, and the second, their analytic formulation, to the Bernoullis, Euler, Clairaut, and d'Alembert. Laplace's was a third stage, that of deriving interactions and interrelations, and reducing or extending the scale, between macrocosmic and microcosmic levels. Even so, the gravitational explanation of the tides belonged to Newton and its subsequent formulation in terms of analytical mechanics to Laplace's predecessors. He cut into the theory by attacking the anomaly of the near equality of consecutive high tides and proceeded to derive from its resolution a relation to precession and nutation. When he corrected predecessors, it was seldom, if ever, over physical data: It was because their analyses did not account for the data. Thus, the existing theory of precession and nutation did not give the wrong result. But it ought to have done so, and discovering why it did not led Laplace to an unsuspected and higher-order relation between the two axial oscillations.

Here, as throughout, his own analyses bespeak an enormous fund of physical knowledge and acuteness of insight into the body of existing theory. He saw into the physical situation for consequences—for example, the asymmetrical gravitational pull on the seas in the two hemispheres—that others either had not noticed or had not been able to calculate. Then he would calculate them, in memoirs that run, like this one, into many pages of differentiation, integration, and approximation. Did he see how to make an analysis come out before deciding which phenomena to handle—or the other way about? That question cannot be answered in the present state of scholarship, and probably the alternatives are two sides of the same coin. However that may have worked, one remark may be ventured with a certain confidence. His instinct for what was true and interesting in a set of physical relations was prompted by its accessibility to the kind of analysis in which he was a virtuoso. His knowledge of physics itself, however, was that of an omnivorous reader, an *érudit*, rather than an inquirer or discoverer.

A further feature of Laplace's mature technique first became fully apparent in these *recherches*, and that is the ingenuity with which he formulated expressions to contain terms representing the separate elements of a composite cycle of physical events. In this instance, he handled serially the several cycles of oscillation—annual, diurnal, and semidiurnal—that, when taken together, produce the gross ebb and flow of the tides. Laplace himself did not use the term harmonic theory that has since been applied to the type of tidal analysis that he started here. The phrase is apt, however. In effect, he did treat the determination of the location of a molecule of seawater relative to its equilibrium position like that of the displacement of a point on a string vibrating so as to produce a beat of tones and overtones. The actual position at any moment is the linear resultant of the respective displacements occasioned in consequence of oscillations of several frequencies.

The technique appears to good advantage in the second installment of this series, submitted to the Academy on 7 October 1778, almost a year after the first. There Laplace went over the same material just summarized, cleaning up and ordering the analysis that he had spun out more or less as it came to him the first time around. Perhaps it will be useful to single out the salient features as they pertain to the elements of tidal oscillation. He set up the problem so as to consider the displacements of a fluid molecule, M, on the surface of the sea. The following are the parameters at the outset:[11]

[11] [1779a], *OC*, **9**, pp. 187–88.

θ Complement of the latitude

ω Longitude taken from a meridian fixed in space

(After time t, θ becomes $\theta + \theta u$; and ω becomes $\omega + nt + \alpha v$, where nt represents the rotational motion of the earth, and α is a very small coefficient)

αy Elevation of M above the surface in the equilibrium state it would have reached without the action of the sun and moon

$\alpha \delta B$ Components of the attraction for M of an aqueous spheroid of

$\alpha \delta C$ radius $(1 + \alpha y)$, decomposed perpendicularly to the radius in the planes of the meridian and to the parallel of latitude

δ Density of seawater

s Mass of the attracting star

v Complement of its declination

ϕ Longitude reckoned on the equator from the fixed meridian

h Distance from the center of the earth, the minor semiaxis of which is taken for unity, and we let $3s/2h^3 = K$

g Gravitational constant

$l\tau$ Depth of the sea, l being very small, and τ being any function of θ.

With these quantities, Laplace restated in terms of polar coordinates three equations that he had formulated in his initial studies of the motion of M, obtaining:[12]

$$y = -\frac{l}{\sin\theta}\frac{\partial(uy\sin\theta)}{\partial\theta} - l\gamma\frac{\partial v}{\partial\omega}, \qquad (32)$$

$$\frac{d^2u}{dt^2} - 2n\frac{dv}{dt}\sin\theta\cos\theta = -g\frac{\partial y}{\partial\theta} + \delta B + \frac{\partial R}{\partial\theta}, \qquad (33)$$

$$\frac{d^2v}{dt^2}\sin^2\theta + 2n\frac{du}{dt}\sin\theta\cos\theta = -g\frac{\partial y}{\partial\omega} + \delta C\sin\theta + \frac{\partial R}{\partial\omega}, \qquad (34)$$

where

$$R = K[\cos\theta\cos v + \sin\theta\sin v\cos(\phi - nt - \omega)]^2. \qquad (35)$$

R is here a force acting in a given direction to disturb the equilibrium of the particle M. Its derivatives are taken with respect to θ and ω, and the terms K, v, and ϕ are given by the law of motion of the attracting star as functions of the time t.[13]

---

[12] Ibid., p. 188.

[13] Ibid., Articles III, XXII, XXIV, *OC*, **9**, pp. 95, 187–89, 198–99. Cf. *Mécanique céleste*, Book IV, Nos. 1–4, *OC*, **2**, pp. 184–95.

Equations (33), (34), and (35) were then combined and transformed by various manipulations into a single equation in rectangular coordinates, setting

$$y = a \cos(it + s\omega + A)$$

$$y' = a' \cos(it + s\omega + A)$$

$$u' = b \cos(it + s\omega + A)$$

$$v = c \sin(it + s\omega + A),$$

(36)

where i and s are any constant coefficients, and a, a', b, and c are functions of θ. Then, when $\sin \theta = 0$, the following equation comprises for Laplace the entire theory of the tides:

$$
ix^3 a(i^2 - 4n^2 + 4n^2 x^2)^2
$$

$$
= -lgzix^2(1 - x^2)(i^2 - 4n^2 + 4n^2 x^2)\frac{\partial^2 a'}{\partial x^2}
$$

$$
+ lgzix^3 \frac{\partial a'}{\partial x}(i^2 + 4n^2 - 4n^2 x^2)
$$

$$
+ lgsza'[(is + 2n)(i^2 - 4n^2 + 4n^2 x^2) + 16n^3 x^2(1 - x^2)]
$$

$$
- lg\frac{\partial z}{\partial x}x(1 - x^2)(i^2 - 4n^2 + 4n^2 x^2)\left(ix\frac{\partial a'}{\partial x} + 2nsa'\right). \quad (37)
$$

Fortunately, it did not have to be solved; it needed only to be satisfied, and that could be done piecemeal, introducing simplifications by dint of reasonable suppositions about the physical factors at work—as to equilibrium, fluid friction, resistance, and so on—and determining the coefficients i and s, and the function a' in terms of a.

To accomplish this task, Laplace grouped the different terms of the expression for R (Eq. 35) according to the cycle that they represent in the total tidal flow. Thus,

$$
R = K[\cos \theta \cos v + \sin \theta \sin v \cos(\phi - nt - \omega)]^2
$$

$$
= K \cos^2 v + \tfrac{1}{2}K \sin^2 \theta(\sin^2 v - 2\cos^2 v)
$$

$$
+ 2K \sin \theta \cos \theta \sin v \cos v \cos(nt + \omega - \phi)
$$

$$
+ \tfrac{1}{2}K \sin^2 \theta \sin^2 v \cos(2nt + 2\omega - 2\phi). \quad (38)
$$

Developing the values for R, Laplace obtained, in place of the three ranks of terms above, three series of the respective forms and periodicities

$$(1°) \quad K' + K'' \sin^2 \theta \cos(it + A),$$

$$(2°) \quad K' \sin \theta \cos \theta \cos(it + \omega + A), \tag{39}$$

$$(3°) \quad K' \sin^2 \theta \cos(it + 2\omega + A),$$

where K', K'', and A are any constant coefficients. These three classes of the expressions for R, and consequently for y, group terms in the following manner:

- (1°) includes terms that are independent of ω. Their period is thus proportional to the time of revolution of the star in its orbit—annual in the case of the sun;
- (2°) includes terms of which the period is approximately one day;
- (3°) includes terms of which the period is half a day.

By the conditions of the problem, i was small compared to n in terms of the first form, and if the orbit is assumed circular, $i = 0$ or $2m$, m being the mean motion. In terms of the second form, i differs very little from n, and this difference is 0 or $2m$ if the orbit is circular. In terms of the third form, i is very little different from $2n$, which difference is again 0 or $2m$ in the case of circular orbits.

Examining now the terms of the first class (1°) of Equation (39), which has the form

$$(K' + K'' \sin^2 \theta) \cos(it + A), \tag{40}$$

the condition that they govern the annual revolution gives values for the coefficients that need to be determined in order to satisfy Equation (37), such that, for this part of the oscillation,

$$y = \frac{K\left(\cos^2 v - \dfrac{1}{2} \sin^2 v\right)}{6g\left(1 - \dfrac{3\delta}{5\delta'}\right)} (1 + 3 \cos 2\theta), \tag{41}$$

where δ' is the mean density of the earth, and $g = (4/3)\pi\delta'$. The expression is exact for the action of the sun, although less so for the moon, since the waters have less time to return to equilibrium in a month than in a year. But any errors are of little importance in the theory of the tides, since they affect only the absolute heights relative to the phases of the moon, and not the difference between high and low tide.

The depth of the sea figures in this analysis, its expression being $(l + q \sin^2 \theta)$, then $z = x + (q/l)x^3$ in rectangular coordinates, q being a very small constant coefficient of the order of l. This value is important for two reasons. Physically, the equilibrium considerations require that the mass of the seawater remain constant. Analytically—and this is more important, although it comes in rather unobtrusively—the determination of the values of a and a' to be substituted in Equation (37) depended on assuming that q is approximately zero, and that the depth of the sea is constant. "This method," Laplace observed, "can thus serve to find the approximate values of a, given this hypothesis concerning the depth [of the sea], which we shall see in the following article is approximately that of nature."[14]

It is the terms of the second class (2°) of Equation (39) that are the most important for the inversion of the tides, their form being

$$K' \sin \theta \cos \theta \cos(it + \omega + A), \qquad (42)$$

and their period approximately a single day. Here the conditions are such that analysis yields the expression for the part of y corresponding to these terms

$$\frac{2q}{2qg\left(1 - \dfrac{3\delta}{5\delta'}\right) - n^2} K' \sin \theta \cos \theta \cos(it + \omega + A). \qquad (43)$$

And now we begin to see the reason for all this. Laplace next sums all terms of this second form given by the development of R and designates by Y the corresponding expression for y:

$$Y = \frac{4Kq \sin \nu \cos \nu \sin \theta \cos \theta}{2qg\left(1 - \dfrac{3\delta}{5\delta'}\right) - n^2} \cos(nt + \omega - \phi). \qquad (44)$$

The value of Y will be a maximum when $nt + \omega - \phi = 0$ or π, which is to say when the star that is the origin of the attracting force passes the meridian. The expression has a negative value when $nt + \omega - \phi = \pi$; and the difference between the maximum positive and negative values, which equals twice the former, will measure the difference between the two full tides of the same day. That difference will be proportional to the unwieldy coefficient of $\cos(nt + \omega - \phi)$ in Equation (44). Since observation shows this difference to be extremely small, the value of q must be nil or almost nil. Finally, when the depth of the sea is

[14] [1779a], *OC*, **9**, p. 202.

$l + q \sin^2\theta$, it follows that "in order to satisfy the phenomena of the ebb and flow of the tides, this depth must be approximately constant." That is, it must be l, which result agrees with what Laplace had already found. Thus he could conclude—"without fear of observable error"—that $Y = 0$, and determine the values needed to satisfy Equation (37).[15]

Laplace went on to set out verbally what he had in mind as a physical picture. For this purpose, he designated by **u** and **U** (where boldface does not signify a vector) the parts of u and v that correspond to the terms of the second class for the expression of R. It will be recalled that u and v are the variations respectively of θ, the complement of latitude, and of ω, the longitude. Thus defined,

$$\mathbf{u} = \frac{2K}{n^2} \sin v \cos v \cos(nt + \omega - \phi), \qquad (45)$$

and

$$\mathbf{U} = -\frac{2K}{n^2} \sin v \cos v \frac{\cos \theta}{\sin \theta} \sin(nt + \omega - \phi). \qquad (46)$$

Confining attention now to expressions of the form

$$2K \sin v \cos v \sin \theta \cos \theta(nt + \omega - \phi), \qquad (47)$$

it is, writes Laplace, "easy to see" that the molecules will slide over each other as if isolated, with no detectable loss from internal collisions in the system.[16] We are to imagine a slice of this fluidity contained between two meridians and two parallels infinitely close together. Since the value of **u** is constant for all molecules located under the same meridian, the length of the slice remains constant in that direction. To the extent that the fluid flows toward the equator, however, the space enclosed between the two meridians increases. Since the slice thus widens along the parallels, the surface of the water would tend to descend if it were not for the latitudinal component of the velocity of the molecules, which tends to squeeze the two meridians together and correspondingly diminish the width. Again "it is easy," we are assured, to conclude that this diminution, resulting from the value for **U**, is compensated by the motion of the slice toward the equator. Therefore, the width actually remains constant along the parallels, and the height of the slice is not affected detectably by the motion of the fluid. That is

[15] Ibid., *MARS* (1776/1779), pp. 199–200. It is preferable to refer to the original printing for these passages, since their republication in *OC* modifies the notation.

[16] Ibid., pp. 200–202.

the reason why in a sea of uniform depth (such as would be constituted by the sum of all such slices taken side by side) the difference between consecutive high tides is almost undetectable.

Now then, whatever may be thought of this argument, it can have referred only to the notion that Laplace had actually formed of physical reality. He was thus enlarging on the phenomenon, he said, because it was very important for the understanding of the tides and entirely contrary to received theory. He went on to illustrate the argument by data drawn from observations made at Brest and turned it against Daniel Bernoulli's idea that the earth rotates too rapidly for the tides to differ conformably with the results of theory. He himself had just shown that whatever the speed of rotation, consecutive high tides would be very unequal if the sea were not everywhere of approximately the same depth. "It seems to me," he concluded, "to result from these considerations that only an explanation founded on a rigorous calculation like this one ... is capable of meeting all the objections that can be lodged against the principle of universal gravitation on this score."[17]

Terms of the form $K' \sin^2 \theta \cos(it + 2\omega + A)$ constitute the third class (3°) of Equation (39) produced by the development of R. Their period is half a day, or that of a single tidal cycle. Examining these terms, Laplace substituted values for the observed tidal range, or difference between low and high tides, in various localities and determined that deriving the law of gravity from the data required assigning the figure of four leagues to the mean depth of the sea "on which only vague and uncertain conjectures have been formed until now."[18] We shall not excerpt that analysis, which is of the same type as the foregoing, nor summarize the calculation that exhibits the proportionality of the slight difference between daily high tides to the influence of the tides on the ratio of nutation to precession. We shall also pass over the third installment of this memoir, in which Laplace calculated much more succinctly that the gravitational effect of the sun and moon on the atmosphere is bound to cause oscillations comparable to tides but too slight to detect.[19] More important, he redeemed the promise of an earlier memoir[20] and showed that these effects cannot be the source of the trade winds, recognizing that oversimplified assumptions, most notably that of constant temperature, reduced his results to a qualitative significance.

---

[17] Ibid., *OC*, **9**, p. 214.
[18] Ibid., p. 216.
[19] See Chapter 28.
[20] [1776a, 2°]; see Chapter 4.

Instead, let us conclude this discussion of Laplace's youthful period with a further memoir, "Sur la précession des équinoxes," which he read before the Academy on 18 August 1779. It opens with a simplified derivation of the results of his previous research on the relation of tidal action to precession and nutation, the consequence of which is now stated as a theorem:[21]

> If the earth be supposed an ellipsoid of revolution covered by the sea, the fluidity of the water in no way interferes with the attraction of the sun and the moon on precession and nutation, so that this effect is just the same as if the sea formed a solid mass with the earth.

This theorem might well have been thought to hold true for any spheroid, observed Laplace, although he had been unable to demonstrate its generality by his previous analysis. The expressions that he formed there to represent the influence of the oscillations of the sea on precession could not be integrated generally. In the meanwhile, he had found a much simpler method, which formed the subject of the present memoir and which met this difficulty. Essentially it consisted of nothing other than the application to the problem of d'Arcy's principle of areas, to which Laplace had also had recourse in analyzing the gravitational interaction of Saturn and Jupiter.[22] Unlike the other basic principles, particularly *vis viva* conservation or least action, which are limited to gradual and continuous changes of motion, d'Arcy's principle holds good in cases of shock or turbulence, such as friction with the bed of the sea and resistance from the coasts in tidal movements. We shall not follow the analysis, except to note its crux, which was that the expression of the variation of the overall motion of the fluids covering a globe contains no terms dependent on the secular passage of time. Thus, both for mechanical and formal reasons, the new method might be extended to what Laplace now, not even a year later, called the case in nature, "in which the figure of the earth and the depth of the sea are very irregular, and the oscillations of the water are modified by a vast number of obstacles."[23]

Now that the seabed had suddenly—and one is tempted to say analytically—become very irregular instead of uniform, Laplace was led to speculate that the spinning earth may not be perfectly symmetrical. In that event, the resulting irregularities in angular momentum, added

---

[21] [1780c], *OC*, **9**, pp. 341–42.
[22] See Chapter 4.
[23] [1780c], *OC*, **9**, p. 342.

to the direct gravitational effects of the sun and the moon and the reaction of the seas to their tidal pull, might change the axis of rotation over very long periods. The poles would then migrate into other regions in the course of time—but this he would leave as a conjecture, worthy of the attention of mathematicians because of its difficulty and importance.

Part II

LAPLACE IN HIS PRIME, 1778–1789

CHAPTER 8

Influence and Reputation

BY THE LATE 1770s Laplace had begun to win a reputation extending beyond the small circle of mathematicians who could understand his work, and by the late 1780s he was recognized as one of the leading figures of the Academy. On 15 May 1788 he was married to Marie-Charlotte de Courty de Romanges, of a Besançon family.[1] She was twenty years younger than he, and they had two children. Laplace's son, Charles-Émile, born in 1789, followed a military career, became a general, and died without issue in 1874. The daughter, Sophie-Suzanne, married the marquis de Portes and died in childbirth in 1813. Her child, a girl, survived and married the comte de Colbert Chabannais. The living descendants derive from that marriage, having taken the name Colbert-Laplace.

We know practically nothing of Laplace's personal life in the years before his marriage, but there are a few indications of his effect on others. Not a single testimonial bespeaking congeniality survives. There are hints that the aged d'Alembert began to resent the regularity with which his recent protégé was relegating his own work to the history of rational mechanics. Overly elaborate tributes by Condorcet and Laplace himself bear the scent of mollification, notably in the prefaces to the "Recherches sur plusieurs points du système du monde," the memoir discussed in the last section. Anders Johan Lexell, a Swedish astronomer who spent part of the winter of 1780–1781 in Paris, wrote in a gossipy account of the Academy that Laplace let it be known that he considered himself the best mathematician in France. (That he was right would have done little to disarm resentment.) He also had extensive knowledge in other sciences, reported Lexell, but presumed too far upon it, "for in the Academy he wanted to pronounce on everything."[2] At about the same time, Laplace fell into a heated dispute with Jacques-Pierre Brissot, the future Girondist leader and a rising scribbler, over the optical experiments of Jean-Paul Marat, who was then seeking scientific recognition and entry to the Academy. In Brissot's dialogue on "academic prejudice" Laplace is the original of the Newtonian idolater, all arrogant in his *fauteuil* and contemptuously spurning

---

[1] Marmottan (1897), p. 7.
[2] Birembaut (1957).

the aspiring "physicist" from the impregnable—and in Brissot's view irrelevant—plane of mathematics.[3]

In 1784 the government appointed Laplace to succeed Bezout as examiner of cadets for the Royal Artillery.[4] The candidates had generally completed secondary school and a year or two of special preparatory school before going on to La Fère or one of the other artillery schools, or in some cases to Mézières for engineering. The responsibility was more serious than might at first glance be supposed. An individual report had to be written on each cadet, all of whom were of good family. The annual scrutiny brought Laplace, as it also did Gaspard Monge, the newly appointed examiner for naval cadets, into regular contact with ministers and high officers; and it introduced them to the practice of recruiting an elite by competitive examination that was later greatly expanded in scale and intensified in mathematical content by the procedures for selecting students to enter the École Polytechnique, founded in 1794. Certain of Laplace's reports survive in the Archives de la Guerre at Vincennes.[5] The government also named him to the most famous of the select commissions through which the Academy investigated and made recommendations on matters of civic concern in the last years of the Old Regime. Laplace was a member of the commission headed by Bailly to investigate the Hôtel Dieu, the major hospital in Paris, as well as hospital care in general. Calculations on the relative probabilities of emerging alive from its wards, and comparisons of the mortality there to other hospitals in France and abroad, must almost certainly have been his contribution to its report.[6]

Laplace was promoted to the senior rank of pensioner in the section of Mechanics on the occasion of the reorganization of the Academy in April 1785. In this, technically the most proficient and productive period of his life, he pressed forward with all the topics that he had begun investigating in his youth, added physics to them, and achieved many of the results for which he is famous. From the published record, it appears that in the early 1780s his emphasis shifted from the problems of attraction, which were occupying him in 1777 and 1778, to probability again, both in its calculus and now in its application to demography, and equally to the experimental and mathematical physics of heat (24, 26,

---

[3] Brissot (1782), p. 335. On Marat's scientific pretensions, see Gillispie (1980), pp. 290–330, and on his life and politics in general, Coquard (1993).

[4] Duveen and Hahn (1957).

[5] Reports on Laplace's examinations for 1784, 1785, and 1786, and his recommendations for reform of the system in July and August 1789, are conserved in the Archives de la Guerre at Vincennes, X^{D}249.

[6] HARS (1785/1788), pp. 44–50. On this and other commissions, see Gillispie (1980), pp. 244–56.

28). In the mid-1780s his interest again centered on attraction and the figure of the earth, and in the latter half of the decade it turned to planetary motions. It will be best to consider these subjects in that sequence, beginning with a pair of relatively brief but important mathematical memoirs and continuing into probability. These dictates of convenience, however, must not obscure the evidence that, to an extraordinary degree, Laplace was able to hold and mature all these matters in his mind concurrently, readily turning from one to another. For even while occupied with probability about 1780, and beginning his collaboration with Lavoisier on the physics of heat, he composed the paper on precession already noticed and in 1781 drafted another, very important, memoir on the determination of cometary orbits, to be discussed in Chapter 13.

CHAPTER 9

Variation of Constants,
Differential Operators

THE EARLIER of the two mathematical papers that preceded Laplace's resumption of probability pertained to astronomy and was entitled "Mémoire sur l'intégration des équations différentielles par approximation" [1780a]. Laplace there simplified the technique of varying arbitrary constants in approximate solutions to differential equations of planetary motion in order to eliminate the troublesome secular terms that crept in and destroyed convergence. Having brought out the method by means of numerous examples in a very lengthy memoir,[1] he now gave rules for applying it generally in problems of theoretical astronomy. Laplace's technical innovations were often thus introduced discursively and then shortened and pointed for restatement.

The latter paper, "Mémoire sur l'usage du calcul aux différences partielles dans la théorie des suites," was the more important mathematically [1780b]. It belongs to the nascent stages of the calculus of operations, as it was called in the nineteenth century, or, more recently, the calculus of differential operators. Laplace submitted it to the Academy on 16 June 1779 (29). Three years previously he had appended Articles XI and XII, "Sur les fonctions," to his memoir on the mean inclination of the comets. He there gave a general demonstration of a theorem that Lagrange had obtained by induction in a very important memoir of 1774.[2] Lagrange had then developed the analogy between positive exponents and indices of differentiation, and reciprocally between negative exponents and indices of integration, and stated the following theorem. We give it in the notation of Laplace, who unlike Lagrange considered only a single variable:

$$\Delta^n u = \left(e^{\alpha \frac{du}{dx}} - 1\right)^n \quad \text{and} \quad \Sigma^n u = \frac{1}{\left(e^{\alpha \frac{du}{dx}} - 1\right)^n}. \quad (48)$$

Uneasy about the legitimacy of Lagrange's analogies, which identified the power of differentiation with its order (i.e., $d^n u/dx^n = (du/dx)^n$),

---

[1] [1776c], Chapter 6.

[2] [1776b]. The Lagrange memoir was "Sur une nouvelle espèce de calcul relatif à la différentiation et à l'intégration des quantités variables," *OL*, **3**, pp. 441–76.

Laplace sought to re-prove Lagrange's results by means more in keeping with the usual methods of the differential calculus. Laplace's 1776 proof depends on showing that in the expression for the developments

$$\alpha^n \frac{d^n u}{dx^n} = \Delta^n u + s\Delta^{n+1}u + s'\Delta^{n+2}u + \cdots \qquad (49)$$

and

$$\frac{1}{\alpha^n} \int^n u \, dx^n = \Sigma^n u + f\Sigma^{n-1}u + f'\Sigma^{n-2}u + \cdots \qquad (50)$$

the coefficients s, s',..., f, f',... are constant and independent of α, thus depending only on n.[3] Lagrange's expressions are obtained if $u = e^x$, and the coefficients can thus be determined by the choice of the function u. Now, in the fullness of an entire memoir, Laplace dealt more generally with the possibility of moving back and forth between powers and indices of differentiation and integration by more orthodox means, giving several alternative demonstrations of the Lagrange theorems and drawing out corollaries [1780b].

According to his own testimony, this research was one of the factors leading to the development of his theory of generating functions, which followed immediately after the memoir on probability that he submitted to the Academy on 31 May 1780.[4] The draft was entitled "Mémoire sur le calcul aux suites appliqué aux probabilités," but as will appear in the next chapter, the treatment is far more comprehensive than that phrase would imply.[5] For with principles or motifs, Laplace's pattern contrasts with that just noticed in the development that he often gave his technical innovations. Whereas the latter were abbreviated and focused after a lengthy initial statement, whole topics such as universal gravitation or inverse probability were first mentioned almost in passing and then, as the ideas continued germinating in his mind, were broadened to embrace entire domains of science and knowledge.

[3] [1776b] *OC*, **8**, pp. 314–19.
[4] [1782a] *OC*, **10**, p. 2.
[5] [1781a], (33).

CHAPTER 10

Probability Matured

THE VERY TITLE OF "Mémoire sur les probabilités" indicates that by July 1780, when Laplace presented the paper to the Academy, he had largely completed the transition in his thinking that led from theory of chance to theory of probability, with his focus on inverse probability.[1] In a letter to Lagrange of 11 August 1780 he wrote that the principal object of this memoir was the "method of going behind events to causes."[2] The preamble is very general. He does not say so, but he had evidently read Bayes by this time, and he proceeds to review the entire status of inverse probability in a Bayesian sense. The approach pertains to what he now calls a "very delicate metaphysics," the use of which is indispensable if the theory of probability is to be applied to life in society. He proposes to treat two distinct but closely related aspects of probability, neither of which had been adequately considered in theory of chance. In the first, the task is to calculate the probability of composite events compounded from elementary events of which the respective possibilities, or (in his terms) prior probabilities, are unknown. In the second, the goal is to determine numerically the influence exerted by past events on the probability of future ones or, as would now be said, to draw statistical inferences. In Laplace's mind, the goal was to uncover the law that reveals the causes.

Before discussing problems of the former class, Laplace refined his epistemology slightly while distinguishing his procedure from previous practice in the theory of chance. In the traditional approach, which assumed the prior probabilities to be given, the respective possibilities of simple events, such as the chance of heads or tails in tossing a coin, had been determined in one of three ways: (1) a priori, by the assumption of equal possibility; (2) a posteriori, by repeated experiment; and (3) by whatever reason there might be to judge the likelihood of their occurrence. In the case of two players, A and B, the calculator in

---

[1] [1781a]. Laplace read a draft under the title "Mémoire sur les suites appliquées au calcul des probabilités" on 31 May 1780 and submitted the final version on 31 July (31). It is illuminating to consult the summary of this memoir that Condorcet prepared in his official capacity of Permanent Secretary and his private role of competitor-colleague, *HARS* (1778–1781), pp. 43–46, reprinted in Bru and Crépel (1994), pp. 159–61.

[2] Laplace to Lagrange, 11 August 1780, *OL*, **14**, p. 95.

approach (1) assumes their skills to be equal; in approach (2) he observes the outcomes of a number of matches; in approach (3) one player is the more adept, but the observer has no grounds for judging which it is, so that in his eyes the bets are even. The a priori assumption (1) gives what Laplace here calls the absolute possibility of simple events. It might now be described as a physical, or logical, probability. Approach (2), as Laplace will argue later, gives the approximate possibility. Approach (3) gives their possibility relative to our knowledge.

Since every event is actually determined by the general laws of the universe and is probable only relative to us, Laplace acknowledged that the distinction between approaches (1) and (3), absolute and relative possibility, may appear to be imaginary. Not so, however. Amid all the factors that produce a composite event, some are different every time, such as the precise movements of the hand in throwing dice. The overall effect of these factors is what we call chance. Other factors are invariant, such as the relative skill of players or the weighting of dice. Taken together, these are what constitute absolute possibility (to which he intended to assimilate approximate possibility, although he did not say so at the outset). It is our greater or lesser knowledge of invariant factors that constitutes "relative possibility." Invariant factors do not suffice to produce the event. They must be joined to the operation of the chance or variable factors first mentioned and only increase the probability of events without determining their occurrence. Thus, at this juncture Laplace considered that the state of knowledge enters into the determination of inverse probability at two levels: in what we know (relative possibility) of the invariant factors (absolute possibility) in simple events, and in our inevitable ignorance of the laws governing events that appear to be subject to chance.

All previous mathematicians had presupposed knowledge of the absolute possibilities, or prior probabilities, in their calculations. Except for the few paragraphs he had devoted in his earliest memoirs to the effect of real but unknown asymmetries in coins and dice, no one had considered, as he now proposed to do, cases in which all we have is relative possibility. Yet many important questions, and most problems in theory of games, are of this sort. Outcomes will be very different depending on whether absolute or relative possibility of simple events (for example, equal or unequal possibility of heads or tails) figures in the calculation.

Laplace began the first main topic, the probability of complex events composed of simple events whose possibility is unknown, by generalizing the treatment that he had already given in the 1774 memoir on probability of cause to what he now calls relative possibility. For slightly asymmetric coins he substitutes two players A and B of differing skills such that $(1 + \alpha)/2$ represents the probability that the stronger will win a move and $(1 - \alpha)/2$ the probability that it will be the weaker. We do not know which is which, but in either case after n moves the probability P that A will win equals

$$\frac{1}{2^{n+1}}[(1 + \alpha)^n + (1 - \alpha)^n].$$

Clearly that is not the same as the probability $1/2^n$, which follows from the assumption that ignorance of the relative skills is equivalent to equipossibility.

Laplace developed the consequences of the supposition of unequal skills, which is to say prior probabilities, in other two-person games and proceeded to generalize the analysis to cases involving any number of players. All that he could conclude from these examples was that the outcome would be more favorable for a particular player than it would be according to the ordinary calculus based on equipossibility. To estimate the amount of the improvement was impossible. If, however, the limits and law of possibility (or prior distribution) of α were known, nothing, Laplace wrote, would be easier than to solve the problem exactly.

As elsewhere, Laplace's idea of easiness is not the reader's. The ensuing calculations were an onerous affair. Allowing q to be the upper limit, he represented the probability of α by the function $\psi(\alpha)$, which since α falls between q and 0 is such that

$$\int d\alpha \, \psi(\alpha) = 1,$$

the integral being taken from $\alpha = 0$ to $\alpha = 1$. The value of the probability P as determined above is thus given by

$$\int d\alpha \, \frac{\psi(\alpha)}{2^{n+1}}[(1 + \alpha)^n + (1 - \alpha)^n],$$

when the integral is taken between $\alpha = 0$ and $\alpha = q$.

The quantity α is a function of the ratio of the skills of the two players. Rather than suppose its a priori distribution known, Laplace set out to derive it from a comparison of the skills of all possible players to that of a single player taken as unity. He imagined a probability curve but did not draw it, as he had done in considering the minimization of error in a series of astronomical observations in the 1774 memoir. The abscissa x represents the ratios of skill. Ordinates y proportional to the number of players whose skill is x are erected at each point of the abscissa. The resulting curve is enclosed between limits h and h' representing the least and greatest skills respectively. The sum of all ordinates, which is to say the area of the curve, then represents the probability that the skill of a given player is between h and h'.

In order to determine the prior distribution of possible values of α, the function being $y = \phi(x)$, its integral, $\int dx\, \phi(x)$ taken from $x = h$ to $x = h'$, is designated a. Further, if x is the skill of the weaker of the two players A and B, and $x + u$ the skill of the stronger, then

$$\frac{x}{x + u} = \frac{1 - \alpha}{1 + \alpha},$$

which when rearranged is

$$x + u = \frac{x(1 + \alpha)}{1 - \alpha}.$$

The probability that when the skill of one player is x, that of the other will be $x + u$, is equal and therefore double the probabilities of x and of $x + u$. It will thus be equal to

$$\frac{2\phi(x)\phi(x + u)}{a^2} = \frac{2\phi(x)\phi\left(\dfrac{1 + \alpha}{1 - \alpha}x\right)}{a^2};$$

giving

$$2\int dx \frac{\phi(x)\phi\left(\dfrac{1 + \alpha}{1 - \alpha}x\right)}{a^2}$$

for the prior probability of α when the integral is taken from $x = h$ to $x = h'(1 - \alpha)/(1 + \alpha)$. As for the limit q of α, it will equal $(h' - h)/(h' + h)$.

These were ungainly expressions. When the function $\phi(x)$ was unknown, moreover, the outcome for the two players A and B could not be calculated exactly, and Laplace proceeded to develop a general method for determining the outcome for any number of players when nothing is known of their skills except the law of their possibility, that is, the prior distribution. The difficulties of analysis were, Laplace now admitted, considerable. He addressed them in the form of a solution to the following problem: "Let there be n variable positive quantities $t, t^1, t^2, \ldots, t^{n-1}$, of which the sum is s, find the sum of the products of each value for a given function $\psi(t, t^1, t^2, \ldots)$ multiplied by the probability corresponding to that value."[3] Laplace treated the problem he here posed himself with great generality and did not assume either equipossibility or continuity in the prior probability distribution.

The simplest cases in which he could employ the very cumbersome formulas he obtained were to the probability that the sum of errors in a given number of observations would fall within certain limits when the probability of error is known for a single observation. The solution was applicable, to Laplace's evident pleasure, to the topic he had already treated in another manner, determination of the probability that the orbits of any number of comets is contained within given limits (Chapter 5). In Stephen Stigler's opinion, the importance of finding the mean in a series of astronomical observations may have been what had attracted Laplace to probability in the first place.[4] It may well be so. The statement of the above methodological problem itself would appear to be a generalization of his procedure in defining the probabilistic mean in the 1774 memoir, where he supposed that the greater the error the lower its probability but did not place limits on its degree. Now he does. Indeed, the systematic assignment of limiting values to prior probability distribution was Laplace's central strategy in deploying the examples he chose from games of chance, cometary orbits, errors of observation, and births of boys and girls throughout the memoir. As in many other instances, it was an investigation by Lagrange that suggested the approach. Laplace acknowledged that Lagrange had already resolved the problem of finding the probability that the sum of errors of observation will be contained within given limits when the prior possibilities are known, but he thought his own method to be more direct and more general.[5]

Turning from error back to games, Laplace next considers the skills of $(n - 1)$ players as the variable quantities of his general methodological problem. In order to simplify the analysis, he supposes that the law of possibility of skills is the same for all players. What that law actually is, however, can be determined only by a long series of observations that it is normally impractical to perform, and there is nothing for it but to choose the likeliest function. Laplace makes no apology for that. "The analysis of chance," he remarked, "which is itself only the art of estimating likelihood, will guide us in the choice."[6] What it teaches is that the limits of skills from the least to the most successful player are

[3] Ibid., *OC*, **8**, p. 396.

[4] Stigler (1986a), p. 101.

[5] Ibid., *OC*, **9**, p. 407. Cf. Lagrange, "Mémoire sur l'utilité de la méthode de prendre le milieu entre les résultats de plusieurs observations," *OL*, **2**, pp. 173–234, originally published in *Miscellanea Taurinensa*, **5** (1776).

[6] Ibid,. *OC*, **8**, p. 409.

much easier to determine by observation than is their distribution. It is likely, Laplace argues, that the probabilities of skill increase from either limit to a mean value and that the distribution is the same on both sides. Again Laplace imagines, but does not draw, a probability curve:

> Let $2a$ be the interval contained between the two limits and x the distance from the midpoint of this interval to any point on either side. If an ordinate y representing the probability of x is erected at that point, a curve will be obtained bounded between two limits. Since the value of x falls within those limits, the surface of the curve will equal unity, so that from the mean to either limit the surface will be $1/2$. The quantity $1/2$ is to be considered as divided into an infinite number of equal parts distributed above the interval a. By the conditions of the problem, this distribution is such that there are fewer parts above each point in proportion to its distance from the mean. Since all the combinations in which that may occur are equally admissible, the mean ordinate for the abscissa x results from taking the sum of all the ordinates y relative to each combination and dividing it by the number of combinations.

For such conditions Laplace's resolution of the general problem yields

$$y = \frac{1}{2a} \log \frac{a}{x}. \tag{51}$$

"Such is the equation," he says in some triumph, "to be used when we have no data for the possibility of values of x except that the larger they are the lower it is. Just that, however, is the situation in a great number of circumstances."[7]

Laplace immediately adduced his favorite illustration of determining the true instant of a phenomenon recorded by a number of observers. Instead of rejecting far-out values as bad, each observer can readily assign the limits of error, both plus and minus, in his own series by taking half the interval (designated $2a$ in the above calculation) between the two at either extreme. Its magnitude depends on the skill of the observer, the quality of his instrument, and the precision with which such measurement can be made. It is to be supposed the same for all observers if there is no reason to think otherwise. It is natural, moreover, to assume that positive and negative errors are equally probable, and that the likelihood of incurring an error decreases with its size. In the absence of any other information about the likelihood of error, the case is the same as that of the previous problem, and Equation (51) is to be employed in seeking the best mean among the results of a series of observations.

[7] Ibid., pp. 410–13.

The argument so far supposes no reason to attribute to one player or observer greater skill or accuracy than to his fellows. In the consideration of games that is clearly true when play begins. As it continues, however, and one move follows another, evidence of the respective skills of the players accumulates, so much so that, as Laplace will demonstrate, they could be determined exactly after an infinite number of moves. More generally, he proceeds to argue, investigation of the causes of events turns on knowledge of their occurrence in ways that it is important to exhibit. Only when the absolute possibility of events is known a priori, such as in the abstract instance of throwing perfectly symmetrical dice, can it be said that the past has no influence on the future. Not so when prior possibilities are unknown, as is normally the case. In real world situations, experience of past events does influence estimates of the probability of future events.

These were the considerations that led Laplace to the second topic of "Mémoire sur les probabilités"—given the events, to determine the probability of causes. The development he gave it broke new ground in the field of application. Indeed, it can be argued that social statistics as a mathematical subject had its start right there. Laplace began the discussion by addressing himself to the first of two difficulties that impeded the application of Bayes's theorem to problems in the real world. (It is interesting that, although Laplace himself still ignored Bayes, whether in the French or English sense of the verb, Condorcet mentioned Bayes and Price in the summary of Laplace's memoir in the preface to the volume in which it appears, saying that they had stated the principle for determining causes from effects, but without any calculations.)[8] The first impediment that Laplace identified was practical: Experience was hardly ever extensive or controlled enough to yield reliable values for the a priori probabilities. The second impediment was analytical: In order to achieve numerical solutions, it was often necessary to integrate differential equations containing terms raised to very high powers.

To overcome the former, experiential difficulty, Laplace turned to the one subject on which statistically significant information had already been assembled—population. Population studies as a science owe much substance to the growing professionalism of eighteenth-century public administration. French parish registers in principle contained records of all births, marriages, and deaths. In 1771 the Controller General of Finance, the abbé Terray, instructed all intendants in the provinces to have the figures for their generalities compiled annually and to report

[8] *HARS* (1778/1781), p. 43. Bru and Crépel reprint Condorcet's "Compte-Rendu" (1994), pp. 159–61.

the results regularly to Paris in order that the government might have accurate information on the entire population.[9] In August 1774 the young Louis XVI appointed Turgot, a statesman close to the scientists in spirit and in program, to be Controller-General. Chief minister, in effect, he took it on himself to attempt to reform the regime at the dawn of the new reign. At Turgot's behest, the Academy of Science addressed itself to problems of population and published a summary of the figures for the city of Paris and the faubourgs covering the years 1709 through 1770.[10] The record showed that in the last twenty-five years of that period, 251,527 boys and 241,945 girls had been born. The ratio of approximately 105 to 101 remained virtually constant year by year. Data also existed for London, and there too, more boys than girls had been born, although in the slightly higher ratio of 19 to 18.

Unlike imaginary black and white balls drawn from a hypothetical urn, the births of real children afforded a genuine numerical example, and Laplace seized on the opportunity to try out his technique for determining the limits of probability of future events based on past experience. Given p male and q female babies, the probability of the birth of a boy is $p/(p + q)$. Laplace now invoked the limit theorem he had given in his original paper on probability of cause [1774c], arguing that if P designates the probability that the chance of the birth of a boy is contained within the limits $p/(p + q) + \theta$ and $p/(p + q) - \theta$, where θ is a very small quantity, the difference between P and unity would vary inversely as the value of p and q. The latter quantity could be increased so that the difference between P and unity would be less than any given magnitude. To that end, he represented the definite integral for P by a rapidly convergent series, which on evaluation reduced to unity when p and q became infinite.

The calculation may be paraphrased as follows in Laplace's notation.[11] If x is the probability of the birth of a boy, and $(1 - x)$ that of a girl, then to determine the probability that x will fall within arbitrary limits became a problem of evaluating between those limits the definite integral

$$\int x^p (1 - x)^q \, dx \qquad (52)$$

[9] On eighteenth-century population studies, see Brian (1994), pp. 153–92.

[10] Jean Morand, "Récapitulation des baptêmes. mariages, mortuaires & enfans trouvés de la ville et des faubourgs de Paris, depuis l'année 1709, jusqu'à & compris l'année 1770," *MARS* (1771/1774), pp. 830–48.

[11] [1781a], Article XVIII, *OC*, **9**, pp. 422–29.

taken from $x = 0$ to $x = 1$, where p and q are very large numbers. Letting

$$y = x^p(1 - x)^q, \tag{53}$$

it followed that

$$y \, dx = \frac{x(1 - x)}{p - (p + q)x} \, dy. \tag{54}$$

If we set $p = 1/\alpha$ and $q = \mu/\alpha$, α being a very small fraction since p and q are very large, then Equation (54) becomes

$$y \, dx = \alpha z \, dy, \tag{55}$$

where

$$z = \frac{x(1 - x)}{1 - (1 + \mu)x}. \tag{56}$$

Thus, whatever the value of z,

$$\int y \, dx = C + ayz \left\{ 1 - \alpha \frac{dz}{dx} + \alpha^2 \frac{d(z \, dz)}{dx^2} - \alpha^3 \frac{d[z \, d(z \, dz)]}{dx^3} + \cdots \right\}, \tag{57}$$

where C is an arbitrary constant depending on the initial value of $\int y \, dx$.

The series (57) would no longer be convergent if the denominator of z in Equation (56) were of the same order of magnitude as α, which would be the case when x differed from $1/(1 + \mu)$ by a quantity of that order. The series was to be employed, therefore, only when that difference was very large with respect to α. But even that did not suffice. Since each differentiation increased the powers of the denominators of z and its derivatives by one, the term of which the coefficient was α^i had for its denominator the term in z raised to the power $2i - 1$.

Thus, for the series to be convergent, α had to be not only much less than the denominator of z, but even much less than the square of the denominator. Under these conditions, the series (57) by very rapid approximations would give the value of the integral $\int y \, dx$ between the limits $x = 0$ and $z = 0$, and in that case, Equation (57) becomes

$$\int y\,dx = \frac{\alpha\mu^{q+1}[1 - (1 + \mu)\theta]^{p+1}\left(1 + \dfrac{1 + \mu}{\mu}\theta\right)^{q+1}}{\theta(1 + \mu)^{p+q+3}}$$

$$\times\left\{1 - \frac{\alpha\left[\mu + (1 + \mu)^2\theta\right]}{\theta^2(1 + \mu)^3} + \cdots\right\}. \tag{58}$$

That series gave the limits between which the value of $\int y\,dx$ was contained, a value less than the first term in the braces and greater than the sum of the first two terms.[12]

A similar demonstration shows that the series (57) also gave values of $\int y\,dx$ from $x = 1/(1 + \mu) + \theta$ to $x = 1$. In showing that such was the case, Laplace proved that the more p and q were increased, the more α was diminished, and that the difference between P and unity was proportional to α, so that by increasing p and q, and thus diminishing α, the difference could be reduced below any given magnitude.

It remained to draw upon the data. Calculating the probability that the possibility of a male birth in Paris was greater than 0.5, Laplace found it to be less than unity by the fraction 1.1521×10^{-42}. He further calculated that the probability that in any year baby boys would fail to exceed baby girls in number was $1/259$ in Paris, whereas in London it was $1/12{,}416$. Although the ratio of male to female births was only slightly higher in London, the probability of male preponderance approached certainty at a drastically increasing rate as the proportion grew. It would have been reasonable to bet that boys would outnumber girls in any of the next 179 years in Paris and in any of the next 8,605 in London.

Here, then, Laplace had worked out a method for approximating the value of definite integrals and thus finding numerical solutions to a type of problem for which the analytical solutions contained terms raised to such high powers that the expressions became impracticable when numbers were substituted in the formulas. Moreover, his method was applicable to practical calculation of future events given the experience of the past. In short, not only did he have a method for statistical inference in hypothetical circumstances, but he really drew inferences in the area of population.

To the modern reader the last three articles of "Mémoire sur les probabilités" may come as something of an anticlimax. Not so to

[12] Ibid., p. 425.

Laplace—he there returned to determination of the mean value among the results of a series of observations. It is, he says, one of the most useful problems pertaining to the probability of cause. Evidently he remained dissatisfied with the solution he had found in the 1774 memoir that had introduced probability of cause (Chapter 3). There, it will be recalled, the analytical difficulties limited its application to the case of three observations. In the meantime Lagrange published the memoir already mentioned (note 5) dealing with the probability distribution of the arithmetical mean among many observations. Laplace had had word of it in 1774. Its appearance in 1776 prompted him to take up mean values once again, this time in a lengthy paper that is important in the background of "Mémoire sur les probabilités." He never published this investigation but read it in two installments before the Academy of Science on 22 February and 8 March 1777.[13]

Though Laplace expressed admiration for Lagrange's virtuosity, as he always did, he also disagreed with the approach. The problem of determining a mean might, he observed at the outset, be regarded from two points of view, depending on whether the observations are considered before or after they are performed. Lagrange and all others who had addressed it undertook an analysis by direct probabilities in order to determine a priori the optimal value to be taken among a set of prospective observations. The alternative method was to consider the observations only after they have been made, taking into account the actual distances between them, and to proceed a posteriori, by probability of cause, in order to determine the mean value with the minimum probability of error. We see easily, Laplace said (a favorite phrase), that the two methods must lead to different results, "but it is at the same time obvious (*visible*) that the second is the only one that should be employed."[14]

The paper consists of a lengthy and laborious series of calculations to bring out why that should be obvious. He began by pointing out that the term "mean result" of a number of observations could be understood in an infinite number of different ways depending on the specified conditions. One might require that the sum of positive errors be equal to the sum of negative errors; or that the sum of positive errors each multiplied by its probability be equal to the sum of negative errors each multiplied by its probability; or that the mean be the point with the maximum probability of being the true point; and so on ad infinitum. The choice among the possibilities was not simply arbitrary, however.

[13] "Recherches sur le milieu qu'il faut choisir entre les résultats de plusieurs observations" (21), published by Gillispie (1979).

[14] Ibid., p. 229.

Only one condition conforms to the nature of the problem, namely that the time of the phenomenon be fixed at a point such that the resulting error in taking a mean be a minimum. The theory of games offered an analogy. Just as expectation might be evaluated by summing the products of each advantage multiplied by the probability of obtaining it, so error is to be estimated by summing the products of each error multiplied by the probability of incurring it. The best mean is the instant for which the sum of those products is a minimum.

Laplace repeated in simpler form the demonstration from his 1774 discussion that this, which he there called the astronomical mean, is the same as the probabilistic mean, defined as the value at which the odds are even that the true instant is before or after it. He had still to specify the corresponding error curve. For that purpose, he considered various cases in which different conditions are attached to the scale and possibility of error, and he proceeded in each instance to calculate the mean first by Lagrange's method of direct probabilities prior to the making of the observations, and then by inverse probabilities in his own posterior approach. Not surprisingly, the latter come out better. In each example Laplace exhibited the function by means of an error curve. He did not, however, arrive at a general curve to replace the expression, $\phi(x) = m/2\, e^{-mx}$, that he had found in his 1774 discussion and been able to evaluate only in the instance of three observations.[15] Instead, he plotted the logarithmic expression he did obtain by means of the following series of curves, one for the probability of each of a^{n-1} observations of the instant of the phenomenon (Figure 8).

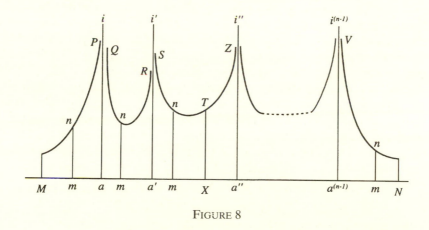

FIGURE 8

[15] See Chapter 3, Equation (12).

In this figure the line *MN* represents time. The points *a*, *a'*, *a"* designate the recorded instant of each observation. The lines *ai*, *a'i'*,... are asymptotes of the different parts of the curve. The line *XT* divides the area of the sum of the functions represented in two equal parts. The ungainliness of that result may have been the reason that this hard-fought paper remained unpublished.[16]

Meanwhile, Laplace charted his way by recourse to limiting values through the argument in "Mémoire sur les probabilités," which culminates in application of probability of cause to the problem of determining the mean value in a series of observations. Article XXX repeats almost word for word the discussion from the 1777 draft leading up to definition of the mean that minimizes the probability of error.[17] The only difference is that Laplace now considers observations of the same phenomenon by a number of different observers (players in the astronomical game, perhaps) rather than a series by the same person. He then gives a more economical demonstration than he had previously done of its equivalence to the probabilistic mean. The laborious calculations and graphic representations from the 1777 draft are not included. They must have found their reward, after Laplace had read it the Academy, in his obtaining Equation (51) in the 1781 memoir. For now he does give the error curve

$$\phi(x) = \frac{1}{2a} \log \frac{a}{x},$$

where the limits of error *x* are $\pm a$. That, it will be recalled, is the equation he proposed for the prior distribution of any quantity when all that is known is that the probability is inversely proportional to the magnitude of *x*. Nothing, Laplace concluded triumphantly, henceforth stood in the way of determining a mean value among a series of observations except "the inevitable difficulties of analysis." These, he had to admit, were severe, and the method he had just outlined correspondingly awkward to use. It might perhaps be employed, he thought, in very delicate investigations requiring the greatest possible precision, such as observations of the passage of Venus across the face of the sun. His purpose in giving it was rather to bring to bear "all the light that the theory of chance can cast on the matter than to offer observers a practical method that was easy to use."[18]

---

[16] For a discussion in modern terms of Laplace's attempt in this manuscript to describe a discrete random probability distribution, see Stigler (1986a), pp. 120–22.

[17] Ibid., *OC*, **9**, 476–78.

[18] Ibid., p. 479.

A further passage from the 1777 paper that Laplace did not repeat in "Mémoire sur les probabilités" was the criticism he had then made of Lagrange and all others who had employed the method of direct probabilities: "their research, though very ingenious, could be of only very little use to observers."[19]

[19] Gillispie (1979), p. 229.

CHAPTER 11

Generating Functions and Definite Integrals

OF THE NEXT two memoirs to be discussed, the first, "Mémoire sur les suites" [1782a], was entirely mathematical. The second, "Mémoire sur les approximations des formules qui sont fonctions de très grands nombres" [1785a], was largely so. Both were motivated mainly, though not exclusively, by probabilistic concerns. The former, which introduced the theory of generating functions, was particularly important to Laplace. Years later, when composing the *Théorie analytique des probabilités*, he would subordinate all the analytical part to the theory of generating functions and represent the whole subject as their field of application. The first part of Book I is in the main a reprinting of this memoir, incorporating clarifications and simplifications in detail and several important additions. He defined generating functions in much the same words in the opening pages of the *Théorie analytique* and of the "Mémoire sur les suites:"

> Let y_x be any function of x. If there be formed the infinite series, $y_0 + y_1t^1 + y_2t^2 + y_3t^3 + \cdots + y_xt^x + y_{x+1}t^{x+1} + \cdots + y_\infty t^\infty$, and u be designated the sum of that series, or (which comes to the same thing) the function of which the development forms the series, this function will be what I call the generating function of the variable y_x.[1]

Thus, he explained, the generating function of a variable y_x is a function of t that, developed in powers of t, has that variable for the coefficient of t^x. Reciprocally, the corresponding variable of a generating function is the coefficient of t^x in the development of the function in powers of t. The exponent expressing the power of t then indicates the place that the variable y_x occupies in the series, which may also be extended indefinitely to the left according to negative powers of t.

Laplace regarded generating functions as something of a panacea for problems involving the development of functions in series and evaluation of the sums. Those procedures embraced many of the possible applications of mathematics to nature. His memoir consists in showing

---

[1] *OC*, **10**, p. 5. Stephen Stigler considers that the origin of generating functions may be found in Abraham de Moivre's method for finding the distribution of the sum of the faces of several dice in *Miscellanea analytica* (1730), and that the approach derived ultimately from Pascal's arithmetic triangle. See Stigler (1986a), pp. 91–92.

how to apply the device to problems of interpolation both in convergent and recurrent series. As a corollary to the latter, he gave a method for solving the linear finite difference equations that express the relation between terms. From finite analysis he proceeded to the comparable infinitesimal expressions and from there to series involving two variables, recurro-recurrent as well as convergent in nature, observing by the way that he himself had initiated the theory of recurro-recurrent series.

In the course of considering the solution of linear partial difference equations on which recurro-recurrent series depended, he had reduced the problem to one of infinitesimal differences by means of definite integrals involving a new variable. In effect, the technique results in finding discontinuous solutions to differential equations, the existence of which had emerged historically in the analysis of sound waves. Laplace reminded the reader that the integral of second-order partial differential equations contains two arbitrary functions of integration. Euler and Lagrange had been the ones to discover these mathematical objects in analyzing acoustical problems in which the movements of the air in transmitting sound were considered three-dimensionally. (This occasion was Laplace's first mention of the theory of sound.)[2] It had been left to him, however, to find a general method for integrating second-order linear partial differential equations in the cases where that was possible and for identifying those cases in which it was not.[3] Now he had found one for integrating many such expressions by means of these definite integrals involving a new variable. The technique was further applicable to problems of vibrating cords, which in turn led to consideration of the employment of discontinuous functions in solutions of partial differential equations.

Thus, many of Laplace's existing analytical and probabilistic interests merged with certain of his future physical concerns in this memoir. What led him to the idea of generating functions may well have been further reflection on the calculus of differential operators (Chapter 9 above). He mentioned at the outset how the relation between a generating function and its independent variable leads directly to the analogy between positive powers and derivatives and between negative powers and integrals. Lagrange had formulated that relation in a theorem that Laplace himself had re-proved by more orthodox means. His calculus of generating functions is, in a sense, a calculus of exponents and characteristics (or operators) as well as of coefficients and variable quantities,

---

[2] [1782a], *OC*, **10**, Section XX, pp. 63–70.
[3] [1777a]; see Chapter 6.

and indeed he devoted Article X to deriving a theorem by Lagrange in the form

$$'\Delta y_x = e^{\alpha \frac{dy_x}{dx}} - 1 \tag{59}$$

by means of generating functions; the symbol $'\Delta y_x$ meaning that x varies by the quantity i.[4]

An example will illustrate the method. Laplace analyzed the series formed by multiplying the terms of a convergent series by the terms of a geometric progression. The general term of the series produced would be $h^x y_x$ (y_x being the general term of the convergent series). Then u would be the sum of the infinite series

$$y_0 + y_1 ht + y_2 h^2 t^2 + y_3 h^3 t^3 + \cdots + y_\infty h^\infty t^\infty. \tag{60}$$

It followed from the general analysis that he had already given of the relation between the coefficients of t and the corresponding powers of the variable that, in this example,

$$u\left(\frac{1}{t^i} - 1\right)^n = u\left[h^i\left(1 + \frac{1}{ht} - 1\right)^i - 1\right]^n. \tag{61}$$

The coefficient of t^x on the left of that equation is the nth finite difference of $h^x y_x$, where x varies by the quantity i. Moreover, if the right-hand side of Equation (61) is developed in powers of $(1/ht - 1)$, the coefficient of t^x in $u(1/ht - 1)^r$ will be $h^x \Delta^r y_x$, whatever the value of r. If the exponent n is negative instead of positive, the analysis will relate to integration instead of differentiation.

On 12 December 1781, some months prior to the publication of "Mémoire sur les suites," Laplace presented a report to the Academy of Science on a paper concerning probability by Adrien-Marie Legendre. It had the effect, evidently, of renewing his interest in the problem he had treated in a special case in concluding his observations on birth data in "Mémoire sur les probabilités," that is, evaluation of formulas that are functions of very large numbers. Not yet a member of the Academy, Legendre was then in the earliest stage of his career. His memoir, which Laplace praised very highly, dealt with three problems, the last of them concerning lotteries. If a lottery consists of a numbers and b are chosen in each draw, what is the probability that in p draws at least n different numbers will be chosen? Laplace had resolved a

---

[4] "Sur une nouvelle espèce de calcul relatif à la différentiation et à l'intégration des quantités variables," (1774), *OL*, **3**, pp. 441–76.

similar problem in his earliest discussion of probability in the memoir on recurro-recurrent series. It had been put to him, he said then, on the occasion of a bet placed on the lottery of the École militaire.[5] Legendre was quite correct, Laplace noted, in observing that when the total number was large, as in the national lottery of France, such problems could be resolved analytically, but not numerically. "It is highly desirable," he concluded his report, "that we should have a general method of approximation by means of which to obtain readily the numerical value of such expressions when the numbers of which they are functions are considerable."[6]

Four years later, in 1785, Laplace responded to his own summons in "Mémoire sur les approximations des formules qui sont fonctions de très grands nombres" [1785a]. It was chiefly, though not exclusively, in the theory of chance that analysis would often result in such unusable formulas. Only in two special cases, the product of the natural numbers $1, 2, 3, 4, \ldots$, and determination of the middle term of a binomial raised to a high power, did techniques exist for readily achieving numerical solutions.

In the latter instance, if the power were supposed to be even and equal to $2s$, that term would be, as everyone knew,

$$\frac{2s(2s - 1)(2s - 2)(2s - 3) \ldots (s + 1)}{1 \cdot 2 \cdot 3 \cdot 4 \ldots s}.$$

Even there, formulating the expression in numbers became difficult once s grew large. Earlier in the century, James Stirling, the Scottish mathematician whose work Laplace consistently admired, had seen how to transform it into a series the sum of which was equivalent to $\sqrt{\pi/2}$, and which converged more rapidly as s grew larger.[7] The transformation was remarkable in that it introduced a transcendental number into the investigation of purely algebraic expressions. It was still applicable only to special cases, however. Laplace himself had given a method in "Mémoire sur les probabilités" for converting integrals of differential functions containing factors raised to very high powers into rapidly

---

[5] [1774b], *OC*, **8**, pp. 17–18.

[6] The referees were Laplace and Vandermonde. The manuscript of the report, in Laplace's hand, is in the Archives of the Academy of Science, where it was discovered by Bru and Crépel. It is among many valuable texts included in their *Condorcet* (1994), pp. 157–59. On 25 March 1783 Laplace gave another, equally favorable report on two further memoirs by Legendre, this time on the attraction of a spheroid. It is printed in Gillispie (1979), pp. 262–64.

[7] James Stirling, *Methodus differentialis: sive tractatus de summatione et interpolatione serierum infinitarum* (London, 1730).

converging series.[8] But he was busy with other matters, and only since then had further reflection shown him the way to extend the method generally to any functions involving very large numbers, thus reducing them to series that, even like the series of Stirling's theorem, become more convergent as the numbers increase.

Since the central difficulty lay in finding numerical solutions for complicated expressions containing many terms, the strategy had to be to transform such formulas into series that converged rapidly enough so that only the first few terms had to be considered. If each of them contained only a few factors, then it would not matter if they were raised to high powers, and resort to logarithms would yield solutions. On the face of it, there seemed no natural way to transform such complicated functions into convergent series. Reflecting, however, that differential expressions simple in form often yield just such functions on integration if the terms have large exponents, Laplace considered that complicated functions of any sort ought to be reducible to integrals of that type, which could then be transformed into convergent series. The problem had two aspects: on the one hand, to integrate by approximation differential equations involving functions that contain very large factors; and on the other, to convert the functions for which approximate values are required into integrals of this type.

Laplace expected that a solution to the former aspect would prove particularly valuable in estimating the probability of causes. The approximation involved series that complement each other: One type was to be employed for points far from the maximum value of the differential function, and the other was to be used for points close to it. The latter contained transcendental functions, usually reducing to the form

$$\int e^{-t^2}\, dt. \tag{62}$$

Since that integral evaluated from $t = 0$ to $t = \infty$ equals $\sqrt{\pi/2}$, Stirling's theorem came out as a special case of the general analysis.

As for transforming the approximate evaluation into the integration of differential expressions multiplied by factors raised to a high power, Laplace outlined a more indirect approach. The method consisted of representing complicated functions of s (where s is a large integer) by y_s, y_s', y_s'',.... These functions are to be supposed given by linear differential or difference equations of which the coefficients are rational functions of s. Equations are then written to set

$$y_s = \int x^s \phi\, dx;\; y_s{}' = \int x^s \phi'\, dx;\ldots, \tag{63}$$

[8] [1781a], *OC*, **9**, p. 444.

and manipulated so that they can be separated after integration by parts, one part coming under the integral sign and the other outside it. Equating the parts under the integral sign to zero then yields as many linear differential equations as there are variables ϕ, ϕ', ϕ'', ..., which can thus be determined as functions of x. The parts outside the integral sign also may be equated to zero so that when the arbitrary constants in the values for ϕ, ϕ', ϕ'', ..., are eliminated, a definitive equation for x is obtained. Its roots then determine the limits between which the integrals $\int x^s \phi \, dx$, $\int x^s \phi' \, dx$, ..., are to be taken. It is very important to notice that the series obtained for y_s, y'_s, ..., hold good generally when the constants they contain change sign, because this circumstance greatly extends the applicability of the method, although as a result of such change of sign the definitive equation in x, which gives the limits of the integrals, no longer has several real roots.

The most serious difficulty to be overcome in applying this analysis arises from the nature of the differential equations in ϕ, ϕ', ϕ'', ..., which often cannot be integrated. That obstacle could normally be surmounted through representing the functions y_s, y'_s, ..., in terms of multiple integrals such as $\int x^s x'^s \phi \, dx \, dx'$, $\int x^s x'^s \phi' \, dx \, dx'$, ... The variables ϕ, ϕ', ... might then be determined by equations of lower order that would be integrable. All these possibilities considered, Laplace claimed that the analysis was capable of being employed generally for very complicated functions represented by ordinary or partial difference or differential equations—or for all the normal uses of analysis.[9]

He reserved his own application for a sequel to this memoir, which had grown very lengthy. Reading the continuation before the Academy of Science on 25 and 28 June 1785, Laplace again reviewed the epistemological status of probability.[10] It was there that he first employed phrasing of which the second sentence is famous from the *Essai philosophique sur les probabilités* thirty years later:

> The word "chance" then expresses only our ignorance of the causes of the phenomena that we observe to occur and to succeed one another in no apparent order.
>
> Probability is relative in part to this ignorance, and in part to our knowledge.[11]

For the first sentence, however, Laplace substituted the following in the *Essai philosophique*: "The curve described by a simple molecule of air or of water vapor is regulated in as certain a manner as are the orbits of

[9] [1785a], *OC*, **10**, p. 213; cf. Chapter 29.
[10] (49), [1786b].
[11] [1785a] *OC*, **10**, p. 296. Cf. *Essai philosophique*, *OC*, **7**, p. viii.

the planets; there is no difference between them except that which our ignorance introduces."[12]

We come now to the final technical factor that drew Laplace to the study of population statistics. In order to appreciate its role in his mathematical development, let us take stock of how far he had prevailed against difficulties still thwarting probabilistic analysis as he had outlined them at the beginning of the "Mémoire sur les probabilités."[13] He had there found in birth data a basis for determining the law according to which a succession of events gives access to knowledge of causes in a real case. In the intervening two memoirs he had developed generating functions and the technique for numerical evaluation of integrals of differential equations containing terms with very large exponents. It remained to put on a footing more realistic than the hypothetical case of lightly loaded dice the class of problems calling for the probability of complex events compounded from simple ones of which the respective possibilities are unknown quantities, and reciprocally for estimating the number of observations necessary in order that a predicted result might have a specified probability. It was in furnishing data for these calculations that the record of births, with its slight but known disproportion between male and female, proved to be such a valuable resource. Laplace devoted the remainder of the memoir to resolving from this point of view, and with the use of his improved techniques, the population problems that he had already handled in "Mémoire sur les probabilités," arriving in most instances at identical values.

[12] Cf. "Notice sur les probabilités," [1810d], p. 100, and Chapter 26.
[13] [1781a]; see Chapter 10.

CHAPTER 12

Population

In 1786 Laplace finally addressed a memoir directly to demography, treated not merely as a convenient repository of problems, examples, and data, but as a subject in its own right.[1] The opening sentences are in a new tone for Laplace. They have the ring of an adviser to government. "Population," he pronounced, "is one of the surest ways to judge of the prosperity of an empire, and the variations it undergoes when compared to the events that preceded them are the most precise measure of the influence of moral and physical causes on the well-being or the hardships of humanity."[2]

Aware of the guidance that such information might provide to those responsible for public policy, the Academy of Science had decided (Laplace continued) to insert each year in its published memoirs the summary of births, marriages, and deaths throughout the kingdom. In consequence, the last volumes of its memoirs in the Old Regime contain six annual installments of an "Essai pour connaître la population du royaume."[3] There was no question of actually counting heads. Instead, late-eighteenth-century demography proposed to reach estimates of the population through determining the factor by which the number of births in an average year was to be multiplied in order to approximate the total. The figure of twenty-six was widely accepted among knowledgeable observers.

Such research, Laplace observed, touched so closely on "the natural history of man" as to be a legitimate subject for the Academy of Science. Accordingly, the goal of the six-part "Essay" published in its *Mémoires* was to estimate the populations of municipalities and regions grouped in accordance with the subdivisions of the Cassini map of France, the nearly completed cartographic enterprise that was also conducted under the aegis of the Academy. Citations to this compilation usually attribute it to the joint authorship of Condorcet, Laplace,

---

[1] [1786c]. Laplace read a draft in the meeting of the Academy on 30 November 1785 (31).

[2] [1786c], *OC*, **11**, p. 35.

[3] *MARS*, (1783/1786) pp. 703–18; (1784/1787), 577–92; (1785/1788), 661–89; (1786/1788), 703–17; (1787/1789), 601–10; (1788/1791), 755–67. On this series, and on its place in late-eighteenth-century demography, see Brian (1994), 272–86.

and Dionis du Séjour, following therein the table of contents of the volumes themselves.

The attribution is misleading. Those three served merely as a commission to receive and communicate to the Academy the work of François de La Michodière, a magistrate who at the behest of the government was continuing research that he had undertaken on his own initiative some thirty years earlier in Auvergne and the Lyonnais, where he was then intendant. La Michodière was not at all mathematical in his attempt to differentiate among causes that led to differing life expectancy in town and in the country and from one region to another. A seasoned administrator and hence an empiricist, he simply observed that a larger than normal proportion of bachelors in such untypical places as Paris and Versailles dictated using a different factor there. In computing those urban—and urbane—populations, he multiplied the average annual births by thirty. His interest, and the interest of the government officials who reviewed his reports, was in the particulars and the variations of population: in the totals of births, deaths, marriages, and religious confirmations city by city and province by province; in the changes year by year and from one place to another in any given year; in the comparison between the figures reached through applying the multiplier to births and through counting the number of households subject to taxation; in the differences in density of population from one town and one region to another. The purpose, in short, was construction of a data bank for the information and guidance of government, national and local.

Laplace's interest, by contrast, was theoretical. Once past the opening paragraphs in his introductory memoir, he dropped the language of political arithmetic, which was proper to Condorcet and others concerned with constructing a social science, and reverted to his own style. Events of the same sort have constant and uniform causes of which the effects may be increased or diminished in the short term by variable factors that we attribute to chance. In consequence of the latter, calculation of the size of the population from the number of births at any juncture can be only a probable result, subject to error. Calculating the probability of error, and the size of the sample to be taken in order that it fall within given limits, pertains to analysis of chance. The procedure depends specifically on a new theory, still little known, that of the probability of future events given observation of their occurence in the past. He had himself worked out the method for solving problems of this sort, involving as they did formulas impossible to calculate directly because of the large numbers involved.

Laplace intended his memoir as a guide to analyzing the data that La Michodière was assembling and that the Academy proposed to publish regularly henceforth. The body of his memoir consists of a study of the vital statistics of Paris from 1771 to 1784, together with an estimate of the total population of France over the two-year period 1781–1782.[4] He regretted that the birth records made no distinction between the sexes and trusted that the administration would correct the oversight in the future. The multiplier 26 was based on a general consensus among students of population and not on any systematic sampling. Applying it to the average annual birth figure for the years 1781 and 1782 gave 25,299,417 for the population of France. In order to reduce to a thousand to one the odds against making an error no greater than half a million in the estimate, a sampling that would statistically justify the factor of 26 would have to be based on an actual counting of 771,469 inhabitants. If the multiplier were taken to be 26 1/2, then the figure for the population would come to 25,785,944, and the sample would need to be 817,219 in order to maintain the same odds against the same error. In view of those results, Laplace recommended that the count be carried to 1,000,000 or even 1,200,000 in order to assure a degree of accuracy compatible with the importance of the information.

It cannot be said that a statistical science of demography stemmed in linear fashion from a synthesis between the respective approaches of Laplace and La Michodière, the mathematician and the administrator. Their interests converged on the same set of problems but did not merge, and they went their not entirely separate ways. When a regular census did begin, in 1801, it did not represent an application of probability to birth records or to anything else. It represented the determination of Bonapartist administrators to base action on facts.

Laplace for his part did not publish further on probability prior to the series of lectures he gave at the short-lived École Normale in 1795 (Chapter 19), one of which was the nucleus of *Essai philosophique sur les probabilités*. Although there are probabilistic considerations in *Mécanique céleste*, notably in Book X, there is no evidence of his resuming sustained work until twenty-five years later, when he composed a memoir that derived the central-limit theorem for the reduction of error from his method for approximating the evaluation of formulas containing terms raised to very high powers ([1810b], Chapter 25). Let us turn back, therefore, to two sets of interests that occupied him concurrently with probability and with each other, cometary theory and his collaboration with Lavoisier in the chemical physics of heat.

---

[4] Ibid., *OC*, **11**, pp. 44–46. The first table had been compiled by Jean Morand in continuation of the data he assembled on the vital statistics of Paris and the suburbs from 1709 to 1770, published by the Academy in 1774 (Chapter 10, n. 10). Morand died in 1784. On his statistical work, see Brian (1994), pp. 260–62, and on other of his enterprises, see Gillispie (1980), pp. 349–50. The second table covering the entire country for 1781–1782 came from La Michodière himself.

CHAPTER 13

Determination of the Orbits of Comets

LAPLACE'S INTEREST in the problem of the curves described by comets antedated the only publication he had yet devoted to these bodies, namely his probabilistic calculation that the cause behind the configuration of the planetary system fails to account for the distribution of the inclinations of the planes of cometary orbits.[1] Earlier in 1776 he had involved himself in an acrimonious dispute with Rudjer Bošković, then resident in Paris. In 1771 Bošković had presented to the Academy of Science a refinement of the method for determining cometary orbits that he had first advanced in 1746 and had advocated ever since.[2]

In the case of comets, uncertainty about the nature of the conic sections they describe, combined with the irregularity of opportunities for observation, left a much wider gap between theoretical and practical astronomy than in the case of planets. In practice, the trajectories of the known comets had been mapped empirically, by fitting the curves to as many observations as could be recorded in the periods of visibility. Newton had handled the problem by supposing that the observable trajectory is a parabola approximately coinciding with a highly eccentric ellipse. On that supposition, widely adopted thereafter, three observations should suffice to determine the curve. The problem was how to arrive at it mathematically. Newton gave an approximate solution that depended on considering a very short element of the trajectory as if it were a straight line. It appears in his "System of the World," appended to the Motte translation of the *Principia*.[3]

Bošković's memoir consists in an analytical development of this, the standard approach, issuing in an elaborate sixth-degree equation of motion. According to Laplace's own account, it was on reading this memoir, published in 1774 in volume 6 of the *Savants étrangers* series—along with two of his own early papers [1774b] and [1774c]—that he realized that the entire method involved a fatal fallacy.[4] Treating the interval between the first and third observations as a first-order infinitesimal entailed neglecting second-order quantities that depend on

---

[1] [1776d]; see Chapter 5.
[2] "De orbitis cometarum determinandis," *SE*, **6** (1774), pp. 198–215.
[3] Ed. Florian Cajori (1934), pp. 619–26. Cf. Cohen (1971), pp. 327–35.
[4] [1784b], *OC*, **10**, p. 93.

the curvature of the orbit and on changes in velocity of the comet. At the same time, the position of the (supposedly) rectilinear fragment of trajectory that the comet is observed to describe has to be determined by second derivatives of its geocentric latitude and longitude. Thus, second-order quantities were both neglected and employed.

From the floor of the Academy, Laplace read out a criticism of the Bošković memoir, castigating all such procedures as "faulty, illusory, and erroneous." We do not know the precise date of his attack. Taking it as a derogation of Newton and an insult to himself, Bošković demanded appointment of a commission to adjudicate the dispute and the exchange with Laplace that ensued. The report of that commission, consisting of Vandermonde, d'Arcy, Bezout, Bossut, and Dionis du Séjour, is recorded in the minute book of the Academy.[5] The commissioners acknowledged that Laplace was right analytically, while deploring the abrasive manner in which he had couched his criticism and regretting that Bošković had taken it personally. Both parties were counseled to bring their findings before the public rather than to quarrel in camera.

Evidently, however, Laplace could not let the matter drop. On 19 June he read a further set of remarks.[6] We do not know what they contained, except that now he also had Lalande in his sights, and another commission was appointed to resolve the affair. Its report has not survived. Years later, in the opening paragraph of the "Mémoire sur la détermination des orbites des comètes," Laplace recalled how he had proved that the method he was criticizing was so faulty that it was capable, in an extreme instance, of reversing the apparent direction of a comet's motion from retrograde to direct.[7]

Having refrained from pressing the investigation at that time, he now resumed it, stimulated by the recent work of Lagrange and Dionis du Séjour. As it was analytically impossible to operate with three widely separated observations, the standard method necessarily depended on three observations of positions that were fairly close together. Inevitably, therefore, small errors of observation would affect the results very considerably. Compensation of errors was then attempted, not by multiplying observations but by increasing the number of terms in the series that expressed the result, in order that it might approximate more closely to the true value. The technique was mathematically laborious and of little practical use. Seeking a simpler way to correct for observational error, Laplace saw that closely contiguous observations could

[5] *PVAS* (5 June 1776), fols. 172–77.
[6] Ibid., fol. 191.
[7] [1784b], *OC*, **10**, pp. 93–94.

serve that purpose if their number were increased. Standard methods of interpolation could then be applied to determine the observational data needed for a solution. The choice of parameters being arbitrary, he preferred to work with the geocentric longitude and latitude of the comet at a given moment and with the first and second derivatives of these quantities with respect to time. These data were the easiest to manipulate analytically, and he could obtain simple formulas that became more precise the larger the number and the greater the accuracy of the observations.

This approach to the determination of cometary orbits had the further advantage that observations separated by as much as 30° or 40° might be employed. In contrast to the established procedure, in which the analysis was an approximation and the observations had to be supposed perfectly exact, Laplace characterized his method as one in which the analysis is rigorous and the observations are acknowledged to be approximations. The second-order differential equations of motion of a comet around the sun at the focus of a conic section yielded directly a seventh-degree equation determining the distance of the comet from the earth:

$$[\rho^2 + 2R\rho \cos \theta \cos(A - \alpha) + R^2]^3 \times (\mu R^2 \rho + 1)^2 = R^6, \quad (64)$$

where ρ is the comet-earth distance, α the latitude, R the radius vector of the earth, r the radius vector of the comet, and A the heliocentric longitude of the earth, and where μ satisfies $\mu\rho = d\rho/dt$.[8] Since the theory holds for any conic section, the supposition that the orbit is a parabola and the major axis infinite yields a new sixth-degree equation for determining the distance of the comet from the earth:

$$[\rho^2 + 2R\rho \cos \theta \cos(A - \alpha) + R^2] \times \left(m\rho^2 + n\rho + \frac{1}{R^2} \right)^2 - 4 = 0$$

$$(65)$$

where, besides the above, m and n are abbreviations representing respectively

$$u^2 + \left(\frac{d\theta}{dt} \right)^2 + \left(\frac{d\alpha}{dt} \right)^2 \cos^2 \theta$$

[8] Ibid., p. 110.

and

$$\left(2u \cos \theta - 2\frac{d\theta}{dt} \sin \theta\right) \times \left[(R' - 1) \cos(A - \alpha) - \frac{\sin(A - \alpha)}{R}\right]$$

$$+ 2\frac{d\alpha}{dt} \cos \theta \left[(R' - 1) \sin(A - \alpha) + \frac{\cos(A - \alpha)}{R}\right],$$

and R' is the radius vector of the earth at longitude $90° + A$.[9] It was possible to combine Equations (64) and (65) and to obtain a linear equation for the distance and an equation of the conditions that the data must satisfy in the case of a parabolic orbit. But the calculation was difficult, and it was more direct to satisfy Equations (64) or (65) by making trials with the data.

Since the problem of determining parabolic cometary orbits could be formulated in a system of equations exceeding the number of unknowns by one, there was a choice of ways to determine the distance of the comet from the earth. The important tactic was to select a method that would minimize the effect of observational error. The second derivatives of the geocentric latitude and longitude ($d^2\theta/dt^2$ and $d^2\alpha/dt^2$) were the quantities most affected. Either but not both might be eliminated, and Laplace formulated two further sets of equations to be used alternatively, according to whether the second derivative of longitude or of latitude was the greater. The former ratio obtained in the case of comets of which the orbital plane was close to the ecliptic, and Laplace was pleased to discover that his first set of equations was nothing other than a translation into the language of analysis of what Newton had demonstrated synthetically.[10]

Although Laplace's method of determining cometary orbits was largely superseded by those of Olbers in the 1790s and of Gauss after 1801, its formulation marks an important stage in his career. It was the first piece of work that brought him into immediate contact with workaday observational astronomers. Following the analytical part of the memoir, Article VIII contains instructions for applying the method to comets themselves and even gives numerical examples.[11] Prior to publication, Laplace sent a copy to the abbé Pingré (34), who immediately applied it to computations in his monumental *Cométographie* (1783–1784). The memoir closes with a contribution by Méchain further illustrating the

[9] Ibid., p. 121.
[10] *Principia Mathematica*, Book III, Proposition XLI.
[11] Ibid., *OC*, **10**, pp. 127–41.

technique in the determination of a comet, the second that he discovered in the year 1781.[12]

Laplace must have requested the favor of that computation from Méchain, who was not yet a member of the Academy, shortly after reading the draft of the analytical parts of the memoir on 21 March 1781 (34). An entry in the *Procès-Verbaux* of the Academy for 2 May (35) probably refers to that. Over three years elapsed before the memoir was published, after a delay rather longer than would be expected from the normal academic lag. The explanation may well be that Laplace became involved with William Herschel's discovery of an object that turned out to be the planet Uranus, initially taken for a comet. Herschel made the observation in Bath on 13 March 1781, and Laplace could scarcely have heard of it—and certainly cannot have been influenced in his treatment even if he had—when he delivered his paper eight days later. On 13 June, however, he read a note (36) to the Academy reporting that he had tried his general method on the new comet but found that neither it nor that of Lagrange—nor any other that he knew—was applicable to the present object, of which the apparent motion in latitude was almost undetectable relative to the motion in longitude. Even so, the notion that he might be dealing with an unknown planet did not then occur to Laplace. He had been trying to find a new method by combining his equations in a different manner and believed that he had succeeded. He planned to read his investigations shortly but needed time to complete the calculations before the comet emerged from the sun, where it was presently hidden. On 28 July (37) he reported those calculations, still taking the object for a comet. Not until a year and a half later, on 22 January 1783 (40), did he make written reference to "Herschel's planet," the ephemerides of which he had been calculating in collaboration with Méchain. Their results showed it to be identical with the supposed star that Mayer had recorded in 1756 and that had mysteriously disappeared (42).

In the meantime, Laplace had become increasingly drawn into another aspect of empirical science, one involving hand as well as mind, and had entered into problems of physics in collaboration with Lavoisier.

[12] Ibid., pp. 141–46.

CHAPTER 14

Lavoisier and Laplace:
Chemical Physics of Heat

THE EARLIEST RECORD of Laplace's interest in a problem of experimental physics is an entry in the *Procès-Verbaux* of the Academy of Science for 9 April 1777 (22). The occasion was the Easter meeting, a session always open to the public. Laplace then read a paper on "the nature of the fluid that remains in the pneumatic machine." No trace of the text remains, but it is evident from the Lavoisier materials that it contained the first fruits of a collaboration between the two colleagues and that the experiments pertained to determinations of the effects of varying degrees of temperature and pressure on the vaporization of water, ether, and alcohol. Henry Guerlac, whose monograph on the subject is the standard treatment, considered it virtually certain that Lavoisier instigated the trials.[1]

Lavoisier and Laplace never published the memoir they intended. They probably never wrote it, but their procedures and conclusion are evident from Lavoisier's other writings. The experiments gave Lavoisier the physical framework he needed in order to come forward in November 1777 with a much rumored criticism of the phlogiston theory coupled with a statement of his own hypothesis on combustion and calcination. In a preface drafted in 1778 for a projected but phantom second volume of his *Opuscules* of 1774, Lavoisier wrote of his intention to practice so far as possible "the methods of mathematicians."[2]

As for Laplace, apart from the intrinsic interest of the work itself, collaboration with Lavoisier, almost six years his senior, offered him the chance to be more than a mathematician. It associated him with the one person who was clearly emerging as the scientific leader of the Academy in their generation, the newly appointed administrator of the Arsenal and reformer of the munitions industry, with influential connections in the worlds of government and finance.[3]

That much transpired in 1777, and although Laplace and Lavoisier served together on occasional committees of the Academy and must

[1] Guerlac (1976), pp. 198–99.
[2] Ibid., pp. 215–16.
[3] Poirier (1993), esp. chapter 9.

have been in each other's company at its semiweekly meetings, they seem to have suspended active collaboration and to have resumed it only in 1781. During that summer they worked together to verify a design for fabricating a barometer with a flat meniscus imagined by Dom Casbois, a Benedictine in Metz.[4] That device was the occasion for Laplace's first recorded scrutiny of the phenomenon of capillary action. Much more immediately, the barometric question led Lavoisier and Laplace during the winter of 1781–1782 to determinations of the thermal expansibility of glass as well as mercury and other metals (38).

The motivation was both instrumental and theoretical. The experience stood them in good stead in the early 1790s, when both were serving on the commission charged with fabricating the standard weights and measures for the revolutionary metric system (see Chapter 18). In their registers they were already using decimal subdivisions of linear and gravimetric units. Along with these practical questions of measurement, the investigation set them both to thinking about heat capacities in general.

Their physical interests transcended heat. Electricity offered the companion example of a subtle fluid for which bodies have characteristic capacities. It was natural for Lavoisier and Laplace to explore the analogy between heat and electricity when Volta came to Paris early in 1782 for an extended visit. He brought with him an electroscope for detecting weak charges and was then harboring the theory that electrical charges in the atmosphere might be produced by vaporization. Lavoisier designed and had constructed a condenser with a marble plate to detect such charges, and Volta tried it in the company of Laplace. Conditions were bad, and the experiment failed. Laplace and Lavoisier then tried it again on their own (39) and published a brief note on what they construed as positive effects [1784c].

Even then they were planning the campaign of their experiments on heat, and Lavoisier must have commissioned construction of the famous ice calorimeter at much the same time. Their joint *Mémoire sur la chaleur* first appeared in a separate printing [1783a]. Laplace's influence was clearly paramount in its theoretical aspect. His was the idea for measuring a quantity of heat by the amount of ice it would melt, and his also the design of many experiments. The memoir consists of four articles. The first discusses the nature of heat and its quantification; the second, the determination of specific heats of selected substances and also certain heats of reaction; the third, theoretical consequences and a program for a chemical physics; and the fourth, the application of the techniques to the study of combustion and respiration.

[4] Guerlac (1976), p. 224.

Article I opens with a frequently paraphrased contrast between the fluid and the mechanical theories of heat. These hypotheses, say the authors, are the only conceivable alternatives. It has generally been supposed that the contrast here was also between their opposing opinions, and there is no doubt that Lavoisier preferred the former, soon to be called the caloric theory after his term for the matter of heat. Moreover, Laplace would almost certainly have been the one to compose the passage elaborating the kinetic theory. According to that school, the quantity of heat in bodies is measured by the sum of the *vis viva* (mv^2) of the vibratory motions of their particles, and the conservation of heat in transfers is a form of the conservation of *vis viva* in gradual changes of motion. Robert Fox, however, questions whether Laplace ever did adopt that position for himself.[5] If that was his view at the time, he certainly changed it for the physics of his later years, when he consistently preferred the caloric model for analysis. In this connection, it is to be remarked that conservation of *vis viva* did not play an important part in his mechanics, any more than it had for Newton; and the body of the memoir consistently and naturally employs the vocabulary of the fluid theory. The contrast concludes with the authors' abstention from choosing between the alternatives. "Perhaps," it is said, "they both obtain at the same time,"[6] an aside about which it may be worth remarking that in Laplace's final view caloric like other subtle fluids is itself particulate. In any case, the only admissible propositions were those that save the phenomena under both theories.

Foremost among those principles were conservation and reversibility in exchanges of heat within a system of bodies. Definitions were needed. *Chaleur libre*, or free heat, is that portion of the total heat contained in a body that may pass to another. Since different quantities of heat are required to raise the temperature of the same mass of different bodies equally, some unit must be designated with which to quantify these comparisons. The heat absorbed in raising one pound of water one degree makes a convenient amount, in terms of which the "specific heats" of other bodies may be expressed. Although probably not invariant with temperature, specific heats may be taken as nearly constant for the range between the freezing and boiling points of water (0°–80° on the Réaumur scale).

It is to be noticed that the notion of specific heat was not quite the modern way of putting it, although it comes to the same thing when reduced to unit mass and referred to the amount for water as unit heat. A more serious reservation must be entered, however, lest this lead to

[5] Fox (1971), p. 30.
[6] [1783a] *OC*, **10**, p. 153.

anachronism. The free heat of a body is only that portion of its absolute quantity that may be exchanged with other bodies in consequence of differences of temperature, change of state, or chemical reaction. It is also the only manifestation of heat that is accessible to measurement, which, moreover, is in degrees of the thermometer. Here the reader must be especially careful to avoid thinking of the absolute quantity in terms of a hypothetical scale of absolute temperature, for there is no anticipation of Kelvin. The conception is that of the eighteenth-century theory of matter, which distinguishes between the portion of electricity, heat, ether, or whatever, that is "fixed" in bodies, and the portion that may be disengaged in natural phenomena or in experiments.

These presuppositions in no way vitiated the design or execution of the experimental program. The problem of measuring specific heats in the simplest case of mixing two miscible substances was formulated algebraically, no doubt by Laplace. He let m and m' be the respective masses, a and a' the initial temperatures, and b the temperature resulting from mixture. Then the ratio of specific heats, q and q', is given by

$$\frac{q}{q'} = \frac{m'(b - a')}{m(a - b)}.$$ (66)

So straightforward an approach was inapplicable, however, to determinations of heat effects involving chemical combination, combustion and respiration, or change of state, the three most important and interesting processes. Faced with this limitation, the authors imagined a method of general applicability. The amount of ice melted in any process involving the evolution of heat could serve to measure the quantity. The notion is introduced by means of an image that bespeaks Laplace and mathematical modeling more obviously than does the elementary algebraic formulation for specific heats in mixture. We are to imagine a hollow sphere of ice with a shell thick enough to insulate the inner surface from the heat of the surroundings. Suppose a warm body were to be introduced in principle into the cavity. Its heat would melt away a portion of the inner surface until it had cooled to zero, and the weight of water would be proportional to the heat required to accomplish that effect.

Lavoisier commissioned the instruments and later named them calorimeters; two were made, each about three feet high. Air could be admitted into one for respiration experiments. A concentric nest of containers like an ice-cream freezer was packed with ice around a central receptacle. A basket could be suspended inside to hold the objects under study. Ice lining the inner shell served for the melting layer of the model. The water was run off through a petcock and weighed. The authors established first that the heat needed to melt a pound of ice will raise the temperature of a pound of water from 0° to 60° R. Here again, it must be emphasized that the concepts of intensity and quantity of heat were not yet differentiated. Ice, as they put it, "absorbs 60 degrees of heat in melting."[7] Thereupon the remainder of their second, entirely experimental article reports the determination of certain specific heats, heats of reaction, and animal heats.

The third article is pure Laplace and exhibits both his capacity to analyze a set of physical phenomena in a highly abstract manner and also the limitations inherent in even the most sophisticated notion of a general theory of heat in the absence of any notion of thermodynamics. Not that Laplace fancied himself in a position to attain such a theory, although what he thought he needed was to know, first, whether specific heat increases with temperature at a uniform rate for all substances; second, the absolute quantity of heat contained in bodies at given temperatures; and third, the quantities of free heat given off or absorbed in chemical reactions. Lacking such data, which only a very elaborate program of investigation could work up, he would simply examine a few problems raised by experiments that the two authors had performed, beginning with the second topic just mentioned, the absolute or total quantity of heat in bodies.

Clearly, its amount is considerable, even at 0° on the thermometer. We wish to know that amount in degrees of the thermometer, but experiments such as those just tried on specific heat could form no basis for calculation, unless it was legitimate to suppose that the specific heats of bodies are proportional to their total heats. That would be a very risky relation to assume without examination. Moreover, there was unfortunately no way to get at it from the values determined by simple mixtures of substances at different temperatures: All that happens there is the exchange of heat. The case is comparable to local exchanges of motion, which tell the investigator nothing of the absolute motion of the earth through space.

There might be a deeper way, however, leading through chemistry. Since the heat of reaction is not the consequence of mere inequality of temperature, it might furnish a basis for relating change of temperature to absolute heat. As usual, Laplace's approach was analytic. He let x be the ratio of the absolute heat contained in water at 0° to the amount that can raise its temperature by 1° (note that he was not yet sufficiently prepared with his own definitions to call the latter "specific heat" and designate it unity). Then, the total heat contained in a pound of water at 0° would melt $x/60$ pounds of ice. Consider any two substances at 0°, m

[7] Ibid., p. 167.

and n being their respective weights, and a and b the ratio of the heat contained in each to the heat contained in a pound of water. They react chemically, and the heat produced when the products are cooled to $0°$ melts g pounds of ice. The heat of reaction alone is sufficient to melt y pounds (y being negative if the reaction is endothermic). Finally, c is the ratio of the heat contained in the mingled products of reaction to that in a pound of water. With these parameters, Laplace formulated and equated two expressions for the quantity of ice that would be melted by the residual free heat:

$$\frac{(ma + nb)x}{60} + y = \frac{(m + n)cx}{60} + g, \qquad (67)$$

whence

$$x = \frac{60(g - y)}{m(a - c) + n(b - c)}. \qquad (68)$$

Here x represents the number of degrees of heat contained by water at $0°$. But how are numbers to be substituted for a, b, c, and y? Two hypotheses made evaluation possible. The first, conservation of free heat in chemical combinations, was generally admitted. The second, that the specific heats of substances are proportional to their absolute heat content, was precisely the risky assumption that could be tested from data tabulated in the preceding article. If it were correct, then $y = 0$, and

$$x = \frac{60g}{m(a - c) + n(b - c)}. \qquad (69)$$

Those data contain values for the specific heats, a, b, and c, in a few selected instances; and if the proportionality of specific heat to total heat was a justifiable hypothesis, all of the different cases ought to give the same value for x. Alas, they did not.

It is true, Laplace immediately went on to argue, that a very small correction in each, no more than $1/40$, would make them all satisfy the relation.[8] Such a correction would be less than the margin of experimental error. Other considerations made such an exercise in curve fitting unpromising, however. Endothermic reactions could not at all be accommodated, and neither could the phenomenon of the dilation of solids on heating. It was very probable that heat is fixed thereby, even as in change of state, although gradually and undetectably; and this

[8] Ibid., p. 178.

reflection gave another reason to surmise that specific heats do increase with temperature but at a different rate for each substance.

Finally, Laplace turned to change of state itself, in order to consider what such episodes can reveal about equilibrium conditions in heat. Again, it was the analogy that is mechanical, rather than the theory of heat itself. Just as there are several positions of equilibrium for, say, a rectangular parallelepiped (resting on its side, balanced on one end, and so on), there may be several conditions around a change of state in which heat is in equilibrium, each involving different physical arrangements of the molecules and different distributions between the portions of the heat that are going into cooling and into freezing. As usual, Laplace formulated analytical expressions. He then adduced the example of super-cooling followed by a sudden crystallization creating a new equilibrium. We need not follow the algebra. The interesting feature is the glimpse that the discussion gives into Laplace's preoccupation with forces and structures at the molecular level. The mutual affinity of molecules of water draws them together on freezing and frees the heat that is keeping them apart. Thus, it seemed probable that their arrangement when frozen is that in which the force of affinity is at its most effective. Hence, it is natural that the surest means of inducing a supercooled sample to freeze is to introduce a bit of ice. The same holds true in all crystallizations.

More generally, and here Laplace laid down a program for research, study of the equilibrium between heat, which tends to separate molecules, and affinity, which draws them together, might well offer a method for comparing the intensities of these forces of affinity. For example, a certain mass of ice plunged into an acid would be melted to the point at which the acid was sufficiently weakened so that its attraction for the molecules of ice would be balanced by their mutual forces of adherence. That point would depend also on the temperature. The further below zero it fell, the higher would be the concentration at which the acid ceases to melt ice. It would thus be possible to construct a statical scale expressing the force of the affinity of the acid for water in terms of degrees of the thermometer. If this procedure were to be followed with solutions of every sort, the relative mutual affinities of all bodies could be stated numerically. But, our author breaks off, this is a major subject; another memoir would be devoted to it.

We shall not consider the final article, in which Lavoisier drew the threads together for the theory of combustion and respiration. Guerlac's monograph shows how it prepared the ground for the "Réflexions sur le phlogistique" (1786) and how, more generally, the further collaboration with Laplace styled the aspiration to make of chemistry in its revolution a mathematical science. Our concern is with Laplace. There are indications that just before undertaking this research, he had been somewhat on again, off again in his commitment. In a letter of 7 March 1782 he asked Lavoisier to release him from their agreement to work together.[9] On 21 August 1783 a covering letter to Lagrange enclosing a finished copy of the memoir was defensive about the time that he had invested in *physique*.[10] Nevertheless, the passages on heat and affinity just discussed are evidence that the work had finally gripped his interest and carried him into the first stages of his physics of interparticulate forces. Only with Laplace, indeed, did chemical affinities begin to be seriously considered as physical forces of attraction. In the preface to the work to be discussed next, *Théorie du mouvement et de la figure elliptique des planètes* [1784a], Laplace referred to the experiments with Lavoisier and repeated the reflection that equilibrium between the attractive force of affinity and the repulsive force of heat might one day furnish analysis with the handle on chemistry that the discovery of gravitation had afforded to the mathematicians who had perfected astronomy since Newton. It also remained to determine the laws of force responsible for the physical effects of solidity in bodies, of crystallization, of the refraction and diffraction of light, and of capillary action.

Laplace continued working with Lavoisier into 1784, when the treatise just mentioned was published. They then began the program of research that he had imagined on affinity. They repeated, largely at Laplace's insistence, measurements of the heat generated by the combustion of charcoal, phosphorus, and other substances. The joint paper reporting those experiments was not written until 1793, however, and appeared after Lavoisier's execution [1793d]. Laplace also worked with Lavoisier on the closely related combustion of hydrogen to produce water. He strengthened Lavoisier's hand in the battle against the phlogistonists by setting him straight on the source of hydrogen when acids act on metals. The gas comes from the acid and is not to be taken for phlogiston escaping from the metal.[11] Thereafter, however, his own affinities in chemistry seem to have drawn him rather toward Berthollet, always more physically oriented than Lavoisier. But we must turn outward now, from particles to stars, from physics and chemistry to astronomy again.

[9] Guerlac (1976), p. 240.
[10] *OL*, **14**, pp. 123–24.
[11] Guerlac (1976), p. 265.

CHAPTER 15

Attraction of Spheroids

IT IS SIGNIFICANT that the paragraph just discussed on physical and chemical forces should have figured in the opening of the *Théorie du mouvement et de la figure elliptique des planètes* [1784a], for that work is very revealing in other important ways of the continuity and persistence of Laplace's interests. It was his first separate publication, and the preface was his first piece of writing suited to the comprehension of laymen. It consists of a summary view of the world in nontechnical language that might easily be taken for a prospectus of the *Exposition du système du monde* (1796). This book has never received the attention it deserves, either among contemporaries or in later generations. It is downright extraordinary that Laplace's first book should have been omitted from both of the two collections of his *Oeuvres*.

At all events, Laplace began by explaining how he came to write the book at the behest of an honorary member of the Academy, Jean-Baptiste-Gaspard Bochart de Saron, a magistrate of the Parlement of Paris and (although Laplace did not put it this way) one of several patrons of science who also contributed to its content. That Bochart de Saron should have commissioned the work and personally subsidized publication is the clearest evidence that Laplace's qualities had come to be recognized in high places. Long ago, he said, he had conceived the notion of drawing into a single work an exposition of the way in which planetary paths and figures follow mathematically from the law of gravity. It seems probable that he was referring here to one of the papers of his youth, "Une théorie générale du mouvement des planètes," which he had submitted to the Academy in November 1771 (9). But he would never have composed the treatise had Bochart de Saron not encouraged him on several occasions to show how the general properties of elliptical and parabolic motion may be derived from the second-order differential equations that determine the motion of celestial bodies.

That exposition occupies Part I of the treatise and, unlike the preface, is addressed to mathematically—and not merely verbally—literate readers. It has the quality of a textbook in the rational mechanics of the solar system limited to the principal motions of the celestial bodies and might appear to consist of a sketch for Books I and II of *Mécanique céleste*.

Besides that, Laplace explained certain of the techniques he had imagined for achieving approximate solutions to equations that defied rigorous integration, and he also included a set of calculations to determine the orbit of Uranus. This topic bulks rather larger than its importance in the solar system would have warranted. It would appear that Laplace took the occasion to publish an analysis that he had submitted to the Academy on 22 January 1783 (40), in which he recognized that the object discovered by William Herschel in March 1781 was a planet and not a comet, as was first thought. Beyond that, he reprinted much of his memoir [1784b] on the determination of cometary orbits.[1] Or rather, he preprinted it, since the memoir, although composed in 1781, came out some months after the treatise. There is no other novelty in this first part, and Laplace simply hoped that the treatment would be pleasing to mathematicians and astronomers.

Part II, subtitled "De la figure des planètes" and addressed "uniquely to mathematicians," is quite another matter.[2] Laplace there resumed investigating the laws of gravitational attraction of spheroids in a far more abstract manner than he had in his earlier pieces on the figure of the earth and on the oscillations of the tides.[3] At the same time, the mathematical problems are of a more specific nature, and it is clear that he had matured his approach since the memoir on the precession of the equinoxes [1780c]. In retrospect, the most interesting feature of the entire treatise is bound to be the emergence of the concept of what is now called potential, although that name was not given to the sum or integral of the action of the elements of an attracting body upon an external point until 1828, when George Green adapted Poisson's application of the expression for it to electrostatic and magnetic effects.[4] There is no reason to suppose that here, at the outset, Laplace thought the notion more signal than other main aspects of his treatise.

In all these researches on attraction, Laplace played leapfrog with Legendre in a manner very like his relation of collaboration and competition with Lagrange over the integral calculus of planetary theory. In the present treatise, he referred to a formulation by Legendre that he had seen in the latter's "Attraction des sphéroïdes homogènes," which was not published until 1785.[5] Legendre for his part there attributed to a communication from Laplace in 1783 the notion that he developed, by way of generating functions, into the polynomials that were later named for Legendre, but that were known as Laplace's

---

[1] See Chapter 13.
[2] [1784a], p. xxi.
[3] [1778a, 1779a, 1779b]; see Chapter 7.
[4] For a discussion of this transfer, see Grattan-Guinness (1995).
[5] [1784a], pp. 96–97. The Legendre memoir is in *SE*, **10** (1785), pp. 411–34.

functions—that was William Whewell's term—through much of the nineteenth century.[6] Legendre was not yet a member of the Academy, and the exchange evidently occurred in connection with Laplace's preparation of a report on the memoir on behalf of Bezout, d'Alembert, and himself, who constituted the committee to which it was referred.[7]

Whatever these priorities may have been, Laplace in the treatise under discussion took as his point of departure the equation for a second-order surface where the origin of coordinates is the center of the spheroid,

$$x^2 + my^2 + nz^2 = k^2, \tag{70}$$

m, n, and k being any constants. For the attraction that the enclosed solid exerts on an external point, he then formulated the integral

$$V = \int \frac{dM}{(a - x')^2 + (b - y')^2 + (c - z')^2}, \tag{71}$$

where dM is the mass of a particle of the solid with coordinates x', y', and z', and a, b, and c are the coordinates of the external point.[8] Then, when the attraction is decomposed along the three principal coordinates,

$$A = -\frac{\partial V}{\partial a}; \qquad B = -\frac{\partial V}{\partial b}; \qquad \text{and } C = -\frac{\partial V}{\partial c}. \tag{72}$$

Applying the analysis, Laplace began by looking backward to Newton's *Principia* rather than forward and by showing that the value of V is the same as if all the mass of the spheroid were concentrated at the center of gravity. In seeking to determine that value in a general manner capable of giving numerical solutions, Laplace turned to polar coordinates and expanded the transformed expression into an infinite series that could be evaluated by approximation. Wishing to achieve a rigorous solution, he in effect reverted to a strategy that he had employed in a fragmentary analysis of 1778, wherein he first investigated the attraction between a spheroid and an external point. At that time he had the rings of Saturn in mind and inverted the problem to consider the attraction of the point for the spheroid.[9]

---

[6] Burkhardt (1908), part 5.

[7] (41). For the text of this report, and a discussion of the relations between Laplace and Legendre, see Gillispie (1979), pp. 257–65.

[8] [1784a], p. 69.

[9] [1778a], *OC*, **9**, pp. 71–87. see Chapter 7.

In the same manner, Laplace now inverted the conditions and made the externally attracted point the origin of polar coordinates. Manipulating the resulting expressions yielded a theorem often called by his name: All ellipsoids with the same foci for their principal sections attract a given external point with a force proportional to their masses. The finding generalized a result already won by Maclaurin for the restricted case of particles located on the extension of the major axis. In modern terms, Laplace's theorem is said to assert that the potentials of confocal ellipsoids at a given point are proportional to their volumes. For the gravitational case, this would presuppose homogeneous density. His conception of the theorem in terms of mass was not intended to obviate that restriction, however, since he did not then envisage any other application. It was simply consistent with his initial motivation, which was to give an up-to-date demonstration of the point-mass gravitational theorem. For the appreciation of Laplace himself, it is more significant to notice that his interest was first mathematical and second physical. His solution compared the attraction of the original ellipsoid to that of a new, confocal ellipsoid containing the attracted point in its surface. That is how he made the problem solvable, for rigorous integration was possible when a point lies on the surface of a figure. Physically, the procedure presented the further advantage of opening the possibility of distinguishing more directly between central attractions and perturbing forces in the motion of planets and their satellites.

Thereupon, Laplace turned to internal particles and found another theorem that surprised him more than anything so far: The attractive force exerted at any point within a homogeneous ellipsoidal shell is equal in all directions. For the component of such an attractive force parallel to the major axis, Laplace obtained the important definite integral.

$$A = \frac{2a\pi}{\sqrt{mn}} \int_0^1 \frac{x^2\,dx}{\sqrt{\left(1 + \frac{1-m}{m}x^2\right) + \left(1 + \frac{1-n}{n}x^2\right)}}, \quad (73)$$

m and n (Equation 70) being positive for bounded surfaces.[10]

The rest of the treatise consists of similar mathematical investigations of particular problems concerning the equilibrium conditions and shape of rotating fluid masses. An important application to the moon shows that the difference in length between the earth-directed axis and the

[10] Ibid., p. 89.

diameter of a spherical body of identical mass is four times the comparable elongation of the orthogonal axis in the orbital plane.[11] The
coincidence of the moon's periods of revolution and rotation is then
deduced from that relation. The dependence of angular velocity on
ellipticity forms another topic, and its application to the earth gives
upper limits for the values of polar flattening. From his earlier essays,
Laplace reiterated in simplified language the finding that, although it
can be determined whether any given form for a solid (or, rather, a
fluid) of revolution satisfies specified forces, it is impossible to determine in a general manner all the forms that do so. All these results
appear in appropriate passages of Book III of *Mécanique céleste*, although he did not incorporate whole articles or sections from this
treatise in his synthesis.

On 11 August 1784, less than a year after finishing the treatise just
discussed (47), Laplace read the draft of a further memoir (48), most of
which is reproduced with little change in *Mécanique céleste*, Book III,
Chapters 1–4. Entitled "Théorie des attractions des sphéroïdes et de la
figure des planètes" [1785b], it contains the basic mathematical theory,
which for Laplace meant formulation, of the subject. The entire memoir
is an exercise in partial differential equations, manipulation of which
enabled Laplace to solve problems by differentiation; by series expansion, especially with Legendre functions and surface harmonics; and by
analysis of coefficients when integration was impossible, as it normally
was. The first of the memoir's five parts contains the simpler and more
direct derivations that this method permitted for the main results of
Part II of the immediately preceding *Théorie du mouvement et de la
figure elliptique des planètes* [1784a]. The rapid sequence fits the pattern
in which Laplace often acted. Dissatisfied with his initial treatment of a
subject as soon as he saw it in print, he would immediately set to work
simplifying the analysis and generalizing the treatment.

That process marks a further stage in the development of what
became the concept of potential. At the outset of his second section,
Laplace gave it a formulation that provided him with a basic equation
from which he could derive the whole theory of spheroidal attraction.
The form will not be immediately recognizable to the modern reader,
however, since the expression is in polar coordinates. He began by
referring to an elementary observation of his earlier treatise, namely
that if $V = \int dM/r$, differentiating V along a direction will give the
attraction of a spheroid in that direction. In the equilibrium case, the
attraction exerted on the particles of a planet takes that form, and

---

[11] Ibid., p. 116.

Laplace proceeded to investigate V. With the origin of coordinates inside the spheroid,

a, b, c, are the coordinates of the point where the attraction is exerted
x, y, z, the coordinates of a particle of the spheroid
$r = \sqrt{a^2 + b^2 + c^2}$ the distance to the origin of the point attracted
θ the angle that the radius r makes with the x-axis
ω the angle formed by the intersection of the invariant plane in which the x-axis and y-axis lie with the plane that passes through the x-axis and the attracted point.

Designating by R the distance $\sqrt{x^2 + y^2 + z^2}$ from the origin to the attracting particle, and by θ' and ω' the values of θ and ω at the point occupied by that particle, Laplace obtained the following expression for V:

$$\int \frac{R^2\, dR\, d\omega'\, d\theta'\, \sin\theta'}{\sqrt{r^2 - 2rR[\cos\theta\cos\theta' + \sin\theta\sin\theta'\cos(\omega - \omega')] + R^2}} \tag{74}$$

in which the integral relative to R is taken from $R = 0$ to the value of R at the surface of the spheroid; that relative to ω' from 0 to 2π, and that relative to θ' from 0 to π.[12]

Laplace's general mathematical virtuosity is nowhere more evident than in this memoir, which also offers a particularly explicit example of the specific advantage that he could draw from his own mathematical innovations. At this point (and this may have been the breakthrough that made the approach possible at all), he applied an important finding of the memoir on generating functions [1782a] and referred also to the early paper on partial differential equations [1777a].[13] He had there shown that integrating second-order partial differential equations was often possible by means of—and only by means of—definite integrals of a sort similar to the expression just given for V. In this instance, he found it easy to show that if $\cos\theta = \mu$, then differentiation produces the following differential equation:

$$0 = \frac{\partial\left[(1 - \mu^2)\dfrac{\partial V}{\partial\mu}\right]}{\partial\mu} + \frac{\dfrac{\partial^2 V}{\partial\omega^2}}{1 - \mu^2} + r\frac{\partial^2(rV)}{\partial r^2}. \tag{75}$$

That equation is in fact equivalent to the modern expression for potential (Equation 94).[14] Laplace never transformed it from spherical

[12] [1785b], *OC*, **10**, p. 362.
[13] *OC*, **10**, pp. 54–60, and *OC*, **9**, pp. 21–24, respectively. See Chapters 6 and 11.
[14] [1785b] *OC*, **10**, p. 362.

polar into Cartesian coordinates in this memoir, however. Rather, he substituted for V in Equation (74), so that he could write Legendre's equation:

$$0 = \left\{ \frac{\partial\left[(1 - \mu^2)\dfrac{\partial U^i}{\partial \mu}\right]}{\partial \mu} \right\} + \frac{\dfrac{\partial^2 U^i}{\partial \omega^2}}{1 - \mu^2} + i(i + 1)U^i, \qquad (76)$$

where U^i is a polynomial function of μ, $\sqrt{1 - \mu^2}\sin\omega$, and $\sqrt{1 - \mu^2}\cos\omega$.[15] In determining that function, and investigating the dependence of the variables on the angles ω and θ, sometimes called Laplace's angles, Laplace arrived at a formula for evaluating auxiliary factors:

$$\gamma\lambda' = 2\frac{1 \cdot 3 \cdot 5 \cdots (2i - 1)}{2 \cdot 4 \cdot 6 \cdots (1 + n)2 \cdot 4 \cdot 6 \cdots (i - n)}. \qquad (77)$$

(An error arising from the assumption that $i + n$ is always even was corrected in *Mécanique céleste* Book III, no. 15.)[16] When the attracted point is internal, the expression for V has to be developed in a series of terms of ascending powers of r; and the analysis compounds the attraction of the sphere, on the surface of which the point lies, with that of the shell constituting the remainder of the spheroid. Laplace did not question the pertinence of Equation (75) for these situations; in 1813, however, his disciple, Poisson, would do so.[17]

Simplifying the problem to the case of almost spherical spheroids, Laplace in a third section achieved a general solution even on the supposition of heterogeneous density. Again he recurred to results he had won in an earlier piece [1776d], namely the equation for the attraction at the surface of a spheroid:

$$-a\frac{\partial V}{\partial r} = \frac{2}{3}\pi a^2 + \frac{1}{2}V, \qquad (78)$$

where a represents half the common diameter of the spheroid and the inscribed sphere, and r is the distance of the attracted point from the center of mass. Laplace reiterated that equation in *Mécanique céleste*

[15] [1785b] *OC*, **10**, p. 375.
[16] *OC*, **2**, pp. 42–43. Cf. Todhunter (1873), **2**, p. 57.
[17] See Grattan-Guinness (1990a) **1**, p. 421.

and always accorded it great importance.[18] The further analysis was now fairly simple. Again, he decomposed the attraction V into two components, the force exerted at the surface of a sphere and that exerted by a further shell of which the outer surface bounds the spheroid. According to the conditions of the problem,

$$r = a(1 + \alpha y) \tag{79}$$

at the surface of the spheroid. Substitution in the basic expansion,

$$V = \frac{U^0}{r} + \frac{U^1}{r^2} + \frac{U^2}{r^3} + \frac{U^3}{r^4} + \dots , \tag{80}$$

and in Equation (78) yielded the equation

$$4\alpha\pi a^2 y = \frac{U'^{(0)}}{a} + \frac{3U^{(1)}}{a^2} + \frac{5U^{(2)}}{a^3} + \dots . \tag{81}$$

Since y is also a polynomial function of μ, $\sqrt{1 - \mu^2}\cos\omega$, and $\sqrt{1 - \mu^2}\sin\omega$, then it could be thought of as a series of functions,

$$y = Y^0 + Y^1 + Y^2 + Y^3 + \dots , \tag{82}$$

where $Y^0, Y^1, Y^2, Y^3, \dots$, like $U^0, U^1, U^2, U^3, \dots$, serve the partial differential equation,

$$0 = \frac{\partial\left[(1 - \mu^2)\dfrac{\partial Y^i}{\partial\mu}\right]}{\partial\mu} + \frac{\dfrac{\partial^2 Y^i}{\partial\omega^2}}{1 - \mu^2} + i(i + 1)Y^i. \tag{83}$$

The functions Y and U are similar in form, and

$$U^i = \frac{4\alpha\pi}{2i + 1}a^{(i+3)}Y^i. \tag{84}$$

Hence

$$V = \frac{4}{3}\pi\frac{a^3}{r} + 4\alpha\pi\frac{a^3}{r}\left(Y^0 + \frac{a}{3r}Y^1 + \frac{a^2}{5r^2}Y^2 + \frac{a^3}{7r^3}Y^3 + \cdots\right). \tag{85}$$

It thus proved possible to determine V by expanding y in a series of functions $Y^0 + Y^1 + Y^2 + Y^3 + \cdots$. Laplace then gave a method for evaluating V by analysis of the coefficients in the case of a spheroid of

[18] [1785b] *OC*, **10**, p. 372; *Mécanique céleste*, Book III, No. 15, *OC*, **2**, p. 30.

whose surface the equation, referred to Cartesian coordinates, is a polynomial function of the coordinates.[19]

Turning to the figure of the planets, Laplace treated the problem as a corollary of the foregoing analysis. In the simplified assumption that these bodies are homogeneous spheroids of revolution, he was finally able to demonstrate the long-sought theorem that, under the law of gravity, they can only be ellipsoids flattened at the poles. Proving that theorem enabled him to derive the law of attraction at the surface by an analysis a priori, which fully confirmed the validity of the above method of expansion of the function V into a series of what have been called Laplace's functions. The form in which he obtained it permitted comparison of the force of gravity, determined by means of the pendulum, with the value calculated for any point where the radius had been determined by geodetic measurements along the meridian. (It should be emphasized that this possibility helps to explain Laplace's strong preference in 1791 for basing the metric system on such a survey, instead of on a standard seconds pendulum [see Chapter 17].) Although existing data were less exact than he could have desired, they nevertheless permitted calculating that the figures assumed by the planets in the course of their rotations, even like their motions in orbit around the sun, confirm the principle of gravity with an overwhelming degree of probability. He had long since calculated that if the law of gravity were to satisfy the phenomena, the ellipticity of the earth had to be between the values 0.001730 and 0.005135.[20] Since observation of the pendulum gives 0.0031171, it is well within the limits.

In the course of this analysis, Laplace demonstrated in passing a theorem, which assumed increasing importance in his later work, with respect to two functions of different orders of the type that he was employing here. If Y^i and $U^{i'}$ are polynomial functions of μ, $\sqrt{1 - \mu^2} \sin \omega$, and $\sqrt{1 - \mu^2} \cos \omega$, and if both satisfy differential equations of the form (83), then they have the property of orthogonality, namely that

$$\int_{-1}^{1} \int_{0}^{2\pi} Y^i U^{i'} \, d\mu \, d\omega = 0, \qquad (86)$$

where i and i' are positive integers.[21] Todhunter points out that although Laplace did not here consider the case in which i and i' are identical, there is an important equation in *Mécanique céleste* that does

[19] Ibid., *OC*, **10**, pp.371–74.
[20] [1779a] *OC*, **9**, p. 269.
[21] [1785b] *OC*, **10**, p. 389.

express all that this entails:[22]

$$\iint Y'^{(i)} \, d\mu' \, d\omega' \, Q^i = \frac{4\pi Y^i}{2i + 1}, \tag{87}$$

where Q^i is a known function of μ and $\sqrt{1 - \mu^2} \cos(\omega' - \omega)$, by means of which U^i may be calculated.[23]

Armed now with a general theory of the attraction of spheroids, which was precisely what he had lacked in his earlier investigation of tidal oscillations [1779a], Laplace returned to that subject and demonstrated conclusively that the equilibrium conditions entail periodicity and that the equilibrium will be stable only if the density of a layer of fluid is less than that of the spheroid it covers. There was nothing new in this, but it served to complete his theoretical investigation of the phenomena of spheroids of revolution serving the law of gravity.

Two short memoirs, on the shape of the earth [1786a] and the rings of Saturn [1789a], apply the theory of gravitational potential, or the attraction of a spheroid as Laplace always called it, to their respective topics. Like the problem of tidal oscillations, both represent earlier interests that Laplace had been unable to resolve in a conclusive manner for lack of an adequate theory. Although both pertain, again like the tides, to application rather than to innovation, both also contain—perhaps for that very reason—important clarifications and simplifications of the mathematical formalism. Thus, in the paper on the figure of the earth, he gave much greater prominence to Equation (86), making it the basis of the analysis.[24]

That paper offers a particularly good example of Laplace's tendency to be preoccupied in the first instance with the analytical representation of the facts and only secondarily with the facts themselves, with how something could be the case, given what the case was. For he was not indifferent to the latter. In navigational and astronomical tables the apparent motions of the sun and moon are referred to the center of gravity of the earth; and it was incumbent, therefore, to know its precise location relative to the point of observation, particularly for the theory of the moon. If the flattening at the poles were 1/178, the lunar parallax would amount to 20″ at certain localities. Thus, it is again clear—to anticipate for a moment—that all these uncertainties about geodetic data were bound to reinforce the motivation to base the metric system on yet another survey of the meridian when the opportunity arose in the Revolution (see Chapter 17).

[22] *Mécanique céleste*, Book III, No. 17, *OC*, **2**, p. 47; Todhunter (1873) **2**, pp. 61–62.
[23] *Mécanique céleste*, Book III, No. 9, *OC*, **2**, pp. 26–27.
[24] [1786a] *OC*, **11**, p. 12.

For the moment, however, the empirical state of the question was what it had been when Laplace in his initial programmatic dual memoir on probability and universal gravitation [1776a] first confronted the discrepancy between the shape of the earth calculated from measurements of the length of a seconds pendulum at different latitudes and that given by the classic surveys of the lengths of the arc in Lapland, Peru, the Cape of Good Hope, and France. From the former data, the ratio of the minor to the major axis came out to be 320:321, and from the latter, 249:250. A brief calculation based on data from these surveys (apparently Laplace's first application of error theory to instrumental data) finds the probability negligible that the departure from the elliptical had resulted from observational error. The survey data also yielded a finite probability that the northern and southern hemispheres are not symmetrical, as well as a possibility that the earth might not be a solid of revolution at all.

As for the lengths of the radii determined by measurements of the seconds pendulum, they too failed to satisfy an ellipsoidal figure. That would have been troublesome to theory only if the earth were assumed to be of homogeneous density, for the differences did follow a regular pattern; and, most important, values for the force of gravity calculated from the length of the seconds pendulum approximated very closely to the law that its variation is proportional to the square of the sine of the latitude. This relation being of the utmost significance for the theory of the earth, Laplace proposed to combine it with the equilibrium conditions for the oscillations of the seas in order to draw from the combination the law of the variation of the radius of the earth.

He had already shown how the expression for the radius of any nearly spherical spheroid may take the form

$$1 + \alpha(Y^0 + Y^2 + Y^3 + Y^4 \cdots). \tag{88}$$

In the case of the earth, the equilibrium conditions for the seas require that $Y^1 = 0$ and αY^0 be a constant.[25] If the equilibrium is to be stable, it is a further condition that the axis of rotation be one of the principal axes of the earth. That requirement entailed the following form for Y^2:

$$H\left(\mu^2 - \frac{1}{3}\right) + H^{IV}(1 - \mu^2)\cos 2\omega, \tag{89}$$

where the constants H and H^{IV}, depending on the physical constitution of the earth, are to be determined by observation.[26]

[25] Ibid., p. 13.
[26] Ibid., p. 16.

These are equilibrium conditions that would hold for any celestial body covered by a fluid. In the case of the earth, measurements of the length of the seconds pendulum make it possible to calculate the values numerically. The constant H comes out approximately equal to 0.003111; H^{IV} is negligible relative to H; the quantity $Y^3 + Y^4 + \cdots$ is very small by comparison to Y^2, and so also is its first derivative by comparison to the first derivative of Y^2. Thus there would be no detectable error in calculating values for the radius of the earth, and also for its first derivative, from the formula

$$1 - 0.003111\left(\mu^2 - \frac{1}{3}\right). \tag{90}$$

What the surveys of the meridian show is that the same approximation is invalid for the second derivatives of the terrestrial radius. The reason for the discrepancy between the results of the two methods for calculating the length of the radius is that the function $Y^3 + Y^4 + \cdots$ becomes significant on a second differentiation. The expression for the radius of the earth has the form

$$1 + \alpha H\left(\mu^2 - \frac{1}{3}\right) + \alpha Y^i, \tag{91}$$

where αY^i is the term representing the variation between the calculated value and the value that satisfies the law of the variation of gravity with the square of the sine of the latitude. Given this expression, the formula for the length of the seconds pendulum is

$$l = L\left[1 + \alpha\left(H + \frac{5}{2}\phi\right)\left(\mu^2 - \frac{1}{3}\right) + (i - 1)\alpha Y^i\right], \tag{92}$$

where l and L are the lengths of the seconds pendulum corresponding to the force of gravity at two different latitudes.[27] The formula for the degree of the meridian is

$$c + \frac{2}{3}\alpha cH - 3\alpha cH\left(\mu^2 - \frac{1}{3}\right) - i(i + 1)\alpha c Y^i + \alpha c\frac{\partial(\mu Y^i)}{\partial\mu}$$

$$- \alpha c\frac{\dfrac{\partial^2 Y^i}{\partial\omega^2}}{1 - \mu^2}. \tag{93}$$

[27] Ibid., p. 18.

We need not follow the calculation in detail in order to seize Laplace's explanation, which is that the term Y^i, representing the variation of the value for the radius calculated from observation by means of the value that satisfies the law of the square of the sine of the latitude, is differentiated once for the pendulum method and twice for the geodetic method. It becomes detectable only in the latter case.

The paper on the rings of Saturn [1789a] was a much more provisional exercise. An "essay for a theory" Laplace called it, pending the development of telescopes powerful enough to reveal the correct number and dimensions of the rings. Like Galileo, Huygens, and many others, Laplace quite evidently found the phenomenon one of the most tantalizing in the whole field of astronomy. In his hands, of course, it was bound to take the form of a gravitational problem. He had first alluded to it in print in a special case discussed at the end of the opening section of his earliest major memoir on spheroid attraction.[28] There the rings are mentioned as the phenomenon instantiating the problem of the (reversed) attraction of an external point for a spheroid. The manuscript draft of that section, which he read on 22 January 1777 (20), closes with an undertaking to devote a future memoir to the problem of the figure of Saturn and the influence on it of the attraction of the rings. That was omitted in the printed text. Now that Laplace was finally redeeming the promise twelve years later, the problem was rather the figure of the rings themselves. There is still some appearance of hesitation. Laplace did not publish the memoir until 1789, five years after submitting the basic theory of spheroidal attraction [1785b] to the Academy (48).

Given the paucity of data, the discussion offers a particularly transparent example of recourse to a mathematical model. Instead of seeking the stability of the ring in any sort of mechanical connection between the particles, Laplace imagined its surface covered with an infinitely thin layer of a fluid in equilibrium under the influence of the forces at work. The shape of the ring would then be determined by the equilibrium conditions of the fluid. The most notable single feature of the memoir is the statement at the outset of the basic equation of spheroidal attraction theory, here referred to the rectangular coordinates that make it immediately recognizable for the first time as the potential function of later physics:[29]

$$0 = \frac{\partial^2 V}{\partial x^2} + \frac{\partial^2 V}{\partial y^2} + \frac{\partial^2 V}{\partial z^2}. \tag{94}$$

[28] [1778a] *OC*, **9**, pp. 86–87.
[29] [1789a] *OC*, **11**, p. 278.

In the case of a spheroid of revolution, the equation becomes

$$0 = \frac{1}{r}\frac{\partial V}{\partial r} + \frac{\partial^2 V}{\partial r^2} + \frac{\partial^2 V}{\partial z^2}, \tag{95}$$

where $r^2 = x^2 + y^2$. If the body is a solid sphere, or a hollow sphere with a homogeneous shell, the equation becomes

$$0 = \frac{2}{r'}\frac{\partial V}{\partial r'} + \frac{\partial^2 V}{\partial r'^2}, \tag{96}$$

where $r' = \sqrt{r^2 + z^2}$.

Investigating the continuity of the ring, Laplace calculated that since the mass of Saturn must be much greater than that of the ring, and since a sphere exerts a much greater force on a particle at its surface than a flattened body of the same mass would do, the relation between the forces of attraction at a point on its inner circumference, its outer circumference, and the surface of Saturn are such that the ring must in fact be a series of concentric rings. He claimed to have been able to predict the discontinuity between the rings from the theory of gravity alone. He then showed how such a ring may be generated mathematically by a very flat ellipse of which the prolongation of the major axis passes through the center of Saturn and which revolves about that center in a plane perpendicular to its own.

There is no point in following the formalism. Todhunter reproduced it more fully than was his wont, and Laplace himself omitted the derivations depending on the most questionable hypotheses from *Mécanique céleste*.[30] It involves the analytic necessity of constructing another ellipse with an infinite major axis on the same equator as the generating ellipse. The two must exert an identical attraction at every point. The ring might then be invested with unequal dimensions in its different parts—and even with double curvature. It had to be so, for if the ring were circular and concentric with Saturn, it would collapse into the planet. The most important findings are that the conditions for stability require that it revolve in the equatorial plane of Saturn, along with the four inner satellites, and that the center of gravity not coincide with the center of rotation. Abstract though the analysis was, Laplace pictured the rings physically as solid bodies, oscillating stably in asym-

[30] Todhunter (1873) **2**, pp. 65–73. The topic occupies Book III, Chapter 6 of *Mécanique céleste*, *OC*, **2**, pp. 166–77, where it formed the point of departure of Maxwell's investigation of the problem. See Harman (1992), pp. 486–87.

metrical rotation around the planet in such a way that their centers of gravity described elliptical orbits about the center of gravity of the planet even like normal satellites. With the delay in publication until 1789, the paper on Saturn's rings comes as an afterthought on the theory of attraction in the midst of the series of memoirs on planetary motion.

CHAPTER 16

Planetary Astronomy

ONLY IN the series on the solar system does vindication of its stability, which became the central feature of the Laplace stereotype through its celebration in the nineteenth century, finally appear to be the central motivation.[1] Laplace himself began insisting upon it in a sequence of five memoirs imparting his main discoveries in planetary theory. He composed them between November 1785 and April 1788 (50, 52, 53, 55, 56). Scientifically, however, they inaugurated an even lengthier preoccupation, for during the next twenty years his attention remained focused on problems of planetary astronomy, practical as well as theoretical, until the completion of Volume IV of *Mécanique céleste* in 1805. A near coincidence marks the start of that orientation. He read the first of these pieces, "Mémoire sur les inégalités séculaires des planètes et des satellites" [1787a], on 23 November 1785 and one week later (51) read the last of his demographic papers ([1786c]; see chapter 13).

The [1787a] paper on the secular inequalities of the planets, a relatively brief piece by Laplace's standards, must be considered, together with "Théorie des attractions des sphéroïdes et de la figure des planètes" [1785b] and "Mémoire sur la probabilité des causes par les événements" [1774c], one of the signal memoirs in his oeuvre. It opens with a serene and comprehensive survey of the history and state of the question of apparently cumulative discrepancies between the theoretical and observed positions of celestial bodies, and largely resolves them with a statement of two of his most famous determinations, the interdependence of the acceleration of Jupiter's mean motion with the deceleration of Saturn's, and the rigorous necessity for the mathematical games played by the Jovian moons, the three inner satellites of Jupiter, in the figure dance of their revolutions around their parent.

The second two memoirs in the sequence [1788a, 1788b], actually a single enormous memoir with a sequel, on the theory of Jupiter and Saturn, contain calculations in detail. Only the further argument, that the apparent acceleration of the mean motion of our own moon over time depends upon the action of the sun compounded by variation of

---

[1] For a history of the most important of these problems, the great inequality of Jupiter and Saturn, and a detailed explanation of Laplace's resolution of it, see the admirable monograph of Curtis Wilson (1985).

the eccentricity of the earth's orbit, would appear to have come to him after writing the covering paper on secular inequalities. He announced it in the preamble to the Jupiter and Saturn memoir, which was presented on 6 May 1786 (52), reserving its development for a further memoir on lunar theory, the fourth of the series [1788c], which he submitted to the Academy of Science on 19 December 1787 (55).

Laplace began the series by explaining the distinction that astronomers habitually made between periodic and secular inequalities in the ellipticity of planetary orbits. The former depend on the positions of the planets in orbit, relative to each other and to their aphelions, and compensate for themselves in a few years time. They are to be considered as very small oscillations on either side of a point in motion on an ellipse described in consequence of the attractive force exerted by the sun alone. Secular inequalities are those that modify the elements of the orbits themselves, their inclinations, eccentricities, and the longitudes of their nodes and aphelia. The changes happen very slowly. The effects on the shape and position of the orbits are undetectable in the course of a single revolution and become manifest only over centuries—hence the term "secular."

The terminology may have become a touch inappropriate, since what Laplace was setting out to demonstrate precisely was that the secular inequalities of the eccentricities and inclinations were themselves periodic, the periods occupying centuries. It is true that the term "secular" did not necessarily imply that very gradual inequalities need to be indefinitely cumulative, but such was certainly the apprehension. He himself spoke of them, perhaps inadvertently, as "accumulating ceaselessly."[2] Also, he was slightly less than candid—or at any rate clear—about the background of his own views on the matter. The most important secular inequalities (if any such there were) would be those that change the mean motion of a planet and with it, by Kepler's third law, the mean distance from the sun. Practical astronomers were thereby constrained to include a correction factor proportional to the square of the time in their expressions for the mean motions of Jupiter and Saturn.

In writing now about his own early investigation of this problem, Laplace claimed to have found that theory admitted no secular inequality in the mean motions or mean distances and to have concluded that such an inequality had been nonexistent or at least unobservable throughout the recorded history of astronomy.[3] What he had actually

---

[2] [1787a], *OC*, **11**, p. 49.
[3] [1776a, 2°]; see chapter 4.

said was not quite that. Rather, it was that theory proves the nonexistence of secular inequalities due to the interaction of any two mutually gravitating planets in historic time. He recalled that he had once thought to invoke the action of the comets to explain the undeniable speeding of Jupiter and slowing of Saturn over the span of historic time. Although the analysis did not preclude such a hypothesis, the physics did when he came to consider how slight the masses of the comets are. Moreover, whether the comets or some other agents were at work, Laplace had not at this early stage discussed or even mentioned the possibility of a long-term periodicity in these greater inequalities.

That strategy for saving the system was suggested by the work of Lagrange, who investigated the problem of planetary perturbations in a series of studies between 1775 and 1785. A crucial memoir of 1783, "Sur les variations séculaires des mouvements moyens des planètes," addressed the question whether, even allowing unlimited time, the perturbing forces of other planets, themselves subject to variations in the elements of their orbits, could cumulatively alter the mean motions and mean distances of a planet and thus disturb its service to Kepler's third law.[4] Lagrange's was a more abstract analysis than Laplace's had been, or would be, and he extended it to investigating the terms involving the squares of the eccentricities. There he did find a secular equation; but its maximum value was one-thousandth of a second, and its effect was negligible over all time, even in the cases of Jupiter and Saturn, which exhibited the largest inequalities. The question, therefore, was not whether those inequalities were to be explained by expressions involving lengthy periodicity, but how to do it. It was unthinkable that they should be random. This was ever the sort of problem at which Laplace excelled, and he set himself to examining the figures.

His first hope was that the mutual gravitation of the two planets might suffice to reveal periodicity. In planetary interaction over long periods, the sum of the masses of each planet divided by the major axis of the respective orbits is approximately constant. Thus, given Kepler's third law, if Saturn is slowed by Jupiter, Jupiter should be accelerated by Saturn. Their masses are respectively $1/1067.195$ and $1/3358.40$ that of the sun. It followed that the ratio of Jupiter's acceleration to Saturn's deceleration should have been approximately 7:3. Halley assumed $9°16'$ for the deceleration of Saturn in two millennia. If that were correct, Jupiter's acceleration should have been $3°58'$, an amount that differed from the tables by only nine minutes of arc. Thus, near equality made it probable that the variations in speed of Jupiter and Saturn are an effect

[4] OL, 5, pp. 381–414. For Lagrange's work on planetary perturbation theory, and its importance to Laplace, see Wilson (1985), pp. 195–221.

of their mutual gravitation. It had been established, however, (by Lagrange and himself) that the action of gravity cannot produce any inequality, whether cumulative, or of a period, even a very long one, that is independent of the positions of the planets. Since their mutual action could produce only inequalities that do depend on their dual configuration, just such an inequality may well be the cause of the observed variations in speed. The problem was to find it in the theory.

That the mean motions were nearly commensurate was well known: Five times the mean motion of Saturn almost equals twice the mean motion of Jupiter. It was characteristic of Laplace that he should have seen how to exploit this fact. It led him to surmise, he explained, that in the differential equations of motion of the two planets, the terms with an argument of $5n' - 2n$ (his designation for the two mean motions) could become detectable on integration, even though they were multiplied by the cubes and products of three dimensions of the eccentricities and inclinations of the orbits. Accordingly, he took the above inequalities to be indeed the probable cause of the variations in speed of the two planets and set out to make the "long and difficult" calculation that would demonstrate the relation. The expressions for Saturn yielded a secular equation of around 47', with a period of approximately 877 years, which depended on $5n' - 2n$; the theory of Jupiter contained an equation of 20' with opposite sign and identical period. The ratio is about 3:7.[5] Recalculating in the full Jupiter-Saturn memoir [1788a], Laplace found 48'44" for Saturn and 20'49" for Jupiter with a period of 929 years.

With these figures, and designating by nt and $n't$ the sidereal motions of Jupiter and Saturn respectively since 1700, the longitudes, reckoned from the equinox of 1700, will be given by the following formulas. For Jupiter:

$$nt + \epsilon + 20' \sin(5n't - 2nt + 49°8'40''), \qquad (97)$$

and for Saturn:

$$n't + \epsilon' - 46'50'' \sin(5n't - 2nt + 49°8'40''). \qquad (98)$$

where ϵ and ϵ' are constants that depend on the longitudes of the two planets on 1 January 1700. In order to evaluate the acceleration and deceleration empirically, the mean motions determined over a very long period could be compared to the values obtained from recent data. The time elapsed between the opposition of Saturn recorded for 228 B.C. and that for 1714 should give the "true" mean motion.

[5] [1787a], OC, **11**, pp. 49–56.

If the deceleration given by the above theory was correct, the value computed for the mean motion using the observations of 1595 and 1715 should have been too small by 16.8″. In fact the deceleration was 16″, so that the fit between observation and experiment was as good as the relative imprecision of sixteenth-century observation could allow. Agreement with Halley's tables was equally good for the acceleration of Jupiter, which came out in the ratio of 7:3 to be approximately 7″. The mean motion of Jupiter was at its maximum and of Saturn at its minimum around A.D. 1580. Since then the apparent mean motions had been coming closer to the "true" values taken over all time.[6] Laplace showed his calculation to be conformable to a set of corrections to be applied over short periods that Lambert had published in 1775.[7] His object had been an empirical determination of the "law of error"—this would appear to be the first time Laplace used the phrase—in Halley's tables.

Also contributing to determination of the inequalities of Jupiter and Saturn were terms due to the ellipticity of the orbits on the assumption that $5n' = 2n$ exactly, along with other terms taking account of perturbations that were due to variations in eccentricity of the orbits. Previous mathematicians had calculated these effects, but with such discordant results that the whole thing needed to be verified. That task Laplace deferred to a further memoir containing the calculations—here he gave only the results—in which all ancient and modern oppositions of both planets would be accommodated and certain strange aberrations of Saturn explained. Centuries would have been required to accomplish that empirically: "Thus, on this point, the theory of gravity has moved ahead of observation."[8]

The present memoir contained the analysis for two other objects that were also intrinsic to the system of the world. The first still concerned the uniformity of mean motion of celestial bodies, in the case of satellites rather than the primary planets, however, and specifically that of the moons of Jupiter. At issue was the invariance of their mean distances from a principal center of force. The second set of problems concerned the other variations of celestial orbits. The methods for determining inclinations and eccentricities were very simple. But were these elements oscillating within certain, presumably narrow limits, or did cumulative changes occur?

---

[6] Ibid., pp. 53–54.

[7] Johann Heinrich Lambert, "Résultats des recherches sur les irregularitès du mouvement de Saturne et de Jupiter," *Mouveaux mémoires de l'Académie Royale de Berlin* (1773/1775), pp. 216–21.

[8] Ibid., *OC*, **11**, p. 56.

The motions of the moons of Jupiter were well-studied phenomena when Laplace took up the question of their stability over the span of astronomical time. They were most famous for the observations that had enabled Römer in 1676 to measure the velocity of light by recording the dependence of variations in the apparent time of Io's eclipses on the distance of Jupiter from the earth. Such observations were of little use to a navigator peering through his telescope on the bridge of a ship pitching and rolling in heavy seas, or even in a light roll. They had, however, been the main resort for cartographical surveyors in determining longitude prior to the availability of reliable chronometers in the late eighteenth century. The Swedish astronomer Pehr Wargentin devoted much of his life to perfecting the tables, which were first published in 1746, and which he revised constantly in correspondence with Lalande. In the theory of these bodies, the most important item by far was a memoir by Lagrange, which had won the prize for 1766 set by the Academy of Science for a study of their inequalities.[9] Lagrange calculated the effects on each of the four satellites of the oblateness of Jupiter, of the action of the sun, and of the action of the other three. He did not investigate the relation of the inequalities in longitude to variations of eccentricity and inclination; and his results, however impressive analytically, were of no help in improving the tables. Moreover, he made a mistake: He assumed that the angle between the planes of Jupiter's equator and orbit was negligible.

Such, in summary, was the state of knowledge when Laplace took the problem in hand. The mean motion of the first satellite is about twice that of the second, which in turn is about twice that of the third. It followed that the difference between the mean motions of the first and second should be twice the difference between the mean motions of the second and third. Curiously enough, however, the tables showed the latter doubling to hold very much more exactly than the former. (The fourth was the odd moon out, its motions being incommensurable with the inner three; and Laplace neglected its influence here, as well as that of the sun.) If the tripartite relation were to prove really exact, so that

$$n + 2n'' = 3n',\qquad(99)$$

it would follow from this simple equation that the three moons could never be in simultaneous eclipse. In point of predictable fact, the data from Wargentin's tables showed that such an eventuality could not occur in the next 1,317,900 years, and it would be entirely precluded by a modification in the annual motion of the second moon smaller than

[9] "Recherches sur les inégalités des satellites de Jupiter," *OL*, **6**, pp. 67–225.

the margin of error in the tables. "Now," Laplace observed,

> it may be laid down as a general rule that, if the result of a long series of precise observations approximates a simple relation so closely that the remaining difference is undetectable by observation and may be attributed to the errors to which they are liable, then this relation is probably that of nature.[10]

The remark offers an especially strategic example of the role of probability, causality, and error theory in his planning of an investigation. The equation (99) of the mean motions of Jupiter's moons could scarcely be thought the effect of chance; yet it was improbable that those bodies had originally been placed at just the distances from Jupiter that it requires. It was natural, therefore, to try out the hypothesis that their mutual attraction is the true cause. Thus, Laplace observed, in anticipation of a later investigation [1788c], the action of the earth on the moon had brought about the equality between its periods of revolution and rotation, which might have been very different at the beginning.

The object of the research, therefore, was to determine whether the mean motions of the first three Jovian satellites became and remained stable by virtue of the law of universal gravitation. Laplace based the analysis on the motion of the second moon. He set up general equations of motion of a system of mutually attracting bodies and applied them to the motions of all three satellites, showing how the principal inequalities of the second depend on the influence of the first and third. In his expressions,

t designates the time

n, n', and n'', the mean motions of the first three satellites

s, the quantity $n - 3n' + 2n''$

$nt + \epsilon$, $n't + \epsilon'$, $n''t + \epsilon''$, the projections onto the same plane of the mean longitudes taken from the x-axis and

V, the angle $(2n'' - 3n' + n)t + 2\epsilon'' - 3\epsilon' + \epsilon$, the mean longitudes being taken from a fixed point in Jupiter's orbit.

The tables called for s and V to be approximately constant in value, but Laplace could find no reason for that condition in any terms that depended on the first powers of the perturbing masses. Accordingly, he proceeded to an examination of terms that depended on the squares and products of the masses. Multiplication of the expressions by the masses of the satellites, taken two at a time, did introduce into the

---

[10] Ibid., *OC*, **11**, p. 57.

values of s and V quantities proportional to time, giving for the variation of s,

$$\delta s = \alpha n^2 t \sin V, \tag{100}$$

where α stands for an immensely complicated function in the equation[11]

$$2\delta n'' - 3\delta n' + \delta n = \frac{3}{2} a n^2 t \sin V \alpha. \tag{101}$$

Laplace then eliminated from Equation (100) the quantities that increase with time, using his method of varying the constants of integration (chapters 7 and 10). He thus obtained two first-order differential equations among s, V, and t:

$$\frac{ds}{dt} = a n^2 \sin V \tag{102}$$

and

$$dV = dt(2n'' - 3n' + n) = s\, dt, \tag{103}$$

whence

$$\frac{d^2 V}{dt^2} = \alpha n^2 \sin V. \tag{104}$$

Multiplying by dV and integrating gives

$$\frac{\pm dV}{\sqrt{\lambda - 2\alpha n^2 \cos V}} = dt, \tag{105}$$

where λ is an arbitrary constant. Of the three possibilities, first that λ is positive and greater than $\pm 2\alpha n^2$, second that α is positive and $\lambda < 2\alpha n^2$, and third that α is is negative and $\lambda < -2\alpha n^2$, the second case entailed a negative value for V oscillating around a mean of 180°; and the data showed this to be the case in nature, with the oscillations of very small amplitude.

Laplace drew four consequences from the analysis pushed thus far. The first was practical. Since both s and V are periodic, the relation (99), $n + 2n'' = 3n'$, is rigorously exact, and Wargentin's tables needed only slight corrections. The second was historical. It was not a necessary condition that the three satellites should originally have been placed at the distances from Jupiter's center of mass that, in service to Kepler's laws, would result in mean motions satisfying the relation (99). If they

---

[11] Ibid., p. 77.

had merely been close to those distances (and Laplace calculated what the limits were), then their mutual attractions would bring them into that relation with each other. The third consequence concerned the stability of the system. There was no need to fear that the tables for the principal equation of the second satellite would be off, even after the lapse of many centuries. The fourth conclusion, and in Laplace's eyes the most important, raised problems for the future. Consideration of the angle V created a second condition that the tables had to satisfy, namely that

$$nt + 3n't - 2n''t = 180° \qquad (106)$$

where nt, $n't$, and $n''t$ are the respective mean longitudes at time t.

The angle V is subject to a periodic inequality that Laplace likened to the oscillations of a pendulum. That inequality affects the motions of the three satellites in varying degrees, according to the ratio of their respective masses and distances from the center of Jupiter. The mass of the second is adequately determined by the inequalities it causes in the motions of the first—that is why Laplace could make its motion the basis of reference. The masses of the first and third were still unknown, however. All that could be said was that a relation exists that accounts for the inequalities of the second. That was how he was able to determine that the libration of V was contained within limits of 4 1/8 and 11 1/3 years. Both its zero point and amplitude were quantities to be determined by observation.

In considering only the interaction of Jupiter's first three satellites, their motion still depended on no fewer than nine second-degree differential equations, the integrals of which contained eighteen arbitrary constants. The eccentricities, orbital inclinations, and positions of nodes and aphelia served to determine twelve of those constants. The mean motions and their epochs would give six others, without the two conditions to which these arbitrary quantities are subject. That reduced the six to four. To answer to that, the expression for V contained two arbitrary quantities. Since the tables do work well without regard to the periodic inequality of V, its amount had to be small. But the uncertainties still prevailing over most of the elements of the theory of Jupiter's satellites made the determination very difficult, and Laplace left that question to the astronomers to resolve. He had shown them the two conditions that the tables must satisfy, namely Equations (99) and (106). Concluding this discussion, he again recurred to the analogy with our own moon. Those conditions would still obtain if, like the moon, the Jovian satellites showed an acceleration in the mean motions such that the mean distances progressively decreased. Even as the satellites drew in on Jupiter, the same relations would be preserved.[12]

Thus, Laplace showed that the mean motions of Jupiter's moons are subject only to periodic inequalities, and he felt justified in drawing conclusions for the whole planetary system from this test case. Considering only the laws of universal gravity, the mean distances of celestial bodies from their principal centers of force are immutable. It by no means followed that the same is true of the other orbital elements, namely eccentricities, inclinations, and the positions of nodes and aphelia, all of which undergo continual change.

Good methods for determining these quantities existed, subject to the hypothesis that the orbits differ little from the circular and that their planes are only slightly inclined to that of the ecliptic. Laplace himself had long since shown [1776c] that the eccentricities and inclinations of the orbits are bound to remain small under the operation of gravity, provided that only two planets are considered. In 1782 Lagrange had further shown that the same is true generally of the location of nodes and aphelia, given plausible hypotheses about the masses (over several of which some uncertainty did linger).[13] But before the entire system of the world could be deduced from the law of gravity it remained to demonstrate that the eccentricities and inclinations of any number of planets are contained within narrow limits. That was the final object of the present memoir. Analytically, what was required was to prove that the expressions for the secular inequalities of eccentricities and inclinations contain neither secular terms nor exponential terms. The analysis is uncharacteristically brief and occupies the final two articles.[14]

In those expressions,

$m, m', m'', \ldots$, designate the relative masses of the planets, referred to the sun as unity

$a, a', a'', \ldots$, the semi-major axes

$ea, e'a', e''a'', \ldots$, orbital eccentricities

$V, V', V'', \ldots$, longitude of aphelions

$\theta, \theta', \theta'', \ldots$, tangent of orbital inclinations, referred to a fixed plane.

$I, I', I'', \ldots$, longitude of ascending nodes.

---

[12] Ibid., *OC*, **11**, pp. 60–61. In a private communication Curtis Wilson notes that in recent years it has been shown that resonances such as those in the mean motions of the first three Jovian satellites are in large part the effect of tidal friction. The rotation of the planet thereby transmits energy to the innermost satellite which, in turn, shares its energy with those further out. The satellites thus fall into resonance dynamically.

[13] "Théorie des variations séculaires des éléments des planètes," (part 2, 1784) *OL*, **5**, pp. 211–344.

[14] [1787a], *OC*, **11**, pp. 88–92.

The quantities $e \sin V$, $e \cos V$, $e' \sin V'$, $e' \cos V'$, ..., $\theta \sin I$, $\theta' \sin I'$, $\theta \cos I$, $\theta' \cos I'$, ..., could then be given by linear differential equations with constant coefficients. Since the eccentricities and inclinations were very small, the system of equations for the eccentricities was independent of the system for the inclinations. The former system was the same as if the orbits were coplanar, and the latter the same as if they were circular. When the former system was integrated, each of the quantities $e \sin V$, $e \cos V$, $e' \sin V'$, $e' \cos V'$, ..., was given by the sum of a finite number of sines and cosines of angles proportional to time t. The numbers by which the times were to be multiplied to give the angles were the roots of an algebraic equation of degree equal to the number of planets. Laplace called that equation k. The same was true for the quantities of the second system, although the equation on which the angles depend was not the same. He called that equation k', referring the reader to his discussion in [1776c] and to Lagrange's "Théorie des variations séculaires des éléments des planètes."[15]

If all the roots of k and k' were real and unequal, the values of the preceding quantities would contain neither secular terms nor exponentials and would thus remain within narrow limits. It would be otherwise if some of the roots were imaginary, for then the sines and cosines would change into secular terms or exponentials. But, in any event, the values of $e \sin V$, $e \cos V$, $e' \sin V'$, $e' \cos V'$, ..., would always take the form shown in the four expressions

$$e \sin V = \alpha f^{it} + \beta f^{i't} + \cdots + \gamma t^r + \lambda t^{r-1} + \cdots + h$$

$$e \cos V = \mu f^{it} + \epsilon f^{i't} + \cdots + \phi t^r + \psi t^{r-1} + \cdots + l$$

$$e' \sin V' = \alpha' f^{it} + \beta f^{i't} + \cdots + \gamma' t^r + \lambda' t^{r-1} + \cdots + h'$$

$$e' \cos V' = \mu' f^{it} + \epsilon' f^{i't} + \cdots + \phi' t^r + \psi' t^{r-1} + \cdots + l'.$$

(107)

(It should be explained that f stands for the e of later notation—that is, the base of the natural system of logarithms—but i is not $\sqrt{-1}$.) In these expressions, the coefficients α, β, μ, ϵ, ..., α', β', μ', ϵ', ..., of the exponential terms are real quantities not subject to exponentials. They may, however, become functions of the arc t and of sines and cosines of angles proportional to that arc. As for the quantities γ, λ, ϕ, ψ, ..., h, l, and γ', λ', ϕ', ψ', ..., h', l', ..., they are real, containing no exponentials or secular terms, and hence are either constant or periodic.

[15] [1776c], *OC*, **8**, p. 406; *OL* 5, pp. 211–344, esp. pp. 249. 255.

Laplace then supposes that, without regard to signs, $i > i'$, $i' > i''$, ..., when $e = (e \sin V)^2 + (e \cos V)^2$. It would then follow that

$$e^2 = (\alpha^2 + \mu^2)f^{2it} + \cdots + (\gamma^2 + \phi^2)t^{2r} + \cdots + h^2 + l^2. \quad (108)$$

Similarly,

$$e'^2 = (\alpha'^2 + \mu'^2)f^{2it} + \cdots + (\gamma'^2 + \phi'^2)t^{2r} + \cdots + h'^2 + l'^2, \quad (109)$$

and so on. These expressions give the values of the eccentricities, which hold good, however, only for a limited period of time. The precondition that they remain small, varying within narrow limits, is valid if—and only if—all the roots of the equation k are in fact real and unequal. It was very difficult to show this directly, and Laplace went back to an expression for the eccentricities that he had obtained in discussing the general equations of motion of a system of mutually attracting bodies under the law of gravity[16]

$$c = m\sqrt{\frac{a(1 - e^2)}{1 + \theta^2}} + m'\sqrt{\frac{a'(1 - e'^2)}{1 + \theta'^2}} + m''\sqrt{\frac{a''(1 - e''^2)}{1 + \theta''^2}} + \cdots ,$$

$$(110)$$

neglecting constant and periodic quantities of the degree m^2, where a, a', a'', ... are the semi-major axes of the orbits of bodies m, m', m''; ea, $e'a'$, $e''a''$, ... are the eccentricities; and θ, θ', θ'', ... are the tangents of the inclinations; and c is an arbitrary constant.

If quantities of the degree e^4, $e^2\theta^2$, and θ^2, are ignored, then Equation (110) becomes

$$c = m\sqrt{a} + m'\sqrt{a'} + \cdots - \frac{1}{2}m(e^2 + \theta^2)\sqrt{a} - \frac{1}{2}m'(e'^2 + \theta'^2)\sqrt{a} - \cdots .$$

$$(111)$$

Furthermore, since the mean distances of the planets from the sun are unaffected by their interactions,

$$m(e^2 + \theta^2)\sqrt{a} + m'(e'^2 + \theta'^2)\sqrt{a'} + \cdots = \text{constant}. \quad (112)$$

It had already been pointed out that the values of e, e', e'', ... are independent of those for θ, θ', θ'', ..., so that they will be the same as if the latter were null. Thus Equation (112) reduces to

$$me^2\sqrt{a} + m'e'^2\sqrt{a'} + \cdots = \text{constant}, \quad (113)$$

---

[16] Ibid., *OC*, **11**, p. 89.

which is the equation that must be satisfied by the values of e, e', e'', ... after the lapse of any amount of time.

Substituting the general expression of those quantities as given in Equation (107) into Equation (113) gives

$$[m\sqrt{a}\,(\alpha^2 + \mu^2) + m'\sqrt{a'}\,(a'^2 + \mu'^2) + \cdots]f^{2it} + \cdots$$

$$+[m\sqrt{a}\,(\gamma^2 + \phi^2) + m'\sqrt{a'}\,(\gamma'^2 + \phi'^2) + \cdots]t^{2r} + \cdots$$

$$+ m\sqrt{a}\,(h^2 + l^2) + m'\sqrt{a'}\,(h'^2 + l'^2) + \cdots = \text{constant.} \quad (114)$$

This equation must hold whatever the value of t, and it is therefore essential to eliminate exponential powers of t and secular terms. To that end, the coefficient of f^{2it} is equated to zero, so that

$$0 = m\sqrt{a}\,(\alpha^2 + \mu^2) + m'\sqrt{a'}\,(a'^2 + \mu'^2) + \cdots . \quad (115)$$

Since $m\sqrt{a}$, $m'\sqrt{a'}$, ... are positive quantities, and since α, μ, a', μ', ..., are real quantities, Equation (115) will hold only on the supposition that $\alpha = 0$, $\mu = 0$, $\alpha' = 0$, $\mu' = 0$,.... It follows that there are no exponential terms in the values of e, e', e'', ...

Returning now to Equation (114) and equating the coefficient of t^{2r} to zero gives

$$0 = m\sqrt{a}\,(\gamma^2 + \phi^2) + m'\sqrt{a'}\,(\gamma'^2 + \phi'^2) + \cdots , \quad (116)$$

whence $\gamma = 0$, $\phi = 0$, $\gamma' = 0$, $\phi' = 0$,.... Thus, neither do the values for e, e', e'', ..., contain secular terms. They reduce, therefore, to periodic quantities of the form $\sqrt{h^2 + l^2}$, $\sqrt{h'^2 + l'^2}$,..., and we know from Equation (114) that these quantities serve the equation

$$\text{constant} = m\sqrt{a}\,(h^2 + l^2) + m'\sqrt{a'}\,(h'^2 + l'^2) + \cdots . \quad (117)$$

When the right-hand side of that equation is expanded in a series of sines and cosines, the coefficients of each sine and cosine vanish automatically.

As for the inclinations. the same reasoning may be applied to the expressions for θ, θ', θ'', ... Laplace thus satisfied himself analytically of the periodicity of variations in the orbital eccentricities and inclinations, thereby completing the mathematical demonstration that "the system of the planets is contained within invariant limits, at least with respect to their action on each other."[17]

[17] Ibid., p. 92.

The calculations comparing the theory of the long-term interdependence of the inequalities of Jupiter and Saturn to observations are enormously lengthier. Laplace deferred their publication to the "Théorie de Jupiter et de Saturne" [1788a, 1788b], presented in two installments. Unlike the analysis in the preceding memoir [1787a], which was mainly a mathematical exercise, they involved him in detailed reference to the corpus of recorded astronomical data in order to reach numerical solutions to his equations. He had tried, he said, to give his results a simple and convenient form. By contrast to predecessors, who considered only inequalities that are independent of the eccentricities and inclinations or that depend on their first powers, he had come to recognize that the resulting approximations were insufficiently exact. What distinguished his analysis was that in the theory of the mutual perturbations, he took into account inequalities that depend on the squares and higher powers of these variations, carrying the approximation up to the fourth power of the eccentricities. He had, moreover, found a method for determining them without entering into excessively lengthy calculation. By a happy coincidence the very necessity of recurring to these inequalities, that is, the very length of the period occasioned by the near commensurability of the mean motions, simplified their determination. For they became detectable in the theory as the effect of extremely small divisors.[18]

The need to consult the most ancient as well as the most recent data led Laplace into the way of certain historical reflections. For when modern observations were compared with ancient ones, Jupiter appeared to have speeded up and Saturn to have slowed down. But when modern observations were compared with each other, the contrary appeared to be happening. It was a curious chance that the turning point, the date at which Jupiter's mean motion was at its maximum and Saturn's at its minimum, should have been c. 1560 (Laplace had said 1580 in the previous memoir [1787a]). The changes had thus occurred at the time of the revival of astronomy in the generations of Copernicus and after. Since that juncture, the apparent values were approaching the true values of the mean motions. If astronomy had been "renewed"—apparently Laplace gave little credit to Arab contributions—three centuries sooner, its practitioners would have observed the contrary phenomena.

The motions that a culture attributes to the two planets might thus serve as an index of the period in which it was founded or at least sufficiently developed to support astronomical observation. Laplace would judge by data from Hindu astronomy that in India the mean

[18] Ibid., pp. 143–50. Cf. Bowditch 1 (1829), pp. 652–53 n.; Wilson (1985), pp. 256–61.

motions had been determined at a time when Saturn was moving at its slowest speed and Jupiter at its fastest. Two of the principal Hindu astronomical eras appeared to fulfill these conditions, one centering around 3103 B.C. and the other around A.D. 1491.[19] For Laplace was as credulous in accepting historical statements as he was critical of astronomical data. He probably garnered the former information from Bailly's *Traité de l'astronomie indienne et orientale* (1787), a work based rather on undigested tradition than on scholarship.

Saturn being the more recalcitrant and less studied of the two major planets, Laplace devoted the first installment of the memoir to calculating its theory in detail. The general strategy for finding its inequalities was to substitute its elements, together with those of Jupiter, into the formulas. Before that could be done, another anomaly besides the long inequality deriving from the action of Jupiter had to be noticed. It was responsible for the slight discrepancy in what was only a near equality between twice the mean motion of Jupiter and five times that of Saturn, and its amount was about ten minutes of arc. If that equality had been exact, it would have coincided with the inequality due to the ellipticity of the orbit. A complete theory of Saturn had to provide for all these inequalities. Working out the mathematical basis in the first section of the memoir required varying the constants in order to eliminate terms introduced in successive integrations that increased with time. These "arcs de cercle" would preclude stability for the system if they referred to something real, and Laplace included a more succinct exposition of his method for variation of constants than he had yet given.[20]

Before Laplace could give his formulas numerical values, which task he carried out in the second section, he had to know what the arbitrary constants represented. In the expressions for the elliptical motion of the planets, these constants were the mean distance, the inclination, the mean longitude at a given moment, the eccentricity, and the positions of aphelion and nodes. They came from the data of observation, which needed to be corrected both for inaccuracies and for the effect of the perturbations to which the bodies are subject. The latter effects had to be predicted from theory. For that purpose, the masses had to be known, taken relative to the sun as unity. In the cases of Jupiter and Saturn, determinations of their inequalities and of the elements of the orbits depended reciprocally one on the other. In a strict sense, therefore, the whole procedure might be thought circular.

Empirically, however, the determinations could be, and had to be, accomplished by successive approximations. Halley's tables were the

---

[19] [1788a], *OC*, **11**, pp. 178–79.
[20] Ibid., pp. 131–40.

main source of data for Saturn. Dating back to 1719, they needed systematic correction, and Laplace had first of all to arrive at a formula, or "law of errors" to be applied in using them.[21] He then perfected the law by trying it on numerous oppositions of the planet. Jupiter presented less of a problem: Wargentin's tables were more up-to-date for that planet than Halley's for Saturn, and their author had allowed for the inequalities that were independent of eccentricity and also for those depending on the first power of that element. He had not known about the effect of the long inequality on the mean motion, eccentricity, and position of aphelion, and he had bundled those effects into his terms for ellipticity, from which they could simply be subtracted in correcting the several elements. Laplace referred his numerical calculations to the beginning of the year 1750, mean time of Paris, choosing that date because it was in the midst of the period of good modern astronomical observations.

It will help visualize his tactics if some numbers for Saturn are reproduced:

Mean longitude ϵ'	$7^s 21°17'20''$
Mean longitude of aphelion ω'	$8^s 28°7'24''$
Eccentricity e'	0.056263
Mean longitude of ascending node	$3^s 21°31'17''$
Inclination of orbit to elliptic (tan θ')	$2°30'20''$
Sidereal motion during 365-day common year, n	$12°12'46.5'' = 43966.5''$
Mean distance to sun (earth-sun = 1)	9.54007.

(118)

The mean sidereal motion of Jupiter (n') was $30°19'42''$, or 109,182''. The problem was then to determine numerically the secular inequalities affecting the orbits of both planets, beginning with those dependent on the angle

$$5n't - 2nt + 5\epsilon' - 2\varepsilon.$$

Calculation from the theory yielded the amount by which to correct the observed value for this class of inequality. Next came the periodic inequalities that are independent of eccentricities and inclinations, followed by those that derive from the eccentricity, and finally, those depending on the squares and higher powers of eccentricities and inclinations. When Laplace included the conversion from mean to true anomaly, he could find the formula for determining the heliocentric longitude of Saturn in its orbit at any time past or future.

[21] Ibid., p. 190.

These measures presupposed, however, that the above elements for Saturn in 1750 were exactly correct, together with the theory by which the inequalities were determined. No doubt the approximation was very close, but the values needed to be checked against the entire body of modern observations. To that end Laplace designated by $\delta\epsilon$, $\delta n'$, $\delta e'$, and $\delta\omega'$ the corrections that would need to be applied to the 1750 values for mean longitude, annual mean motion, eccentricity, and position of aphelion, respectively. In determining these quantities empirically, he drew on twenty-four observations of Saturn, choosing instances well distributed in time from 1591 to 1785 and that his prior investigation of the law of error in Halley's tables had shown to be tolerably precise. Not that these could be supposed to represent the "true" longitudes—Halley had made his calculations on the basis of Flamsteed's observations and had been unaware of the phenomena of aberration and nutation, and also of more recent data on the location of certain reference stars. It would be well to calculate them all. Pending that immense task, however, Laplace's strategy compensated for the lack of precision of individual observations by the number he employed.

His procedure thus required minimizing the error, and to that end Laplace wrote twenty-four equations of condition for the theory of the motion of Saturn. Here, for example, is the equation for the opposition observed on 23 December 1679 at 10:39 P.M. Paris time, when the longitude was $34°54°0''$:

$$0 = 3'9.9'' + \delta\epsilon' - 70.01\,\delta n' + 2\delta e' 0.12591 - 2e'(\delta\omega' - \delta\epsilon')0.99204.$$

$$(119)$$

Stephen Stigler has characterized the solution of this system of twenty-four equations as an early example of a multiple linear regression.[22] For Laplace combined them into a set of four equations by (1) summing all twenty-four; (2) subtracting the sum of the second twelve from the sum of the first twelve; (3) combining one set of twelve in the order $-1 + 3 + 4 - 7 + 10 + 11 - 14 + 17 + 18 - 20 + 23 + 24$; and (4) combining the remaining twelve in the order $+2 - 5 - 6 + 8 + 9 - 12 - 13 + 15 + 16 - 19 + 21 + 22$. He then solved these four equations to get

$$\delta\epsilon' = 3'23.544''$$
$$\delta\omega' = 5'45''$$
$$2\delta e' = 30.094''$$
$$\delta n' = 0.11793'' \qquad (120)$$

[22] Stigler (1975).

Applied to the elements for 1750, these corrections gave

$$\epsilon' = 7^s 21°20'44''$$

$$\omega' = 8^s 28°13'19''$$

$$e' = 0.056336, \tag{121}$$

and the value of $\delta n'$ showed that the mean sidereal motion of Saturn had to be increased by 1/9 second.

Armed with these corrections, Laplace gave simple formulas for finding the location of Saturn at any instant, having now brought theory and practice into conformity. Thus, the mean longitude $n't + \epsilon'$ is given by taking the corrected mean longitude for 1 January 1750, $7^s 21°20'44''$ (121), and adding to it the angular distance traversed in the mean sidereal motion at a rate of 43966.6'' per year of 365 days for the number of days elapsed, instead of 43966.5'', as supposed in the data for Saturn (118).

Laplace employed a similar procedure with the other elements of the orbit and constructed a table, not only of the twenty-four oppositions used in refining the data, but of some forty-three extending from 1582 to 1786. In it he gave observed heliocentric longitudes together with the excess or residual differences between the observed value, the value that Halley had calculated in 1719, and the value that he had just calculated. Stigler points out that Laplace's solution—for the twenty-four equations that he linearized—included a residual sum of squares only 11 percent greater than a least-squares solution would have yielded, and that the residual sum of squares from Halley's calculation was eighty times greater. The increase of Laplace's powers is no less striking in the contrast between this memoir and his early essays in error theory (see chapter 4) than it is in astronomy itself. He concluded this part with an application of his formulas to the Ptolemaic data, the most ancient observation being that reported from the Babylonian legacy for 1 March 228 B.C. The memoir had grown so lengthy that Laplace postponed Jupiter to the sequel. There he again turned the crank to accomplish the mutual reconciliation of the observations with the theory of the largest planet, in its interaction with Saturn as in its lesser inequalities.[23]

More interesting for the history of Laplace's career is the evidence of his involvement in the activities of routine instrumental and positional astronomy. That had already begun, to be sure, in 1781 with the memoir on determination of cometary orbits and the discussion that followed Herschel's discovery of Uranus (chapter 13). With the Saturn memoir

---

[23] [1788b] *OC*, **11**, pp. 226–39.

[1788a] the association became systematic. Laplace read the first install-ment to the Academy on 10 May and the second on 15 July 1786 (52, 53). In the preamble to the second, he recognized that the theory of Saturn still contained three very small though detectable inequalities. Their sum came to less than one second of arc, but they would need to be taken into account in the most exact and rigorous possible calcula-tion. What was needed was that an astronomer familiar with such computations should rework all the oppositions of Jupiter and Saturn in the seventeenth and eighteenth centuries, calculating them directly from the perfected theory in each instance.

Jean-Baptiste Delambre thereupon came forward and volunteered to undertake that task—a *délicate et pénible* discussion, Laplace called it.[24] The word "drudgery" might occur to others. For his part, he made the small corrections just mentioned in the theory, and Delambre proceeded to compare the formulas to a very large number of observa-tions and to compile the results. A young scholar seeking a foothold, Delambre had moved from classical studies into astronomy by way of following Lalande's course in the Collège de France and had become his mentor's assistant. He became known through his dramatic success in observing the final moments of the transit of Mercury across the disk of the sun on 3 May 1786 after bad weather had led others to give up. It was exactly one week later, 10 May, that Laplace presented the first installment of his "Théorie de Jupiter et de Saturne" to the Academy (52). The opportunity Delambre seized to review the whole corpus of the data in the light of Laplace's theory opened the way to what became a distinguished career. He presented his *Tables de Jupiter et de Saturne* to the Academy on 26 April 1789 (57) and published them later in the year accompanied by an explanation of his methods. In 1792 Lalande included them in the third edition of his *Astronomie*, the standard manual of practical astronomy. Laplace characterized the tables as the first ever to be based "on the law of gravity alone," since they depended on observation only for the data needed to determine the arbitrary constants introduced by the integration of the differential equations of planetary motion. One particular advantage of that austerity was the possibility that they afforded of deciding whether causes of any sort external to the solar system could disturb its motions. More mundanely, wrote Laplace of Delambre, "I acknowledge with pleasure that, if my research is useful to astronomers, the merit is owing principally to him."[25] From then on, Laplace in his astronomy was never without a calculator at his beck and call.

[24] Ibid., p. 211.

[25] Ibid., pp. 211–12. For an exact estimate of the improvement in accuracy achieved by Delambre's review of the observational record of oppositions of Jupiter and Saturn on the basis of Laplace's theory of their perturbations, see Wilson (1985), pp. 277–86.

There remained the moon, the last member of the solar family whose apparent behavior failed to conform in all respects to the rule of universal gravity. Halley had discovered the acceleration of its mean motion, and since then astronomers corrected for it in their tables by adding to values for the mean longitude at a given date a quantity proportional to the number of centuries that would have elapsed after 1700. There was some disagreement among them on the exact rate of the increase, but none on its overall effect. Delambre had just confirmed that the secular motion was three or four minutes greater than in Babylonian times. As to the cause, the Academy had offered several prizes, but no one had been able to identify anything in the configuration or the motion of the earth or its satellite that would explain these variations in a manner conformable to the law of gravity. As in the case of planetary anomalies, various ad hoc hypotheses had been invoked—resistance of the ether, the action of comets, the transmission of gravity with finite velocity.

It seemed a pity to Laplace that the moon should still be a renegade, and he had decided to investigate the phenomenon, failing in his first few attempts and telling of his eventual success in "Sur l'équation séculaire de la lune" [1788c], the final substantive memoir in the seminal series on planetary motion. He had been led to his formula for the moon—so he recalled in later years—by the application that he had been able to make of his theory of Jupiter and Saturn to the problem of the moons of Jupiter.[26]

The cause, in a word, lies in the action of the sun combined with variations in the eccentricity of the earth's orbit owing to the action of the planets. The action of the sun—to take that first in a qualitative summary of the finding—tends to diminish the gravitational attraction of the earth for the moon and thus to expand its orbit. In itself that effect would slow the angular velocity and appear as a deceleration in the lunar mean motion. The solar action is strongest when the sun is in perigee, and the lunar orbit is then at its maximum, decreasing as the sun moves toward apogee. Thus, in the motion of the moon there is an annual equation that is identical in period with that of the apparent motion of the center of the sun, though opposite in sign. But the gravitational force exerted by the sun on the moon also varies with the changes in the terrestrial orbit that are caused by the resultant of the influence of the other planets. Over the long term, the major axis is fixed and the other elements—eccentricity, inclination, position of nodes, and aphelion—change incessantly. The mean force exerted by the sun on the moon varies as the square of the eccentricity of the terrestrial orbit. This effect produces contrary variations in the motion

[26] [1799a] *OC*, **12**, p. 193.

of the moon, the periods of which are enormously longer, extending over centuries. Currently the eccentricity of the earth's orbit was decreasing, and the consequence was an acceleration of the moon. The motions of the lunar nodes and apogee were also subject to secular equations of a sign opposite to that of the equation of the mean motion: for the nodes the ratio to the equation of mean motion was 1:4 and for the apogee, 7:4.

It might also have been expected that variations in the position of the ecliptic would influence the action of the sun on the moon by changing the angle between the lunar orbit and the ecliptic. The sun keeps the angle between the lunar and terrestrial orbits constant, however, so that the declinations of sun and moon serve the same periodic law. Finally, Laplace calculated that neither the direct influence of planets on the moon, nor consequences arising from its shape, had any detectable effect on the mean motion. To Laplace, the most surprising finding was that the decrease in the eccentricity of the sun's apparent orbit should have had an effect so much more evident in the motion of the moon than in the solar theory itself. The decrease in eccentricity of the solar orbit had modified the time that lunar eclipse occurs by less than 4 minutes since ancient times, while the mean motion of the moon had increased by more than 1 and 1/2 degrees. The influences causing these variations clearly affect the lunar motion of revolution differently from the motion of rotation, and it might have been expected that they would disturb the identity of period between the two so that in time the moon would show the earth its other hemisphere.

In fact, however—and this effect was the subject of Laplace's final calculation—the period of the moon's secular variation in mean motion of revolution was so long, and the earth's gravity by comparison so powerful a force, that the major axis of the lunar equator is always drawn toward the center of the earth, subject only to the libration that shows a tiny rim of the hidden hemisphere, now on one side and now on the other.

The period of the moon's long inequality was the greatest that Laplace had yet studied, amounting to millions of years. Yet there could be no doubt of its existence, and he assured his readers that the acceleration would be reversed one day, also by the force of gravity, and that the moon would retreat from the earth on its next beat. It would never come crashing down upon us, therefore, as it surely would do if ether resistance or a temporal transmission of gravity were the cause of its present approach. Acknowledging that his numerical results in these calculations were much less secure than in the theory of Jupiter and Saturn, he left it to posterity to perfect them. Posterity has obliged. In 1853 John Couch Adams estimated that barely half of the lunar acceleration can be explained by decrease in the earth's orbital eccentricity, and in 1865 Charles Delaunay attributed the remaining increase to slowing of the earth's rotation by tidal friction.

Laplace himself returned to the theory of the moon on several occasions [1798b], [1799a] prior to the publication of *Mécanique céleste* (see chapter 21). He there modified this analysis and its application more fully than he needed to do for his treatment of the other topics in what he had already developed into a body of celestial mechanics, dispersed amid these memoirs and lacking only the name. Since the last, and briefest, in the series, "Mémoire sur les variations séculaires des orbites des planètes" [1789b], merely gives a more general proof of the final item that he had demonstrated in the first [1787a], namely the periodicity of the variation of orbital eccentricities and inclinations, perhaps it will be fitting to conclude this account of the work that Laplace accomplished at the height of his powers with the peroration to the lunar memoir, presented in April 1788 (56), just one year prior to the onset of the French Revolution:

> Thus the system of the world only oscillates around a mean state from which it never departs except by a very small quantity. By virtue of its constitution and the law of gravity, it enjoys a stability that can be destroyed only by foreign causes, and we are certain that their action is undetectable from the time of the most ancient observations until our own day. This stability in the system of the world, which assures its duration, is one of the most notable among all phenomena, in that it exhibits in the heavens the same intention to maintain order in the universe that nature has so admirably observed on earth for the sake of preserving individuals and perpetuating species.[27]

[27] Ibid., *OC*, **11**, pp. 248–49.

Part III _____

SYNTHESIS AND SCIENTIFIC
STATESMANSHIP

CHAPTER 17

The Revolution and the Metric System

THERE IS no more instructive example of the effect that institutional developments may have, even in the most mathematical reaches of science, even upon the most insensitive of political temperaments, than the modification in the pattern of Laplace's career in consequence of the events following out of the French Revolution. The earliest record of a reforming initiative on his part is a proposal that he advanced "anew," according to the minutes of the Academy on 4 July 1789. It went dead against the current, not to say the torrent, of the times. Laplace wished the Academy to require an elementary knowledge of mathematics and physics of artisans qualified for a *brevet*, or license, in its officially recognized corps. Discussion was postponed to the next session. His colleagues then voted by a large majority to "require" nothing and resolved only that they would "prefer" to license those who knew these subjects.[1]

On 18 July, four days after the fall of the Bastille, Laplace read a paper on the inclination of the ecliptic (58). That autumn the Academy began considering a liberalization of its own structure and procedures intended to bring its regime into congruence with the emerging constitutional order. Laplace was appointed to a committee to review suggestions along these lines advanced by the duc de La Rochefoucauld, and he joined with Condorcet, Borda, Bossut, and Tillet in composing a memoir on the subject submitted on 1 March 1790.[2] There is no evidence that he took further part in the reformation or political defense of the Academy, however, the leadership of which was left largely to Lavoisier, whose efforts proved increasingly forlorn as the extremists gained in power.

In the early years of the Revolution, the Academy was inundated with proposals for inventions and technical schemes of many sorts, and Laplace's committee work, like that of all his colleagues, became much more demanding. On 2 November 1791 he was elected to the panel of fifteen academicians who, together with a like number of representatives of inventors' societies, constituted a new Bureau de Consultation

---

[1] *PVAS*, **108** (4 and 8 July 1789), fols. 184, 190–91.
[2] Ibid., **108** (23 Nov. 1789); **109** (1790–1792), folio 75.

des Arts et Métiers intended to relieve the Academy itself of responsibility for advising government agencies on patents and technological
policy in general. Prior to that, Laplace had taken one other initiative,
together with Condorcet and Dionis du Séjour. Jointly, they framed a
petition, subscribed to by others of their colleagues, and addressed it to
the National Assembly, urging that body to instruct departmental authorities under the new system of local government to continue the
population inquest begun by provincial intendants and published by the
Academy.[3]

The one revolutionary enterprise that consistently engaged Laplace's
interest, however, was the preparation of the metric system, and there
his part in the design was more decisive than has been appreciated.[4]
Like other aspects of modern polity, the standardization of weights and
measures in France had been urged in the programs of the Turgot
ministry (1774–1776) and then frustrated until the Revolution made
reform not only possible but imperative. It was also the sector in which
science and civil life overlapped most extensively. We have seen the
interest that Laplace was already taking in units during his collaboration with Lavoisier on the measurement of heat (see chapter 14), and
also in geodetic surveying during his analysis of the figures of the
planets (chapter 15). In the matter of the metric system, the scientists
moved before the politicians did. On 27 June 1789 the Academy
appointed a commission "for a piece of work on weights and measures";
the members were Lavoisier, Laplace, Brisson, Tillet, and Le Roy. No
record remains of their deliberations or of a memoir of 14 April 1790
drawn up by Brisson on the relation of linear units to units of capacity
and weight.[5] It is probable, however, that these early proposals revived
a suggestion advanced by La Condamine on returning from the
1735–1736 expedition to Peru. His thought had been to take the length
of the seconds pendulum at 45° latitude for the basic linear unit and to
decimalize its subdivisions and multiples. A pamphlet by one recalcitrant commissioner, Tillet, written together with Louis-Paul Abeille on
behalf of the Society of Agriculture, attacked precisely such a scheme.[6]
The authors urged limiting the reform to the standardization of existing
units, pointing out that with a duodecimal basis the shopkeeper and
engineer can manage quarters, thirds, and halves, and warning against

---

[3] Ibid., **109** (11 Dec. 1790), fol. 257; see chapter 12.

[4] For a chronology of his participation, and that of Condorcet, Lagrange, Legendre,
and Monge, see Champagne (1979), and for an overview of the history of weights and
measures, Kula (1986).

[5] *PVAS*, **108** (1789), fol. 170; **109** (1790–1793), fol. 83.

[6] *Observations ... sur l'uniformité des poids et mesures* (Paris 1790).

sacrificing the daily convenience of farmer, merchant, and builder to the exigencies of a perfectionist science.

Alarmed, the scientists evidently felt the need for a spokesman skilled in the ways of the political world. The report that formed the basis of the initial metric law, enacted on 8 May 1790, was presented to the Constituent Assmbly by Charles-Maurice de Talleyrand, bishop of Autun. Published separately, his proposal overwhelmed petty anxieties about habit with the grandeur of a universal reform drawing its units from nature.[7] The seconds pendulum was to be the linear basis, and gravimetric units were to be related to linear through the volume occupied by a unit weight of water at a given temperature. The last value had just been determined by Lavoisier, who almost surely had coached Talleyrand on these matters.

Thereupon, the Academy appointed a new commission on weights and measures consisting now of Laplace, Lagrange, Monge, Borda, and Condorcet. Less than a year later, on 25 March 1791, they made a different recommendation, enacted into law the next day.[8] It provided for a new geodetic survey of the meridian of Paris to be measured from Dunkerque to Barcelona, and for defining the still unnamed meter as the ten-millionth part of the quadrant. Its subdivisions and multiples would be decimal, and its length would be close to that of the seconds pendulum. The reason given for the change was that determining the length of the seconds pendulum depended upon a parameter in time, which was arbitrary, and upon the force of gravity at the surface of the earth, which was extraneous to the determination of a truly natural unit. In the words of Condorcet, who drafted the report:

> If it is possible to have a linear unit that depends on no other quantity, it would seem natural to prefer it. Moreover, a mensural unit taken from the earth itself offers another advantage, that of being perfectly analogous to all the real measurements that in ordinary usage are also made upon the earth, such as the distance between two places or the area of some tract, for example. It is far more natural in practice to refer geographical distances to a quadrant of a great circle than to the length of a pendulum.[9]

It has become conventional wisdom to dismiss this reasoning as a piece of grantsmanship.[10] The argument is that eminent mathematicians and engineers must surely have understood that there is no such thing as a naturalistic metric and that any unit is based on a mere

---

[7] Talleyrand, *Proposition ... sur les poids et mesures* (Paris 1790). For the text of the law, see *Archives parlementaires*. 1st series, **15** (1790), pp. 438–43.

[8] *HMARS*, (1788/1791),"Histoire," pp. 7–16.

[9] Ibid., pp. 9–10.

[10] For example, Favre (1931), Heilbron (1990), Dhombres (1992).

convention, the true meter being an agreed-upon stick. Their real motivation, it is said, was to mount another elaborate and costly scientific traverse at the expense of the State, with the immediate purpose of establishing the reputation of a new surveying instrument, an ingenious repeating circle, invented by one of their number, Borda.

These dark thoughts were also noised at the time and were probably unfair, then as now. It seems likely that Laplace would have been instrumental in changing the commission's recommendation. He and Borda were the only members with any background in geodesy—Borda as a navigator and engineer, and Laplace as a theorist analyzing the figure of the earth. The law of 8 May 1790 had been passed in haste to forestall Tillet and the defenders of routine. It was enabling legislation never meant to be the last word. The draft left it to the Academy to recommend an appropriate scale of subdivision. The importance of that was further emphasized in a companion decree passed the same day, also calling for recommendations from the Academy for a new monetary system.[11]

Evidently the vision of a universal decimal system, embracing not only ordinary weights and measures but also money, navigation, cartography, and land registry, unfolded before the commissioners as they explored the prospect in the summer and autumn of 1790. In such a system, it would be possible to move from the angular observations of astronomy to linear measurements of the earth's surface by a simple interchange of units involving no numerical conversions; from these linear units to units of area and capacity by squaring and cubing; from these to units of weight by taking advantage of the specific gravity of water taken as unity; and finally from weight to price by virtue of the value of gold and silver in alloys held invariant in composition through a rigorous fiscal policy. The seconds pendulum could never anchor that. The earth is round, and we fix our position by astronomical observation. The crux was thus that linear and angular units should be numerically interchangeable. At ordinary dimensions the curvature of the earth's surface would introduce no detectable error. In April 1795 Laplace put the point very simply in a lecture (see chapter 19) before the École Normale of the Year III:

> It is natural for man to relate the units of distance by which he travels to the dimensions of the globe that he inhabits. Thus, in moving about the earth, he may know by the simple denomination of distance its proportion to the whole circuit of the earth. This has the further advantage of making nautical and celestial measurements correspond. The navigator often needs

[11] *Archives parlementaires*, 1st series, **15** (1790), p. 443.

to determine, one from the other, the distance he has traversed from the celestial arc lying between the zeniths at his point of departure and at his destination. It is important, therefore, that one of these magnitudes should be the expression of the other, with no difference except in the units. But to that end, the fundamental linear unit must be an aliquot part of the terrestrial meridian.... Thus, the choice of the meter was reduced to that of the unity of angles.[12]

Laplace went on to justify taking decimal parts of the quadrant rather than of the whole circumference by reason of the role of the right angle in trigonometry. All the rhetoric about the universality of units taken from the earth itself thus turns out to have a perfectly sensible foundation. He incorporated the substance of this lecture in *Exposition du système du monde* (1796).[13] That only Laplace should have explained it clearly—then or now—gives further reason for thinking that he must have been the moving spirit in the design. It is certainly indicative, moreover, that when the two surveying teams took the field in the summer of 1792, one for the southern and the other for the far more extensive northern sector of the meridian, those chosen to head them should have been Laplace's first two calculators, Pierre-François Méchain for cometary data (see chapter 13), and Jean-Baptiste Delambre, for the tables of Jupiter and Saturn (see chapter 16).[14]

A related project linked the metric reform to the fiscal system. The abolition of feudal landholding and the reorganization of local government made it essential to create a new land registry, or *cadastre*, for determinations of title and assessments of tax.[15] The engineer in charge, Gaspard Riche de Prony, was mandated by law to base the task upon the Cassini map of France. That enormous cartographical achievement had been fleshed out on the skeleton of triangles run by the abbé Nicolas-Louis de Lacaille in his 1739–40 survey of the meridian, to which Delambre's and Méchain's results were to be compared. In planning the cadaster, Prony proposed to concert his efforts with those of the metric commission, construct his instruments on a decimal scale, and convert his units to the new basis as soon as it might be accurately known. His "Instruction" to that effect was submitted to the Academy and was largely approved in two reports, both drafted by Laplace. The

[12] Dhombres (1992), in which Laplace's lectures, as well as those of Monge and Lagrange, are reprinted with an admirable introduction by the editor.

[13] Book I, Chapter 14 *OC*, **6**, pp. 64–86.

[14] Delambre (1806–1810). With the immense labor of completing the project behind him, Delambre confirmed the cogency of Laplace's reasoning; (**3**), p. 304.

[15] *Recueil de documents concernant le cadastre*, Bibliothèque Nationale, Lf158.236. On Prony's direction of the cadastre, and particularly his construction of log and trig tables for decimalized angles, see Grattan-Guinness (1991).

first, which concerned methods, was also signed by Borda, Lagrange, and Delambre. In the second, Laplace took up units and proposed the names meter, decimeter, centimeter, and millimeter. The square on one hundred meters was to have been an "-are," divided into deciare, centiare, and milliare.[16]

Laplace had an uncomfortable moment in connection with the design of the Revolutionary calendar, represented by its sponsors as an extension of the metric system to the realm of history and the arts.[17] Together with Lagrange and Lalande, he was consulted by the enthusiast who conceived it, Gilbert Romme. Lalande's notes of the encounter are in the archives of the Paris Observatory. According to him, Laplace refrained from bringing home to the zealots the incompatibility between their desire for a calendar that would embed the civil year in nature and the incommensurability between the day and the year entailing an unavoidable irregularity in the number of leap years per hundred years in centuries to come.[18] One of Laplace's political detractors later called it apostasy in an astronomer that Laplace should have consented to draft the recommendation on the strength of which Napoleon restored the Gregorian calendar effective 1 January 1806.[19] In fact, astronomers require, not a natural, but an arbitrary and universal system of intercalation, and the Republican calendar conjured up instead the confusion of ancient Greek chronology, when particular cities intercalated a day or month named after some hero or victory. It may be true, however, that Laplace lacked the temerity to voice this sentiment in what was soon to be called the Year II of the Republic.

As for the metric system itself, there is no doubt about his fidelity. In his subsequent writings Laplace made a point of expressing angles in the form of decimal subdivisions of a right angle defined to be 100° even though that provision of the original legislation fell into general disuse when decimilization of time was officially abandoned as impractical in 1795. It is perhaps significant that his son in later years should have made himself something of a watchdog over the integrity of the metric system, as he did of the form of his father's work in general.[20]

Laplace took no part in the affairs of the Revolution during the phase of radical republicanism. He had been among the scientists vilified by Marat in the diatribe Les charlatans modernes (1791), although not as

[16] PVAS, 109 (12 May 1792), fols. 147–51; (11 July 1792), fols. 205–97.

[17] "Rapport sur l'ère de la République," Procès-Verbaux du Comité d'Instruction Publique, 2 (1894), pp. 440–50.

[18] Bibliothèque de l'Observatoire de Paris, B-5, 7.

[19] E. Merlieux, "Laplace" in Nouvelle Biographie Générale, ed. F. Hoefer (46 vols., 1855–1866), apropos of [1805].

[20] "Notice sur le Général marquis de Laplace," OC, 1, pp. v–viii.

virulently as Lavoisier.[21] As power shifted to the left in the spring and summer of 1793, the Academy came under increasing pressure from the radicals, and it was suppressed by a decree of the National Convention on 8 August. Laplace appears to have withdrawn from participation as its political vulnerability increased. His attendance became increasingly infrequent in the latter part of 1792. He was present for the last time on 21 December.[22]

A provisional commission was left in charge of the metric system after the abolition of the Academy. On 23 December 1793 Laplace was purged from its membership, along with Lavoisier, Borda, Brisson, Coulomb, and Delambre, on the grounds that such responsibilities were to be entrusted only to those worthy of confidence "by their Republican virtues and hatred of kings."[23] The curve of the Terror was then rising toward its climax. Some time previously (we do not know precisely when) Laplace had decided to remove from Paris. He had been married five years previously, and his two children were infants. He and his wife took a lodging in Melun, thirty miles southeast of Paris, and remained there until he was recalled to the capital to participate in the reorganization of science that followed the fall of Robespierre and the Jacobin dictatorship in July 1794 (see chapter 19).

The Thermidorean Convention thereupon placed the metric system, the Observatory of Paris, and all matters pertaining to navigation and official astronomy under the administration of the new Bureau des Longitudes, created in accordance with a law of 25 June 1795.[24] Laplace served regularly as a member, often chose its meetings as the forum in which to present appropriate papers, and published frequently in its journal, the *Connaissance des temps*.

---

[21] On Marat and science, see Gillispie (1980), pp. 290–330; and on his life and importance in the Revolution, Coquard (1993).

[22] *PVAS*, **109** (21 Dec. 1792), fol. 325.

[23] *Procès-Verbaux du Comité d'Instruction Publique*, **3** (1897), p. 397.

[24] *Procès-Verbaux du Comité d'Instruction Publique*, **6** (1907), pp. 321–27.

CHAPTER 18

Scientific Work in the Early Revolution

IT IS SOMETIMES said that Laplace began writing *Exposition du système du monde* and *Mécanique céleste* during this retreat at Melun. It may be so; there is no way of knowing. We do know that he had presented three further memoirs to the Academy before its situation deteriorated to the point that such works could scarcely have been received, even if they could have been composed. In July 1789 he communicated a memoir on the obliquity of the ecliptic (58), later combined with a miscellany of other topics [1793a]; in April 1790 a further memoir on the satellites of Jupiter (59), to be followed by a sequel [1791a] and [1793b]; and in December 1790 a study of tidal phenomena (60), largely in the port of Brest [1797a]. The disparity of publication dates bespeaks the confusion of the circumstances.

Laplace was forty years old in 1789. All three memoirs exhibit the pattern of the investigations that Laplace put in hand during this second half of his career, while also composing the magisterial treatises that remain its monument. The genesis of the latter, as will appear in the next section, owed much to the educational milieu. The former, his specialized studies, he henceforth conducted in regular and continuing interaction with practical astronomers, physicists, geodesists, meteorologists, and civil officials. On the whole the topics were not new to him, except for those that, in addition to heat, he took up in physics (see chapters 21–23, 26); neither, with the same exceptions, were the results. They contain no great surprises, nothing like the period of the long inequality of Jupiter and Saturn or the potential function. Instead, he returned to phenomena that he had already dealt with, usually with one or both of two purposes in mind: first, to give them a more detailed and general analysis; second, where possible, to give the analysis numerical expression in actual instances. There was precedent, of course. The theory of Jupiter and Saturn had already issued in Delambre's tables (see chapter 16), the cometary theory in the abbé Pingré's *Cométographie* (chapter 13), and the probability of cause in population studies (chapter 12). It would, therefore, be difficult to say to what extent external pressure and opportunity favored the shift in emphasis that was occurring anyway in the natural evolution of his lifework. There can be no doubt about the reality of the pressures or the opportunities, however, and no reason to question their efficacy in this, as in any evolutionary process.

In the opening sentence of the voluminous "Théorie des satellites de Jupiter" [1791a], Laplace responded himself to the challenge he had posed astronomers four years previously in concluding his discussion [1787a] of the pendulum-like libration of the three inner moons (see chapter 16). He now intended to give "a complete theory of the perturbations that the satellites experience and to place before astronomers the resources that analysis can provide to perfect the tables of these stars."[1] Cosmology had motivated the earlier analysis. The moons of Jupiter were a test case for the stability of the planetary system, and the proof was limited to a relatively brief demonstration that, for the inner three, the two relations,

$$n - 3n' + 2n'' = 0 \quad \text{and} \quad nt - 3n't + 2n''t = 180°,$$

are rigorously exact. A more complete demonstration confirming those theorems occupies Articles XIII–XIV of the present memoir concerning inequalities that depend on the squares and products of the perturbing forces.[2]

For the rest, Laplace no longer restricted the problem to the three inner satellites, nearly coplanar and concentric with Jupiter and bound in their triune libration. He now included the outer moon, the orbit of which is more eccentric and a bit more inclined to the plane of Jupiter's equator. The calculations involve the interactions of all four together with the effects exerted by two other perturbing forces, those due to the oblateness of Jupiter and to the gravity of the sun, which is affected in its incidence by the angle between Jupiter's orbit and equator. The latter two factors made themselves felt mainly in the inequalities of motion of the fourth satellite.

Handling these complications one after another, Laplace drew successively on each of the main investigations that he had completed in planetary astronomy. The memoir amounts to a reprise of the entire subject put directly into practice within the compass of the Jovian system. In the overall strategy, he followed the model of the Jupiter-Saturn memoir [1788a], [1788b]. After first setting up general equations of motion for satellites, he then went through the calculation of the effect of each class of inequality to which they are subject: those independent of eccentricity and inclination, those depending on eccentricity of orbit, those depending on the action of the sun, those appearing in the squares and products of expressions for the perturbing forces,

[1] [1791a] *OC*, **11**, p. 309.
[2] Ibid., pp. 369–87.

and those depending on motion in latitude, where the angle of the planes of Jupiter's orbit and equator is significant. For the theory of the figure of Jupiter, he drew on his work on the attraction of spheroids of revolution [1785b] and on its application to the figure of the earth [1786a].[3] For the determination of inequalities independent of eccentricity and inclination, he drew on the formula of the comparable section of the Jupiter-Saturn memoir [1788a].[4] For the action of the sun on the motion of the satellites, he drew on his discovery of the effect of variations in the eccentricity of the earth's orbit on the motions of our moon [1788c].[5] In an interesting aside he explained how the data for Jupiter's moons were calculated from observation of the times and duration of eclipses.[6] The first three satellites disappear behind Jupiter on every revolution, and the fourth intermittently. All that the observer needs to report are the instants of disappearance and reappearance, and the information thus obtained is far more accurate than could be yielded by tracking the actual motions. The further possibility of multiplying observations indefinitely reduces the already small risk of instrumental error to zero.

Armed with this information from the Wargentin tables (see chapter 16), Laplace could compute provisional numerical values for the mean motions and the mean distances from the center of Jupiter. For the distances, the most accurate method consisted in using Kepler's laws to derive the values for the three inner satellites from observed positions of the fourth. With these quantities, he could calculate the flattening of Jupiter and the masses of the four satellites. Theoretically, determination of those unknowns was the main object of the investigation. Practically, its purpose was to permit Delambre to construct tables for the satellites that, like his tables for Jupiter and Saturn, would be derived theoretically from the law of gravity. Observation would confirm the values and not be their source, as in the older tables. Laplace himself published a sample in the *Connaissance des temps* [1790a], the first of a series of contributions to practical astronomy in that journal. Delambre incorporated a more fully developed set, drawing on an enormous body of observations, in the third (1792) edition of Lalande's *Astronomie*.

Solving for the unknowns required manipulating a formidable array of equations. The three differential equations obtained at the outset govern the motion of each satellite. For four satellites there are twelve

[3] Ibid., p. 317.
[4] Ibid., p. 329.
[5] Ibid., p. 346.
[6] Ibid., pp. 361–69.

equations. Integration introduced twenty-four arbitrary constants, to be determined by deriving the elements of the orbits from observations of the eclipses. (Actually, the two relations of longitudes and mean motions among the inner three reduced the number of arbitrary constants to twenty-two, but the need to include terms for their libration added two others.) The indeterminate quantities—that is, the flattening of Jupiter and the masses of the four satellites—raised the number to twenty-nine. Five further items of observational data were required to make their determination possible: the principal inequality of the first satellite, the principal inequality of the second, the annual and sidereal motion of the apside of the fourth, the equation of the center of the third relative to the apside of the fourth, and finally the annual and sidereal motion of the orbital node of the second.

It is a measure of Laplace's insight into the conditions of the problem that he could seize on these pieces of information as practically obtainable and analytically sufficient for a solution. Substituting those values in the analysis, he fixed the masses at 0.184113, 0.258325, 0.865185, and 0.5590808, each multiplied by 10^{-4}, the mass of Jupiter being taken for unity. As for the figure of Jupiter itself, the ratio of the minor to major axis came out to be 67:72.[7] Formulas to compute the motion of each satellite in orbit then followed readily.

The first time that Laplace had run through these calculations, which appeared in a sequel [1793b] to the parent memoir, it had been with reference to a relatively small body of data, and he regarded his results as a first approximation to be corrected by a process of successive approximations that he set forth at the same time. All this he handed along to Delambre. The original plan was that the latter would set to work preparing yet a further stage in the evolution of the tables toward perfection.[8] Delambre's preemption to run the survey of the meridian prevented that, and Laplace took back the material himself, substituting in the analysis the more exact and fuller data from the tables just mentioned in the Lalande compilation, which although still imperfect were far fuller than his own sketchy figures.[9] Since Delambre began work on the metric system in May 1792, readying the calculations for publication must therefore have been occupying Laplace in the latter part of that year and perhaps into 1793.

The reflections with which he concluded the investigation are predictable in one respect. He held that the magnitude of the effect that the flattening of Jupiter has upon the inequalities of motion in its

[7] [1793b], *OC*, **11**, p. 421.
[8] Ibid., pp. 477–81.
[9] Ibid., pp. 415–16.

satellites proves that the attraction is compounded of the gravitational force exerted by every particle of the planet, that hypothesis having been assumed in the formulation of the equations of motion. A second rumination may be no more surprising in itself. It concerns the velocity of light, for which topic the moons of Jupiter had been instrumental long before Laplace's interest in them. He had Delambre calculate a value for aberration from figures for the eclipses of the first satellite. The results exactly confirmed Bradley's value drawn from the well-known method of direct observation of the fixed stars. That it should be so confirmed the uniformity of the velocity of light, at least within the dimensions of the diameter of the earth's orbit. Considering the configuration of his career, the conclusion that Laplace drew from this assertion is significant in a way different from the remark on gravity. Rather than celebrating a victory, it anticipates a battle that he would lose:

> That uniformity is a new reason for thinking that the light of the sun is an emanation from that body; for, if it were produced by the vibrations of an elastic fluid, there would be every reason to think that this fluid would be more elastic and denser on approaching closer to the sun, and that the velocity of its vibrations would thus not be uniform.[10]

One further paper—or, more accurately, collection of short papers— appeared in the volume that the Academy managed to get printed in 1793 prior to its demise. Entitled "Sur quelques points du système du monde," it obviously consists of odds and ends from Laplace's worktable [1793a]. The opening article, which concerns the investigation just discussed, was evidently written after completion of the analytical part [1791a] and before Laplace had decided to publish the numerical application himself [1793b]. Thereafter, the two topics discussed most extensively are the variation in the obliquity of the ecliptic and geodetic data bearing on the figure of the earth.

Laplace had read a draft on the former topic in July 1789 (58). He reminds readers that the decrease in obliquity of the ecliptic was one of the best studied of celestial phenomena, that it occurs in consequence of the action of the other planets on the earth, and that in overall rate and period it is independent of the shape of the earth. Nevertheless, flattening at the poles and bulging at the equator do affect the action on the earth of the sun and moon. The present analysis investigates the magnitude of these secondary effects in the variation of the plane of the ecliptic. They appear in the rate of precession of the equinoxes and in

---

[10] Ibid., p. 473.

the length of the year, and the chief purpose of the discussion was to correct the equations that astronomers employed for precession.

Two main points are to be noted. First of all, the results formed part of a work that Laplace intended to publish on what he still called *astronomie physique*.[11] This remark is the first indication since the general treatise on planetary motion [1784a] (see chapter 15) that *Traité de mécanique céleste* was in gestation. Second, the mode of analysis recalls a technique he had introduced in deriving general equations of motion of a system of mutually attracting bodies in the covering memoir on secular inequalities (chapter 16), and that he employed again in considering the motions of Jupiter's moons in latitude.[12] The orbits of planetary bodies were projected onto a plane fixed in space that passed through the center of the sun and served as the locus of a coordinate system for calculating angular momentum. It seems possible, and even probable, that this approach derives from his early enthusiasm for the application of d'Arcy's principle of areas to problems of planetary motion (chapters 5 and 8). At any rate, Articles XXI–XXII of the memoir under discussion develop it preparatory to a much more extensive use in *Mécanique céleste*.[13]

Laplace goes on to inquire whether the results of many meridional surveys, and also of determinations of the length of the seconds pendulum at various latitudes, could "without doing too much violence to the observations" be reconciled with the hypothesis that the earth is an ellipsoid.[14] The method was simply to compare the data in the literature with the ellipsoidal requirement that the force of gravity at the surface vary with the square of the sine of the latitude. The answer in both cases was negative, although the data from measurements of the pendulum were about eight times closer to satisfying an ellipsoidal figure than were those from meridional surveying. It cannot be said that this finding carried Laplace much farther than his full-scale memoir [1786a] on the figure of the earth (chapter 15), and perhaps the chief interest that it affords for the development of his work is the evidence that error theory was occupying a growing place in his thinking about physical problems. The form of the question was whether the discrepancies between the observed and calculated values exceeded what might be attributed to observational error. Calculating those limits represented another step in error theory itself.[15] It is worth mentioning that the

[11] Ibid., p. 483.

[12] [1787a], *OC*, **11**, pp. 69–70; [1791a], *OC*, **11**, 347–61.

[13] [1793a], *OC*, **11**, pp. 547–53; *Mécanique céleste*, Book II, Chapter 1, *OC*, **1**, pp. 57–73, 309–45.

[14] [1793a], *OC*, **11**, p. 493.

[15] Ibid., pp. 506–9.

degree measured by Mason and Dixon in Pennsylvania and Maryland figures in these data.

For the rest, the memoir is a grab bag. One article points out that the earth cannot have taken form in the fluid state, since if it had, it could have assumed only an ellipsoidal figure under gravity. Another observes that the stability of the seas tolerates disturbances sufficiently great so that occasionally the highest mountains are submerged, a consideration that explains certain curious facts of natural history. There is new proof from the conservation of vis viva that the long-term equilibrium of the sea is stable, whatever its depth and whatever the law of rotation of the earth. Laplace gives yet another simplified and generalized demonstration of the variation of constants to eliminate secular terms introduced into the solution of integrals by the standard methods of approximation. Finally, there is a new and purely analytic formulation of the laws of motion of a system of any number of bodies attracting each other by any law whatever. Like the material on the variation of the ecliptic, much of this found its way into appropriate passages of *Mécanique céleste*.

Laplace returned to the problem of the tides in "Mémoire sur le flux et le reflux de la mer," first read in December 1790 (60).[16] After remaining in manuscript for seven years, it was published with other *Nachlässe* from the Academy in its posthumous volume. He may probably have touched it up during that long interval, for although none of the data refers to observations more recent than 1790, in recurring at the outset to the difficulty of the problem he called it "the thorniest in all of celestial mechanics."[17] Thus, he casually launched that phrase in print two years before the appearance of the first two volumes that bear it for their title. The memoir has a very different quality from the enormous and intricate mathematical model that he had constructed some twenty years previously, which smells of the lamp and not of the brine (chapter 7).[18] In the meantime, he had fixed his attention on the most considerable existing body of data, a corpus of tidal observations in the port of Brest dating from the early eighteenth century, which Jean-Dominique Cassini had found at the Paris Observatory among the papers of his grandfather, Jacques, intendants of that establishment each in his turn.

In 1771 Lalande published this find in the second edition of his *Astronomie*. Jacques Cassini himself had drawn on the data for memoirs of no lasting value. Laplace took the occasion to remark how important

---

[16] [1797a].
[17] Ibid., *OC*, **12**, p. 4.
[18] [1778a], [1779a], [1779b].

it was in serious research on such topics to publish the original observations. The whole mass needed to be available before patterns could be discerned, trivial or accidental effects distinguished from fundamental rhythms, and causes assigned to the latter. When he examined the tables, he found one essential item of information missing; the collection contained no observations on the rate of the rise and fall of the tides at Brest. He therefore requested that detailed observations be made, evidently in the year 1790. He does not say who had carried out that commission; presumably, the naval authorities would have been responsible. It was a stroke of good fortune that Cassini had fixed on Brest. Brittany juts into the sea, and the harbor itself had a long, narrow entrance to a large protected basin so that wave action and other irregular oscillations were damped. Few other locations could have been equally advantageous.

This memoir reads differently, somehow, from any of Laplace's previous works. It would abstract the discussion too far from his own career to say that he did not need to be the mathematician he was in order to compose it, but it can be said that the formulas were not beyond the grasp of anyone capable of doing astronomy or geodesy. It is tempting to infer that so direct a contact with the facts about something as tangible as the tides, all laden with seaweed washing in and out of a working harbor, had a chastening effect, at least upon the writing. Whatever the truth may be, however, it is more enigmatic than that. For Laplace did mention, in a somewhat subdued manner, that although his earlier theory had predicted certain of the phenomena, it had been deepened by others that he had thought irreconcilable on first learning of them.[19] The difficulty is that when the two memoirs are compared, the theory in question—unless Laplace meant the theory of gravity, a claim so broad as to be empty—appears to have been changed in certain respects that appear significant to the outside observer two centuries later.

It is natural simply to suppose that by 1790 he had learned to be clear about a physical picture that had been obscured in a thicket of calculation the first time around. Certainly his verbal account of the conclusions could hardly be more lucid. Even now, it would be difficult to think of a better place to send a reader for a qualitative explanation of how the sun and moon contribute to the motions of the tides and how the magnitude and incidence vary with the seasons and relative positions of those bodies and of the spinning earth. Why does the tide never fall to the lowest point called for in gravitational theory? Why does it always take a little less time to rise than to fall? Why is the magnitude

[19] Ibid., *OC*, **12**, p. 21.

of tidal effects greater the shallower and more extensive the oceanic area? How may local circumstances affect height and times? Why do tides as a rule run most swiftly in shallow bays and narrow passages? The reasons for these and many other effects are expounded with admirable clarity.

Once again, as in the youthful analysis, there are three systems of tidal oscillation affecting the seas concurrently and superimposing their effects on the motions of individual particles of water. There, however, the identity of the two accounts ceases and merges into a resemblance wherein differences are at least as striking as similarities. In the earlier paper, the three cycles were accorded equal attention and were discussed in the order of the length of their annual, daily, and semidiurnal period. Now the order is reversed. The emphasis is also different, and—what is more surprising—so are important effects attributed to each system. In 1777, an account of the near equality of successive high tides (23), which the canonical Newtonian approach failed to give, was the starting point, if not quite the motivation, of the entire investigation. Moreover, Laplace then discussed the phenomenon by means of a mathematical analysis of the terms in his expressions governing the middle set of oscillations, the period of which is one day. Physically, that analysis presupposed a uniform depth of the sea, and its success was said to confirm that hypothesis. Now, in 1790, nothing is said of the latter argument. Even more curious, the near equality of consecutive high tides is attributed to the physical conditions governing the first cycle discussed, the semidiurnal oscillation. What is attributed to the daily cycle is the small difference, rather than the near equality, between the two tides of the same day in times of syzygy.

Neither of these effects is now deduced in the first instance from the mathematics. The return of consecutive high tides to almost the same level is explained in terms of the equilibrium conditions affecting a single particle of seawater. Suppose the sun (or moon) is acting in the plane of the equator. The gravitational force that it exerts on a particle of water directly underneath will be slightly stronger than the force exerted on the center of the earth. Hence, the action of the sun will tend to draw the particle away from the center of the earth. Twelve hours later, the particle is in opposition, and the sun attracts the center of the earth more strongly than it does the particle. It will then tend to separate the center of the earth from the particle. Since the radius of the orbit is enormously greater than that of the earth, the two effects are virtually identical in magnitude. Thus it happens, generalizing over the whole ocean, that the seas return to the same state every twelve hours.

Oscillations of the second sort, with a period of one day, are also given a physical explanation. They arise because the attracting body does not normally act in the plane of the equator, and their amplitude is proportional to the product of the sine and cosine of the declination. This daily variation is now held responsible for the small difference in consecutive high tides. At Brest, the tide in the morning was about seven inches higher than in the evening during syzygies at the winter solstice. The magnitude was small in European latitudes. There might be places, however, where geographic conditions would be such that, in the semidiurnal cycle, a tidal crest coming from one direction would coincide with a trough coming from the other, so that the normal tides would annul each other. In such localities, the second system would produce the only tidal motion, and there would be one tide a day. Laplace understood the port of "Batsha" (Badong?) in Tonkin (Vietnam) to be such an instance.[20]

Finally, Laplace discussed very briefly a system of oscillations like those that he had treated first, and at equal length with the others, in the early memoirs. They are no longer given the period of a year, however, but are simply said to be independent of the rotation of the earth and to result from the sharing of the seas in its other motions. Hence their period is very long though still finite, and their amplitude at Brest very small though still detectable. Centuries of observation would be needed to determine them precisely, after which time the values could be relied on to afford a valuable means of calculating the ratio of the mean density of the earth to that of the seas.

To what extent the theory that Laplace confirmed by the tidal data from Brest remained the theory that he had conceived in his youth is, therefore, problematic. Fortunately the point is not one that needs to be settled. The burden of the new exercise would appear to have been descriptive rather than theoretical anyway. It culminates in a formula for finding the level of the tides at Brest at any instant by means of astronomical data.[21] The concluding table was of a type that could have served the operations of any enlightened harbor master.

[20] Ibid., p. 20.
[21] Ibid., p. 112.

CHAPTER 19

Exposition du système du monde

FROM THE FALL of Robespierre and the Jacobin regime on 27 July 1794 (9 thermidor an II) until 26 October 1795, France was governed by the Revolutionary Convention, purged of its radical elements and often said to have been reactionary. From then until the coup d'état of 9 November 1799 (18 brumaire an VIII) that brought Napoleon Bonaparte to power, executive authority was vested in the collective hands of a Directory that never achieved stability, confronted as it was by the survival of Jacobinism on the left and the revival of royalism on the right. Many judgments, mostly adverse, have been passed on the political tone of the period, but there is general agreement among historians about its importance in the institutionalization of modern French society. For science, the most signal instances are the first École Normale, which held classes for three months beginning on 21 January 1795; the École Polytechnique, which was given that name on 1 September 1795, having started classes on 21 December 1794 as the École Centrale des Travaux Publics; and the Institut de France, the scientific division of which began regular meetings on 27 December 1795. They were intended to be the apex of a system of primary and secondary education, trade schools, and medical schools in which science and systematic knowledge would largely replace the classics as the staple subject matter.

Conceptually, the dominant influence was the school of *Idéologie*, certain of whose adepts had become administrators of science and culture in the government, and who thought to implement a philosophy of science deriving from the Enlightenment. In their outlook the moral and civic function of science was to educate citizens in the order of nature and, by extension, of society. In practice, the Institute never became the quasi ministry of education originally imagined. The planning for the École Normale was inadequate, and it closed after three months, not to reopen until 1812. Napoleon dispersed the *idéologues*, and positivism predominated at the École Polytechnique after 1800. The practical effects were nonetheless decisive in the long run and even in the short. The center of activity in science moved from academies, its home since the seventeenth century, to institutions of higher education, where it instilled the spirit of research. Scientists became educators and professors.[1]

There is no evidence that Laplace was among the organizers or promoters of these enterprises but every indication that when called on to participate, he did so with alacrity and enthusiasm, naturally assuming a leading role. The first, or scientific, class of the Institute amounted to a reincarnation of the Academy decked out in national and republican garb, elitist rather than privileged, civic rather than royal. It so commanded attention in the world of learning that reference to the Institute brings to mind its doings rather than those of its fellow, and in effect lesser, divisions concerned with social science and humanistic culture. At the organizational meeting on 27 December 1795, Laplace was elected vice president and, on 26 April 1796, president. The office was mainly honorific. More significant was his presence from the outset on a host of committees where policies were formulated and decisions preempted. To name only the most influential, he was a member of the committees that dealt with by-laws, weights and measures, finances, and the specification of prize contests in physics as well as mathematics.[2] At the end of the Institute's first year, Laplace was chosen to present before the joint meeting of the two legislative councils a formal address and résumé of the work accomplished.[3] He took the occasion to exhort the legislators to support the implementation of the metric system.

Laplace's activities at the Institute were a case of new or increased prominence in a set of revised procedures, whereas his involvement with the École Polytechnique marked a new departure, for him and for scientific education at large. There, physics first came to be taught systematically as a mathematical subject to well-qualified and highly selected students. True, the graduates were intended to be engineers, but it cannot have been an accident that the most famous of them in the early nineteenth century were engaged in what is now called mathematical physics—Ampère, Arago, Biot, Sadi Carnot, Fresnel, Malus, and Poisson, to name only the best known. Laplace never taught a course at the École Polytechnique. The initial curriculum emphasizing descriptive geometry was mainly designed by Gaspard Monge, the prime mover in the first foundation. Monge built upon his experience in the Old Regime at the former Royal Engineering School at Mézières, where he had established his career.

Laplace's post was that of one of two principal examiners, which gave him power over content as well as standards. From 1797 until the end of

---

[1] Lacroix (1828); Dhombres (1992), editor's introduction.
[2] *PVIF*, **1** (1910), pp. 30, 46, 410.
[3] [P1796].

1799, Monge and Berthollet were away from Paris with Bonaparte, first in the Italian campaign and then in Egypt. In their absence, Laplace became the predominant personality in the affairs of the École Polytechnique. His, indeed, was the primary influence in shifting the basis of the curriculum from geometry to analysis, from the concrete to the abstract, from formulations that could be pictured to formulations that could be generalized.[4]

Only at the École Normale did Laplace actually teach a course, albeit briefly. Design of an educational system had high priority among the idéologues and would-be technocrats who drafted the Constitution of the Year III founding the Directory. Creation of a national network of schools presupposed the preparation of a cadre of teachers. To fill the gap in short order the Committee of Public Instruction of the Convention resolved to follow the model of the "revolutionary" courses in production of saltpeter, gunpowder, and weapons mounted by leading scientists—Monge, Fourcroy, Guyton de Morveau, Berthollet—in the military crisis of 1793–1794. Just so would future teachers be selected from all departments of the Republic to assemble in Paris, there to sit at the feet of the foremost men of science and thought lecturing in the amphitheater of the Muséum National d'Histoire Naturelle. In addition to Laplace the professors of exact science were Lagrange, Monge, Haüy, and Berthollet. Laplace served on the committee that selected students from the Paris region.[5]

The significance of the experiment is not to be measured by the naïveté of the conception pedagogically nor by its inevitable failure in the short term. For the first time in history, science and learning were in principle to feed directly into public education. For the first time in history, young people were to be formed by new knowledge imparted by those who make it, not by old knowledge retailed by clerical pedants. For the first time in history, leading scientists and humanists were to be more than researchers and writers. They were also to be teachers at the highest level. The foremost mathematicians, Lagrange, Monge, and Laplace, thus found themselves obliged to explain their thought to whoever would make the effort of understanding it.

Laplace rose to the occasion in a series of ten lectures, delivered one each decade—or ten-day week—beginning on 1 pluviôse an III (21 January 1795). Originally published from stenographic transcripts in 1800, along with those of his colleagues, they have recently been gathered in an admirable edition of the mathematical courses.[6] Laplace

---

[4] Gillispie (1994); see also Hahn (1994).
[5] *Procès-Verbaux du Comité d'Instruction Publique* **5** (1904), p. 546.
[6] Dhombres (1992). For the original printing, see [1800d].

led his listeners in the classical sequence from arithmetic through the operations of algebra and the formulations of geometry and trigonometry to analytic geometry and the theory of curves and surfaces. The subject matter was elementary but the treatment modern, though only in the last two lectures did Laplace reach topics of intrinsic importance in his own career. The new ground he chose to break, significantly enough, was not so much mathematical as civic in importance. The ninth lecture explains the rationale and advantages of the metric system, while the last treats the theory of probability in nontechnical terms. It amounts, indeed, to an outline of *Essai philosophique sur les probabilités*.

Laplace intended to round out his course with a comparable précis of the results he had reached in his astronomical investigations, which he was even then gathering into the treatise that became *Mécanique céleste*. The schedule, or rather the disintegration, of the École Normale did not permit. Only a small proportion of the fourteen humdred pupils assembled in haste to overcrowd the auditorium of the Museum could have followed the lectures. His listeners ranged in age from extreme youth to near senility and in preparedness from virtual illiteracy to the genius of the young Fourier, who moved to the staff of the École Polytechnique in 1795 as assistant lecturer. As often happens, the courses proved more important to the teachers than to the students. Attendance was spotty. By early April 1795 many pupils had drifted away. The École Normale closed at the end of April, and Laplace never gave the lectures his program envisaged on rational mechanics, the differential and integral calculus, and astronomy. Instead, he referred his auditors to a book he had in preparation, to be entitled *Description du système du monde*, in which he would give a nonmathematical account of all that had been discovered in these subjects.[7]

Whatever the comprehension of Laplace's lectures on the overcrowded benches of a noisy auditorium, the promised book proved to be one of the most successful works of science ever composed. *Exposition du système du monde*—Laplace made a slight change in title—appeared in two volumes in 1796. The impression it made worldwide is conveyed by an autograph on the flyleaf of a copy of the first edition presented by a graduate of the College of New Jersey in 1798 to what is now the library of Princeton University:

> This treatise, considering its object and extent, unites (in a much higher degree than any other work on the same subject that we ever saw) clearness, order and accuracy. It is familiar without being vague; it is

[7] *Essai philosophique, OC,* **14**, p. 146.

precise but not abstruse; its matter seems drawn from a vast stock deposited in the mind of the author; and this matter is impregnated with the true spirit of philosophy.

Calling the book a work of popularization, as is often done, misrepresents the uniqueness it shares with *Essai philosophique sur les probabilitès*. Laplace never writes down to his readers. He employs nothing in the way of merely literary artifice. Not for him Fontenelle's witty conversations about a plurality of worlds with an entrancing marquise; nor yet Euler's letters to a princess sugarcoating a course on natural philosophy; nor, finally, Einstein's fictitious physicists performing thought experiments in relativistic elevators.

Laplace spares his reader nothing but the mathematics, and of that only the calculations and formulations, not the vocabulary or the reasoning. He does not pause to define sines, cosines, tangents, radii, ellipses, epicycles, logarithms, square roots, cube roots, derivatives, and integrals. True, he does introduce each topic with an elementary account of the phenomena, but the level of discourse rises, and rapidly, as he gets into the concepts that explain them. When he reaches his own results and those of his contemporaries, he writes as he would for his colleagues. In certain passages the very phrasing is the same as in prefatory or concluding sections of his academic memoirs.

This is a work of science, then, not a work about science, written for an educated public, not for a general public. Laplace kept his book alive and abreast of his thinking and of developments in astronomy and physics throughout his life. A second edition was published in 1799 to accompany the first two volumes of *Mécanique céleste* (67) as a companion volume to his magnum opus and in identical format. The third edition appeared in 1808. For the commentary on the Republican calendar originally included in deference to the political reality of the 1790s, Laplace substituted an explanation of why the Gregorian, of which Napoleon decreed the restoration in 1806, was more practical after all, despite its imperfections.[8] Laplace was by then deeply involved with the Arcueil group in matters of physics, most notably capillary action, the speed of sound, and the refraction of light, problems that they sought to resolve through analysis of a micro-astronomical model of central forces of attraction acting on point masses (see chapters 22 and 23). The fourth edition (1813) preceded by a few months the first printing (1814) of *Essai philosophique sur les probabilités*. The fifth (1824) announced Laplace's intention to make molecular forces the basis of a theory of heat and gases. Faced with the successes of Fresnel in optics,

---

[8] *OC* **6**, pp. 21–22.

Ampère in electrodynamics, and Fourier in theory of heat, a now aged Laplace apparently abandoned molecular forces before preparing the sixth edition for a publication that was delayed until 1835, eight years after his death.

Let us concentrate our account on the contents of the first edition. The material is organized in five books. Book I begins with what any attentive observer may see if he will open his eyes to the spectacle of the heavens on a clear night with a view of the whole horizon. The most demanding chapter deals with the oblate shape of the earth, the consequent variation of the force of gravity at different points on the surface, and the rationale of basing the metric system on its dimensions (see chapter 17). A passing observation is reminiscent of a saying of Einstein, to the effect that it is the laws of nature that are simple, not nature itself, which on the contrary is very complicated:

> The simplicity of nature is not to be measured by that of our conceptions. Infinitely varied in its effects, nature is simple only in its causes, and its economy consists in producing a great number of phenomena, often very complicated, by means of a small number of general laws.[9]

The remark is a reminder that Laplace, too, was a thinker about the world and not merely an indefatigable calculator or an overbearing dogmatist, although in the mix that made his personality those aspects may also have been combined.

Book II sets out the "real" motions of planets, satellites, and comets and gives the dimensions of the solar system. Such is the clarity and simplicity of the writing that those parts might serve a modern reader who is not conversant with the elements of planetary astronomy in exactly the way Laplace had in mind in writing for contemporaries, except that no one will need to feel edified. By now it may be equally useful in a retrospective way to the scientifically initiated. It is a handbook of what was known of cosmology at the end of the eighteenth century.

Book III is a verbal précis of the laws of motion as understood in eighteenth-century rational mechanics, with special reference to astronomy and hydrostatics. The discussion of the motion of a material point is a reminder that throughout the eighteenth century, force was taken to be proportional not to acceleration but to velocity, and that when Laplace and others spoke of the "force of a body" they meant the quantity of motion, mass times velocity, later called momentum.[10] For

---

[9] Book I, Chapter 14, *OC*, **6**, p. 65.
[10] Book III, Chapters 2, 3, *OC*, **6**, pp. 155–61, 173.

the quantity mass times acceleration, the more restricted term accelerative force was used. The importance to Laplace of d'Arcy's principle of areas (see chapter 4) and its equivalence to conservation of angular momentum emerges very explicitly in Book III, Chapter 5.[11] An especially felicitous analogy between the secondary and tertiary oscillations of a pendulum and the perturbations experienced by the planets makes it easier to see how he was envisioning and formulating problems of the latter sort.[12]

In Book IV, Laplace summarized his own work in gravitational mechanics. Much of it consists of simplifications of the prefatory sections to the published memoirs. The topics are perturbations in planetary motions, with emphasis on the great inequality of Jupiter and Saturn, the shape of the earth, the attraction of spheroids and the rings of Saturn, the dynamics of tidal and atmospheric motion, the motions of the moons of Jupiter, precession of the equinoxes, and lunar theory—all vindicating the stability of the solar system under the operation of the law of gravity. A chapter of "Reflections" recapitulates the basic assumptions about the operation of the force of gravity that Laplace had stated as his point of departure in the dual probability-gravitation memoir.[13] Now they are five instead of four, since the assumption that gravity is indifferent to the state of motion or rest of bodies, and that it acts instantaneously, has become two principles. An interesting passage explains how his analysis of secular inequalities of the moon's motion had led him to change his mind on the latter point.[14]

Only in Book V did Laplace deal with matters on which he had not already published mathematical investigations. It consists of six chapters, the first five on the history of astronomy. He must have taken his history seriously, for he included it in all editions and published a revision separately ([1821e]). Laplace, however, was not a scholar of the caliber of Delambre, a classicist by training, whose histories of ancient and modern astronomy continue to be valuable. Laplace's remarks on the great discoveries of the past are further evidence—if any is needed—that inventiveness in one discipline can accompany banality in another. His treatment of the place of astronomy in the growth of knowledge is not so much warmed-over as it is cooled-down Condorcet, whose *Esquisse d'un tableau historique des progrès de l'esprit humain* had been posthumously published the previous year.

In the sixth, and last, chapter Laplace introduced a speculation on the origin of the solar system and another on the nature of the universe.

[11] *OC*, **6**, pp. 195–96.
[12] Book III, Chapter 5, *OC*, **6**, pp. 190–93.
[13] Book IV, Chapter 17; cf. [1776a, 2°], see chapter 4.
[14] *OC*, **6**, p. 346.

These concluding nineteen pages, written for a wide public, have sustained a more continuing, although not a better informed, commentary than all of Laplace's other pages put together. He presented the former speculation, which has quite generally come to be misnamed the "nebular hypothesis," with all the "misgivings" (*défiance*) that anything should arouse that is in no way the product of observation or calculation.[15] Perhaps it will not strain analogy (one of Laplace's favorite modes) too far to liken it to the Queries with which Newton raised questions he had not answered at the end of the *Opticks*. It makes a curious commentary upon the history of science that the indulgence of exact minds in such flights of fancy should excite so much more interest—to be sure, it is a human interest—even among scientists in later times, than does the content, let alone the detail, of the work that gives them a claim on our attention in the first place. S. L. Jaki has reviewed the successive modifications of Laplace's cosmogony.[16] We shall confine our attention to the first rendition and attempt to situate it in the context of Laplace's own thought.

In the century or more since the emergence of evolutionary modes of analysis and explanation, the Laplacian cosmogony along with the Kantian—it is very unlikely that Laplace knew of Kant in 1796—has conventionally been cited as an early instance, perhaps as marking the introduction, of a historical dimension into physical science. That attribution, indeed, has been its chief attraction. Unfortunately, however, it is also quite anachronistic. If the text itself is allowed to speak for Laplace, it will be altogether evident that evolutionary considerations in the nineteenth-century sense formed no part of his mentality. The conclusions he had reached concerned stability; the evidence for that he had calculated, many and many a time. Although that was not the main burden of these passages, he again referred to it as a warranty for the care that nature has taken to ensure the duration of the physical universe, just as it has the conservation of organic species, to which he alluded in terms clearly taken from Cuvier. It was not about the development of the solar system that he was thinking. It was about the birth.

If we were to find a phrase that would characterize what Laplace had in mind about that event, it would not be "nebular hypothesis." It would be "atmospheric hypothesis." And if, further, we were to identify the context in which he raised the question at all, it would not be the evolution or history of nature. It would be the probability of cause. The motifs are altogether familiar to the student of Laplace's own development. The reader is summoned to contemplate the whole disposition of

---

[15] *Exposition du système du monde*, (1796), **2**, p. 303.
[16] Jaki (1976).

the solar system. At the center spins the sun, turning on its axis every twenty-five and a half days. Its surface is covered by an "ocean" of luminous matter spotted with dark patches some of which are the size of the earth. Above that zone is a vast atmosphere; how far it extends into space cannot be told. Beyond it turn the planets, seven of them, in almost circular orbits, with fourteen known satellites among them, all revolving almost in the same plane and in the same direction. Those whose rotation is observable—the sun, moon, five planets, and the rings and outer satellites of Saturn—also turn west to east on their respective axes.

The question is, can such an arrangement be the effect of chance, or is the existence of a cause to be inferred? There are twenty-nine discrete movements in addition to the revolution of the earth around the sun. The earth's orbital plane serves as reference for determining whether the motion of other bodies is direct or retrograde. If any of the orbits fell outside a quadrant centered on the earth's orbital plane, the motions would appear retrograde. Now then, if the arrangement of the solar system were due to chance, the probability that at least one such inclination would exceed the quadrant is $1 - 1/2^{29}$. Since that value amounts to virtual certainty, and since in fact no orbit does fall outside the quadrant, the arrangement cannot be the result of chance and must therefore bespeak a cause. Other appearances are no less remarkable, notably the very slight eccentricity of the orbits of all planets and satellites. Comets, on the other hand, are highly eccentric. They travel into regions still of the sun's dominion but far beyond the planetary sphere, in orbits inclined at all angles to the plane of the ecliptic.

What, then, can the cause be? It would need to explain five distinct sets of phenomena: (1) motion of the planets in the same direction; (2) motion of satellites in the same direction; (3) motion of rotation in the same direction; (4) small eccentricity of orbits for all the above; and (5) extreme eccentricity of cometary orbits.

The only modern writer Laplace had read who had tried to think seriously about the origin of the planets and satellites was Buffon. In his scheme, a comet had struck the sun and released incandescent matter that cooled and coalesced to become the planetary system. That hypothesis satisfied only the first among the above sets of phenomena. Laplace proposed to rise above that to the "true cause."

Whatever it was, it had to have included all the bodies. It had, therefore, to have been originally in the fluid state in order to have been expansible to the dimensions of the planetary system. It must, in a word, have surrounded the sun like an atmosphere. Might it not, indeed, have been the atmosphere of the sun, which in the course of contracting formed the planets by condensations in the plane of the solar equator at the successive limits of its extension? Similar processes centered on the planets could equally have produced the satellites. Such a mechanism would also account for the cometary appearances. The clue was in the absence of gradation between the near circularity of planetary and the extreme elongation of cometary orbits. The less eccentric comets, which would exhibit such a progression, had been drawn into the sun with the contracting atmosphere, leaving behind those describing the extreme trajectories. Hence, the appearance of chance in the distribution of their inclinations. Contraction of the solar atmosphere did not explain it—which was not to say that there was no other cause. It is often said that Laplace was mistaken in ruling the comets out of the solar system; only in respect to causality did he really do so, not physically.

So much for the origins of the solar system. It is true that Laplace mentioned nebulae in this chapter, but he did so in the course of the second speculation about the immensity of the universe beyond the solar system, and not in connection with contraction of the sun's atmosphere to form the latter. Large telescopes reveal great patches of undifferentiated light in the heavens. It is plausible to suppose that these "*nébuleuses*" are really groups of very distant stars.

This passage contains another conjecture that has recently been picked up in the light (or the dark) of black holes, rather than of evolution. The gravitational attraction of a star with a diameter 250 times that of the sun and comparable in density to the earth would be so great that no light could escape from its surface. The largest bodies in the universe may thus be invisible by reason of their very magnitude. Laplace stated this possibility in a merely qualitative way, almost in passing. He obliged, however, when F. X. von Zach requested the calculations for publications in the latter's journal.[17] Laplace omitted the speculation in later editions, and seems never to have alluded to it again.

The first edition of *Exposition du système du monde* closes with a panegyric of astronomy—and a political statement. The great merit of the science is that it

> dissipates errors born of ignorance about our true relations with nature, errors the more damaging in that the social order should rest only on those relations. TRUTH! JUSTICE! Those are the immutable laws. Let us banish the dangerous maxim that it is sometimes useful to depart from them and to deceive or to enslave mankind to assure its happiness.[18]

[17] [1799d]. The proof is translated as appendix A in Hawking and Ellis (1973), pp. 365–68. For a detailed discussion of Laplace's conjecture, see Eisenstaedt (1991), pp. 365–71.

[18] *Exposition du système du monde*, (1796), **2**, p. 312.

CHAPTER 20

A Scientific Eminence

BOTH TIMING and circumstance associate publication of *Traité de mécanique céleste* with the beginning of the Napoleonic regime. Laplace had first encountered the young Bonaparte at the École Militaire in Paris in September 1785 among the artillery cadets whom he examined that year in mathematics. On 25 December 1797 the Institute elected General Bonaparte, fresh from victories in Italy, to the vacancy in the section of mechanics created by the exile of Lazare Carnot following the coup d'état of fructidor.[1] Laplace accompanied Berthollet in the ceremony of escorting the young general to take his seat. In October 1799, three weeks before the coup d'état of 18 brumaire (9 November) that brought Bonaparte to power as First Consul, Laplace presented him with copies of the first two volumes of *Mécanique céleste*. The acknowledgement is famous. Bonaparte promised to read them "in the first six months I have free" and invited Laplace and his wife to dine the next day, "if you have nothing better to do."[2]

Laplace and Bonaparte were then serving on a commission together with Lacroix to report on an early mathematical memoir of Biot (69). Bonaparte never made the personal favorite of Laplace that he did of Monge and Berthollet, but in 1807 and 1808 his sister, Elisa, having been elevated to the rank of princess, took up Madame de Laplace and attached her as lady-in-waiting to her court in Lucca. Their correspondence offers a glimpse into the Napoleonic world of fashion.[3] On seizing power, Napoleon named Laplace Minister of the Interior. That ministry had responsibility for most aspects of domestic administration other than finance and police. Laplace lasted six weeks in the government, to be replaced by Bonaparte's brother, Lucien. Napoleon's reminiscence at St. Helena is also famous. Laplace, he said, could never "get a grasp on any question in its true significance; he sought everywhere for subtleties, had only problematic ideas, and in short carried the spirit of the infinitesimal into administration."[4]

---

[1] *PVIF* (21 brumaire an VI), **1**, p. 296.
[2] *Correspondance de Napoléon Ier*, 27 vendémiaire an VIII [19 October 1799], no. 4384, **6** (1861), p. 1.
[3] Marmottan (1897).
[4] *Correspondance de Napoléon Ier*, **30** (1870), p. 330.

Thereupon, Napoleon saw value in Laplace as an ornament, though not as an instrument, of state. He appointed him to the Senate and made him chancellor of that body in 1803, an office that Laplace enjoyed throughout the Consulate and Empire at an annual income of 72,000 francs. With other emoluments and honors, he "touched" (as the French has it) well over 100,000 francs a year and became a rich man. In 1805 Napoleon further named Laplace to the Legion of Honor, ennobled him the following year with the title of count of the empire, and in 1813 conferred on him the Order of La Réunion. Laplace in return dedicated the third volume of *Mécanique céleste* (1802) and *Théorie analytique des probabilités* (1812) to Napoleon. The dedication in the latter is adulatory, even by sycophantic standards, and is not reproduced in the *Oeuvres complètes*, where the third edition occupies Volume VII.

Reorganization of the École Polytechnique was the one important accomplishment that marked Laplace's tenure of the Ministry of the Interior. A law promulgated on 25 frimaire an VIII (16 December 1799) established a Conseil de Perfectionnement to oversee the curriculum and standards. The course was cut from three to two years and was made preparatory to the specialist schools, the École des Ponts et Chaussées, École des Mines, and École d'Artillerie, which became essentially professional schools at what would now be called the graduate level. Bonaparte, well disposed at first toward the École Polytechnique, was persuaded to provide adequate financing. Three members of the Council were to be delegated from the Institute. Laplace, Berthollet, and Monge were chosen and reelected annually until Lagrange replaced Monge in October 1805.[5] Laplace and Berthollet continued to serve throughout the Napoleonic regime, and, indeed, Laplace was commissioned by the government of the restored monarchy to oversee a further reorganization in 1816. His report at the end of the first year of the council's responsibility amounts to a catalog of courses and requirements and a prospectus of services to be expected of science by the state.[6]

From this, the period of Laplace's greatest prominence, testimony remains of friendships no less than of enmities. They clustered around his work, naturally enough, and if the element of discipleship was predominant, that is not unusual in the lives of scientists. The most sympathetic personal recollection comes from Biot, who made it the subject of a reminiscence half a century later before the Académie

---

[5] *PVIF* (15 vendémiaire an XIV), **3**, p. 261.

[6] 24 December 1800, [P1800]. For the influence of Laplace in the École Polytechnique, see Hahn (1994).

Française. Biot had graduated from Polytechnique with the first class in 1797 and had received a post teaching mathematics at the École Centrale in Beauvais. It was common knowledge in scientific circles that Laplace was preparing *Mécanique céleste* for publication. Wishing to study the great work in advance, Biot offered to read proof. When he returned the sheets, he would often ask Laplace to explain some of the many steps that had been skipped over with the famous phrase "It is easy to see." Sometimes, Biot said, Laplace himself would not remember how he had worked something out and would have difficulty reconstructing it. He was always patient in going back over these deductions and equally so with Biot's own early efforts. He encouraged him to present before the Institute the memoir (69) on the general method that Biot had conceived, in the isolation of Beauvais, for solving difference-differential equations. Only some time afterward did Laplace show Biot a paper he had put away in a drawer, a paper in which Laplace had himself arrived at much the same method years before.

Biot would often stay to lunch along with others of his age. After a morning of work, Laplace liked to relax in the company of students and young men at the beginning of their careers. In their mature years, they remained an entourage grouped around him like—Biot's phrase may be more revealing than he intended—"so many adopted children of his thought."[7] Madame de Laplace, still young and beautiful, treated them like a mother who could have been a sister. Lunch was frugality itself—milk, coffee, fruit. They would talk science for hours on end. Laplace would often ask them about their own studies and research and tell them what he would like to see them undertake. He was equally concerned with practicalities of their future prospects and would point out opportunities. "He looked after us so actively," said Biot, "that we did not have to think of it ourselves."[8] In 1800 Biot himself was appointed to a chair of mathematical physics at the Collège de France.

Among Laplace's contemporaries, his friendship with Berthollet was the closest and most enduring of which record remains. They had begun to draw together in the mid-1780s, attracted to each other scientifically by their mutual interest in the physics of chemical forces. Both enjoyed greater prestige and influence at the Institute than either had achieved in the last years of the Academy, where Lavoisier had predominated. They were also close scientifically, at least after Berthollet's return from Egypt with Napoleon in 1799. Laplace contributed two notes on pressure-temperature relations in an enclosed gas to Berthollet's master treatise *Essai de statique chimique.*[9]

---

[7] Biot (1850), p. 68.
[8] Ibid., p. 69; cf. Lacroix (1828).
[9] [1803e].

Berthollet then had a country house in the village of Arcueil, five miles south of Paris. He installed a chemical laboratory there and also a physical laboratory, and gathered around his work a younger group, the most notable among them being Gay-Lussac and Thenard. In 1806 Laplace bought the neighboring property. The transfer of his salon there, and their collaboration with Berthollet and his group, created a circle of mathematically and experimentally capable people under strong leadership who soon began informally calling themselves the Société d'Arcueil; its institutional history has been written by Maurice Crosland.[10] Laplace's part in the work of the group, which also included Bérard, Descotils, Biot, Arago, Malus, and Poisson on the side of physical sciences, occupies chapters 22–24 of the present biography. One chronological fact is important to emphasize here. The activities of the Arcueil group clearly postdated Laplace's completion of his astronomical system with the publication of Volume IV of *Mécanique céleste* in 1805 (83). He did do further astronomical work, but it was of an occasional nature, and Volume V, assembled much later between 1823 and 1825, comprises a series of addenda.

Before we proceed to a discussion of that treatise (see chapter 21), two further memoirs on particular topics need to be noticed briefly. In January 1796 Laplace read before the Institute the draft of "Un mémoire sur les mouvements des corps célestes autour de leurs centres de gravité" and had it revised and ready for publication less than a year later (61).[11] His summary reflections on gravity in *Exposition du système du monde* grouped the problems that the law presented into three categories—the motion of centers of gravity of celestial bodies about centers of force, the figures of the planets and the oscillations of the fluids that cover them, and the motion of bodies about their own centers of gravity.[12] He did not say so, but his own work clearly had been addressed largely to problems of the first two types, and he had dealt with rotation only incidentally to analysis of precession, tidal motion, and the coincidence between the lunar periods of revolution and rotation. Now he proposed to give a complete analysis of motions of the last type.

In fact, the memoir is both more and less than that. The most important example, he says, is the earth. Precession of the equinoxes is produced by one of its motions, and we tell time by its rotation. First, however, he digressed to give an application that he had just developed of the generalized gravitational function to the theory of perturbations in planetary motion.[13] The approach came out of his studies of the

---

[10] Crosland (1967).

[11] [1798a].

[12] *Exposition du système du monde*, Book IV, Chapter 17, *OC*, **6**, p. 341.

[13] Chapter 15, Equation (94).

moon. In analyzing its motion around the earth, he now treated its mass as infinitesimal and attributed to the earth a mass equal to the sum of the two masses. The new, and Laplace thought quite remarkable, equation of condition that described this effect gave a direct relation between parallactic inequality and inequalities of lunar motion in longitude and latitude.[14] Moreover, it was easy to verify the theoretical values by observation, since the constants were given by the mean longitudes of the moon and of its perigee and nodes at a given time. More generally, the same equation was applicable to verifying the calculation of the perturbing influence exerted on one planet by another, whose own perturbation is ignored, which procedure was standard in astronomical practice. Laplace promised to develop this first-order theory further and kept his word in *Mécanique céleste*.[15]

Coming back to the theory of rotation about centers of gravity, Laplace considered that the equations Euler had formulated in his *Mechanica* were the simplest and most convenient that he could use. In order to integrate them, terms needed to be expanded in series, and the whole art consisted in identifying those that on integration produced detectable quantities. The finding for the earth was that the only periodic variation in the position of the axis that needed to be considered was the so-called nutation, which depends on the longitude of the nodes of the lunar orbit. Two other axial wobbles, one much smaller and the other of much longer period, might be disregarded. Motions of the axis (and they are the main subject of this memoir) depend upon the shape of the earth, and the analysis led Laplace back to a review of his general memoir on attraction and the shape of spheroids.[16]

He now found that the phenomena of nutation and precession confirmed the figure for the flattening, namely 1/320, given by the measurements of the length of the seconds pendulum at different latitudes. These results were in agreement with each other and were much closer to satisfying an ellipsoidal figure than was the curve constructed by the various surveys of arcs of the meridian. Although Laplace was modifying his opinion about the primacy of geodetic data, as usual when he changed his mind, he did not say so expressly. He said only that in the expression for the radius of the earth, terms that appear to show degrees measured along the meridian as deviating from the elliptical have much less influence on its magnitude and on variations in the force of gravity. An ellipsoidal figure with a flattening of 1/320 may thus be supposed in calculations of parallax, of the length of the

[14] [1798a], *OC*, **12**, p. 136.
[15] Book II, Nos. 14–15, *OC*, **1**, pp. 163–70.
[16] [1785b]; see chapter 15.

pendulum, and of axial motions.[17] He further gave a direct proof of the theorem that he had long since found indirectly, to the effect that precession and nutation have the same quantity as if the seas and the earth formed a solid mass.[18] In the preliminary remarks, Laplace said that he was also extending the analysis to the variations in the direction of the lunar axis and in the inclination of the rings of Saturn. In fact, however, the memoir breaks off with an apology for its length and refers the reader to a further volume of the *Mémoires de l'Institut* for a continuation. That second installment never appeared, although he incorporated very similar material in appropriate passages of *Mécanique céleste*.[19]

One other feature of this memoir is noteworthy. Laplace was now being assisted in his research by Alexis Bouvard, who had succeeded Delambre in the role of calculator, and who performed all the work of calculation for *Mécanique céleste*. Bouvard also made the calculations for a further investigation of lunar variations, the results of which Laplace read before the Institute on 20 April 1797 (63). The seminal series on planetary motions in the 1780s (see chapter 16) had culminated in a paper on the secular inequalities of the moon [1788c], which Laplace had arrived at by means of applying his approach to the theory of Jupiter and Saturn to the moons of Jupiter.

The 1788 investigation obtained a formula for determining the variations in mean motion of the moon, which in Laplace's finding depended on variations of opposite sign in the eccentricity of the earth's orbit. The motions of nodes and apogee of the lunar orbit also exhibit secular inequalities. Laplace had then restricted his determination of those values to terms given by the first power of the perturbing force, although well aware that this was only half the story for the motion of the lunar apogee. The other half was expressed in terms dependent on the square of the perturbing force. Clairaut had discovered that this part was the resultant of the two large inequalities called variation and evection. The secular equation Laplace found for the motion of apogee, added to that for mean motion, gave a secular equation of the anomaly equal to 4.3 times that for mean motion. In like fashion, when he included terms depending on the square of the perturbing force in the secular equation of the nodes, he found its value to be 0.7 that of the mean motion, which amount was to be added to their mean longitudes. Thus, the motions of nodes and apogee decelerated when the mean

---

[17] Ibid., *OC*, **12**, p. 131.
[18] [1780c]; see chapter 7.
[19] Book V, Chapters 2 and 3, *OC*, **2**, pp. 375–402.

motion of the moon accelerated, and the secular equations of the three effects were in the ratio 7:33:10.[20]

These were large inequalities. One day they would produce changes in the secular motion of the moon equal to 1/40 of the circumference and up to 1/12 of the circumference in the secular motion of the apogee. Like the variations of the eccentricity of the earth's orbit, on which they depended, they were periodic but the periods, enormously longer than any others Laplace had yet identified, occupied millions of years. Slow though the changes were, they were sufficiently important to be incorporated in the tables and to appear in the comparison of ancient to modern observations. Laplace had Bouvard compare some twenty-seven eclipses recorded in antiquity, by Ptolemy and by the Arabs, to the figures in the tables, and they made no doubt of the importance of the acceleration of the motion of the lunar anomaly. He took the occasion to review the ancient corpus of lunar data as calculated by Ptolemy on the basis of the observations of Hipparchus and corrected by the further observations of al-Battani.

Laplace's presentation of this research differed from his previous practice. He published the results in a brief paper in the *Connaissance des temps* [1798b], the almanac for practical astronomy and navigation, roundly recommending that astronomers increase the motion of the lunar anomaly by 8′30″ per century and apply a correction to its secular equation equal to 4.3 times the mean motion.[21] The details of the analysis he kept for publication in the *Mémoires* of the Institute. There he chided mathematicians for having been insufficiently scrupulous in examining which of the terms they might legitimately neglect in successive integration of astronomical expressions.[22] In this investigation, he found it best to follow d'Alembert's example and express the lunar coordinates in series of sines and cosines of angles depending on the true motion, a method that had originated with Euler. In those expressions, Laplace made the true longitude the independent variable, rather than the time, as he always did in his planetary theory. There would be an advantage, he thought, in constructing tables that would give time as a function of the true motion of the moon, since terrestrial longitudes were determined in practice on the basis of the time at which the moon was observed to be at some certain position in its motion in longitude.

The papers of this period are indicative of the pattern of Laplace's later work. Henceforth, he tended to divide his efforts between short communications giving the results or applications of his current investi-

---

[20] *OC*, **12**, p. 194.
[21] [1798b], *OC*, **13**, p. 11.
[22] [1799a], *OC*, **12**, pp. 191–92.

gations and the great treatises still to be compiled and issued. These brief reports appeared in the *Connaissance des temps* when they were astronomical; otherwise they were published in one of the other journals started in the 1790s concomitantly with the movement toward specialization in the sciences. Often, as will appear in the bibliography, he published the same paper in several journals, sometimes with slight modifications. The day of the communication of scientific investigations through the medium of monographic research memoirs was, in any case, almost over. Those that Laplace had yet to publish show the tendency, already evident in the 1790s, to explore ever finer points of his earlier investigations. The first two volumes of *Mécanique céleste* were published in the same year, 1799, as the lunar analysis just discussed. In thus drawing together his science into the form of a treatise, Laplace like many of his colleagues was answering to another aspect of the evolution of science, the creation, actual or potential, by the new system of higher education of a truly scientific public within the larger audience that could be expected for his *Exposition du système du monde*.

CHAPTER 21

Traité de mécanique céleste

THOUGH CALLED a treatise, *Traité de mécanique céleste* (1799–1805) is a composite work. It has the aspects of a textbook, a collection of research papers, a reference book, and an almanac and contains both theoretical and applied science. The first two volumes form a largely theoretical unit. Methodologically, their purpose is to reduce astronomy to a problem in rational mechanics, in which the elements of planetary motions become the arbitrary quantities. Phenomenologically, the purpose is to derive all the observed data from the law of gravity. The textbook character of the work is most apparent in Book I and, to a lesser degree, in Book II, which occupy the first volume.

Book I is a mathematical exposition of the laws of statics and dynamics in a development adapted to the formulation of astronomical problems. Laplace's normal practice in those investigations had been to open each memoir with a derivation of the laws of motion in a form suited to the particular set of problems. Here he arranged the same material systematically. The sequence was canonical: first the statics and dynamics of mass points, second of systems of bodies, and third of fluids; the point of view is d'Alembert's. Dynamical laws are derived from equilibrium conditions.

Apart from the motivation, only two features appear to be distinctively Laplacian. In Chapter 5, which is concerned with the general principles of mechanics, Laplace incorporated his concept of an invariant plane into the discussion of the principle of conservation of areas. His introduction of that idea (see chapter 18 here) had been the first published statement that a general work on physical astronomy was in preparation.[1] He there emphasized the utility of such a plane in providing a frame of reference, fixed in space, to which calculations of planetary motion could be reduced in centuries to come. In *Exposition du système du monde* he had specified that a plane "that would always be parallel to itself" would pass through the center of the sun perpendicularly to the plane in which the sum of projected angular momentums of the bodies in the solar system is a maximum.[2] In later terminology, the reference plane is normal to the total angular momentum

---

[1] [1793a], *OC*, **11**, pp. 547–53.
[2] Book III, Chapter 5, Book IV, Chapter 2, *OC*, **6**, pp. 198–99, 218–19.

vector of the system. Laplace had given the mathematical rule for finding it in the *Journal de l'École Polytechnique* [1798c]. Now, in *Mécanique céleste*, he moved the origin of coordinates from the sun to the center of the earth, no doubt because in practice astronomers refer their observations of the motion of celestial bodies to the plane of the earth's orbit.[3]

The second feature that one would not expect to find in a textbook of rational mechanics is the discussion in Chapter 6 of the laws of motion of a system of bodies given any mathematically possible hypothesis concerning the relation of force to velocity.[4] In that apparent digression, Laplace may have been following Newton's example in certain propositions of Book II of *Principia mathematica*. A completely abstract and general system of dynamics might be imagined in which the number of such relations involving no contradiction would be infinite. There are two laws of nature, however, that hold good as principles of dynamics only in the simplest case, which is that of force directly proportional to velocity. The first is the principle of rectilinear inertia; the second is the conservation of areas in angular motion. It is in this discussion that it becomes clearest how, in the astronomical application, Kepler's law of equal times in equal areas had become for Laplace a special case of the principle of conservation of areas, or of angular momentum. He was usually careful to point out, however, that an accurate determination of the masses of all the planets had yet to be achieved.

Taken in isolation, Book II might also appear to have been conceived as a manual in which the mathematically qualified student learns the analysis required for theoretical astronomy. In it the laws of motion are applied to deriving the law of gravity from phenomena and to calculating the displacements of celestial bodies. Here also Laplace is more concerned to impart techniques than results. As he moved beyond the differential equations of gravitational attraction (Chapter 2) and of elliptical motion (Chapter 3), however, the techniques became increasingly his own. Chapter 6, for example, on perturbation theory, generalizes the combination of perturbations in coordinates and in the orbital elements that he had begun working out before 1774 for his first gravitational memoir.[5] Two bodies are assumed to move in coplanar, circular orbits with radii equal to the semi-major axis of the planetary orbits. A disturbing function is developed in sine and cosine series of the longitudinal difference between the bodies in orbit. The coefficients of the series are functions of the ratio of the semi-major axes. Laplace

[3] Book I, No. 21, *OC*, **1**, pp. 63–69; cf. Book I, No. 60, *OC*, **1**, pp. 337–38.
[4] See Vuillemin (1958).
[5] [1776a, 2°], Article LXIII, *OC*, **8**, pp. 241–46.

established the analytical relationships among them, and among their derivatives, with respect to that ratio, finding expressions for which he could later give numerical values for all possible pairs of planets.

The sequence of topics is also distinctively Laplacian and is, indeed, identical to that in *Exposition du système du monde*. At the outset of Book II, Laplace says that he intends to give the mathematization of the phenomena that he had there described in detail. Even in this book, however, the treatment grows more specialized as he continues, and already he was incorporating blocs of material from earlier researches in the exposition. Passages from the memoir on cometary orbits [1784b], for example, reappear in Chapter 4 dealing with motion in very eccentric and parabolic orbits. Similarly, the reciprocity of the acceleration of Jupiter and deceleration of Saturn, and also the libration of the inner three satellites of Jupiter, are introduced to illustrate methodological points. The Jupiter-Saturn relation (Chapter 8, no. 65) exemplifies the method of approximating periodic inequalities that appear in elliptical motion when it is legitimate to neglect terms involving squares or products of perturbing forces; the libration of the Jovian moons (Chapter 8, no. 66) depends on inequalities that appear only in terms of the order of the squares of the perturbing masses.

Volume II continues and completes the mathematical analysis of the three main categories of phenomena outlined in *Exposition du système du monde*. Having handled the motion of celestial bodies in translation in Book II, Laplace turned to the figure of the planets in Book III, to the motions of the seas and atmosphere in Book IV, and to rotational motion in Book V. Book III, nos. 8–15, on the attraction of spheroids, is a systematic reprinting, with some simplification of the mathematical development, of the material from his memoirs on the subject [1785b], [1786a], [1793a]. He had promoted the statement of the most important equation, which gives the potential function (see chapter 15, Equation 94), to the passage developing the basic differential equations governing the motions of mutually attracting bodies in Book II, no. 11.[6] For the rest, he repeated his discussion of the attraction exerted by spheroids of revolution on internal and external points, restated his theorem on the attractions of confocal ellipsoids, showed how to expand the expressions in series, and considered the cases of homogeneous and variable density (chapter 16). The third chapter brings in the demonstration that a liquid mass rotating under the influence of gravitational force will satisfy an ellipsoidal figure and that its axis of rotation will be in the direction that at the outset would have given it the maximum angular momentum.[7]

---

[6] *OC*, **1**, p. 153.
[7] Cf. [1785b], [1793a], Article XV.

The fourth chapter considers the spheroid covered with a layer of fluid and analyzes the equilibrium conditions.[8] The sixth and seventh discuss respectively the shape of the rings of Saturn and an equation governing the atmospheres of celestial bodies applied to the sun.[9] The main novelty is the comparison (Chapter 5) of spheroidal attraction theory with the results of geodetic surveys of meridional arcs. Laplace had introduced that topic in the miscellany published and largely lost to view in the waning days of the Academy.[10]. In *Mécanique céleste* he could and did draw on the data, not previously analyzed, from the Delambre-Méchain survey of the meridian from Dunkerque to Barcelona, on which the metric system was to be based (see chapter 18). He also, and perhaps more importantly, went further than he had previously done in applying error theory to the investigation of physical phenomena. An initial theoretical article (Book III, no. 38) develops the analytic geometry of geodetic lines and results in the following expressions applicable to the case of the earth: for the radius vector of an osculatory ellipsoid,

$$1 - \alpha \sin^2 \psi \{1 + h \cos 2(\phi + \beta)\}; \qquad (122)$$

for the length of a meridional arc,

$$\epsilon - \frac{\alpha \epsilon}{2} \{1 + h \cos 2(\phi + \beta)\}\{1 + 3 \cos 2\psi - 3\epsilon \sin 2\psi\}; \quad (123)$$

and for the degree measured orthogonally to the meridian,

$$1° + 1°\alpha\{1 + h \cos 2(\phi + \beta)\} \sin^2 \psi + 4°\alpha h \tan^2 \psi \cos 2(\phi + \beta), \qquad (124)$$

where ϵ is the difference in latitude between the two extreme points of the arc, ψ is the latitude of the longitudinal degree, ϕ is the angle formed by intersection of the plane xz with the plane that includes the radius vector and the z-axis, and β is a correction for the deviation of the true figure of the earth from an ellipsoid.[11] In his translation Nathaniel Bowditch had to point out that Laplace erred in the calculation, and that his numerical application suffered from this as well as from several arithmetical mistakes.[12] Nevertheless, it was in the derivation of these expressions that spheroidal analysis was brought to bear on actual geodetic measurements. The science of geodesy was thereby

[8] Cf. [1779a], Article XXVIII; [1785b], Article XV.

[9] Cf. [1789a] and *Exposition du système du monde*, Book V, Chapter 6.

[10] [1793a], no. 9; see chapter 18.

[11] *OC*, **2**, pp. 133–34.

[12] Bowditch (1829–39), **2**, pp. 394, 412–16, 447, 459, 471.

moved a significant distance along the scale from the observational to the mathematical.

The method itself is more interesting than the results. It had two stages. The first (Book III, no. 39) involved estimates of observational error. The quantities $a^{(1)}$, $a^{(2)}$, $a^{(3)}$, ... represent the lengths measured for a degree of the meridian in different latitudes; and $p^{(1)}$, $p^{(2)}$, $p^{(3)}$, ... are the squares of the sines of the respective latitudes. On the supposition that the meridian describes an ellipse, the formula for a degree will be $z + py$. Designating the observational errors $x^{(1)}$, $x^{(2)}$, $x^{(3)}$, ... Laplace wrote the following series of equations:

$$a^{(1)} - z - p^{(1)}y = x^{(1)},$$

$$a^{(2)} - z - p^{(2)}y = x^{(2)},$$

$$a^{(3)} - z - p^{(3)}y = x^{(3)},$$

$$\dots\dots\dots\dots\dots\dots,$$

$$a^{(n)} - z - p^{(n)}y = x^{(n)}, \tag{125}$$

where n is the number of meridional degrees measured.[13] The purpose is to determine y and z by the condition that the greatest of the quantities $x^{(1)}$, $x^{(2)}$, $x^{(3)}$,..., $x^{(n)}$ shall have the least possible value. Laplace gave solutions for the cases of two, three, or any number of such equations of condition, pointing out that the method was applicable to any problem of the same type. He mentioned specifically the example of n observations of a comet, from which it would be required to determine (1) the parabolic orbit for which the largest error is a minimum, and (2) whether the hypothesis of a parabolic trajectory can be reconciled with the observations in question. In the present, geodetic case the problem is to determine the ellipse for which the greatest deviation from measured values is a minimum.

The solution would reveal whether the hypothesis of an elliptical figure was contained within the limits of observational error. It would not, however, give the ellipse that the measured values themselves showed to be the most probable. Determining that figure was the object of the second stage of the analysis (ibid., no. 40). Two conditions had to be satisfied, first that the sum of all the errors made in the surveys of entire arcs should be zero, and second that the sum of all the errors taken positively should be a minimum. In the preliminary version of this analysis [1793a] Laplace had attributed the idea for this approach to

[13] *OC*, **2**, p. 135.

Bošković, of whom he made no mention in *Mécanique céleste*.[14] Having developed it, he proceeded to numerical calculations of the ellipticity of the earth, concluding from the data of the metric survey that it cannot be an ellipsoid and that the ratio of flattening of an osculatory ellipsoid is 1/250. The remainder of Chapter 5 consists of calculations of the probable degree of error in the results of other surveys (Lapland, Peru, Cape of Good Hope, Pennsylvania), of the flattening of Jupiter, and of the length of the seconds pendulum at various latitudes.

A propos of the last topic, it is perhaps worth noting that as Laplace came to consider that measurements of the length of the seconds pendulum might be reconciled with an ellipsoidal figure, his interest in them appears to have slackened.[15] Instead, he increasingly turned his attention to the data from direct geodetic surveys, determining how much of the deviation of that figure was owing to observational error and how much to nature. For what the second stage in this investigation finds him estimating is how far nature itself departed from theoretically determined forms. In other words, Laplace was now applying error theory to an investigation of phenomena and not merely to the probability of cause. He had not yet arrived at the least squares rule, and Quetelet's notion that errors of observation and errors of nature may follow an identical distribution would have been foreign to Laplace.[16] Both lay not far in the future along the same path, however.

Book IV, on the oscillations of the sea and the atmosphere (the last four of the 144 pages concern the atmosphere), contains less novelty in principle. Here, too, he first develops theory, which occupies three chapters, and compares it to the observations in a fourth. In his memoir on the tides at Brest [1797a] he had already revised the approach of his youthful investigations of the ebb and flow of the tides [1778a], [1779a], [1779b]. The first chapter now gives the mathematical treatment which that revision had summarized. The second restates his two theorems, to the effect that the seas are in stable equilibrium if their density is less than the mean density of the earth, and conversely. The last chapters are largely a repetition of the Brest memoir on the influence of local conditions, illustrated by the same early-eighteeenth-century data, the use of which is now tempered by incidental consideration of probable error in the observations. The second volume ends with Book V, on the rotation of celestial bodies. It is one of the shorter books of *Mécanique céleste*; and, like the memoir [1798a] hurried into print before publication of the work, has the appearance of an afterthought included for completeness.

[14] Todhunter (1873), no. 962, **2**, p. 134.

[15] [1798a], *OC*, **12**, p. 131.

[16] Gillispie (1963), p. 449.

Laplace had Volume III ready to present to the Institute in December 1802 (80). Three years of the Napoleonic consulate had elapsed since the publication of the first two volumes, which he had designated as Part I. The main purpose of Part II, he announced in the preface, was to improve the precision of astronomical tables. That motivation is consistent with the overall configuration of his career, at least in astronomy and probability. In both, the emphasis shifts to application, and only in physics did new theoretical problems engage his interest. The tendency was already evident in the internal sequence of particular investigations, and it would not force matters unduly to describe the first two volumes as representing largely the work of the early Laplace, and the second two that of the later Laplace. The transition occurred somewhere in the interval between 1790 and 1795, the years of the revolutionary liquidation of the old Academy and the quasi-technocratic reorganization of science at the Institute and related bodies. Nothing more than the environmental conditioning that accompanies change of circumstance can be claimed for the coincidence, but influence is nonetheless important for being felt pervasively.

The third volume is entirely occupied by the theory of the planets in Book VI and of the moon in Book VII. In developing the general formulas and methods for planetary astronomy in Book II, Laplace had limited himself to expressions for inequalities in the motions that are independent of orbital eccentricities and inclinations or that depend only on the first power of those quantities. The precision was insufficient for accurate positional astronomy, however, and Book VI applied to all the planets the method employed for the theory of Saturn in the great Jupiter-Saturn memoir [1788a]. Approximations were carried to the terms involving the squares and higher powers of these quantities and also to those depending on the squares of perturbing forces. Laplace then had Bouvard substitute the numerical values for each planet in these formulas, combined with the general formulas from Book II. The successive chapters then give numerical expressions for the radius vector and for its motion in longitude and latitude. It is in this book, and later in Book VIII (Volume IV) on the moons of Jupiter, that *Mécanique céleste* could serve the practical navigator and observer as the basis for an astronomical almanac. Bouvard was responsible for the enormous labor of numerical computation, for comparing the results with the findings of other astronomers, and for pinpointing the sources of disagreement. There might still be errors, Laplace acknowledged, but he was confident that they were too inconsiderable to vitiate the tables that might now be compiled.

Laplace himself had not previously worked on theories for Mercury, Venus, the earth, and Mars, for all the periodic inequalities are small and are now precisely given. Chapters 12 and 13, on Jupiter and Saturn, mainly repeat his classic work on their long periodic inequality. Although he had investigated the motion of Uranus as early as January 1783 (40), Chapter 14 is his first theoretical account of its motion. Chapter 15 formulates equations of condition for long-term periodic inequalities produced by the mutual perturbations of pairs of planets other than Jupiter and Saturn—that is, earth-Venus, Mars-earth, Uranus-Saturn, Jupiter-earth—and shows how they corroborate the respective planetary theories. Finally, the masses of the planets and of the moon are calculated relative to the sun, the values for Saturn and Uranus still needing considerable refinement. In the preface, he mentioned the discovery of Ceres on the first day (Gregorian style) of the new century, followed by that of Pallas, but gave no detail.

Book VII is devoted to lunar theory. Its object is to exhibit in numerical detail the finding of the initial memoir on the moon [1788c] that all the inequalities of lunar motion, namely variation, evection, and the annual equation, result from the operation of universal gravity, and then to deduce from that law further explanations concerning finer points of the motion, and also of the parallax of the sun and moon and of the flattening of the earth. Laplace followed the practice of his second memoir [1799a) in taking true longitude rather than time for the independent variable in his differential equations of motion, which he had taken the precaution of adapting for the purpose in Book II.[17]

It will be recalled that he had published this paper only a few weeks before the first two volumes of *Mécanique céleste* itself, having worked out the analysis early in 1797 (63). In the meantime, the Austrian astronomer Johann Tobias Burg had been investigating what appeared to be a periodic inequality in the motion of the lunar nodes with an interval of about seventeen years between the maximum positive values and about nineteen years between the maximum negative values. In 1800 the Institute awarded Burg a prize for this research.[18] He had already asked Laplace to investigate what the cause of these effects might be. Employing the appropriate equations from *Mécanique céleste*, Laplace analyzed the data in a memoir [1801b] read before the Institute in June 1800 (71).[19]

The episode illustrates that from the outset, *Mécanique céleste* was furnishing the apparatus for further research and calculation in both practical and theoretical astronomy. Laplace found that the effects result from a nutation in the lunar axis created by a variation in the

---

[17] No. 15, Equation K.
[18] *PVIF* (11 germinal an VIII), **2**, p. 129.
[19] Book II, no. 14; Book III, no. 35.

inclination of the lunar orbit to the plane of the ecliptic. Its inclination is constant with respect to another plane passing through the equinoxes between the equator and the ecliptic. That angle would amount to 6.5″ on the assumption that the flattening of the earth is 1/334. Further comparison by Bouvard of Burg's observations with those of Maskelyne indicated rather a figure of 1/314. In any case, the value was far from the fraction of 1/230 that spheroidal theory predicted for an earth of homogeneous density. Laplace was delighted that so minuscule an anomaly in the position of lunar nodes could thus confirm the direct measurements of geodesy on the shape of the earth and on conclusions to be drawn concerning its internal constitution. He incorporated the material in Book VII of *Mécanique céleste*, where it formed the major novelty, the bulk of the discussion being a recapitulation of his earlier research fortified by Bouvard's indefatigable calculation of numerical values for the formulas to serve in compiling precise tables.

For Volume IV, presented to the Institute in May 1805, over two years after Volume III, there remained the practical theory of the satellites of the outer planets, and also of the comets. Book VIII is almost entirely devoted to the moons of Jupiter. It consists of a revision of the calculations of the memoirs of the early 1790s [1791a], [1793b], which had issued in values for the masses of the four satellites relative to that of Jupiter and for the flattening of the latter. Laplace now gives greater numerical detail on the inequalities of the three inner satellites, concealed in the invariance of the libration that he had discovered in his first memoir on these bodies [1787a]. He likens their lockstep to the observable libration of our own moon and compares other particularities of the motion of the Jovian satellites to the lunar evection, annual equation, and variation in latitude discussed in Book VII. (It is interesting that, having been led to his explanation of the apparent lunar acceleration by his first work on the moons of Jupiter, he was now illuminating finer points in their theory by analogy to earth's moon.) Chapter 7, giving numerical values for the various inequalities, is new.

In deriving formulas for the variations in radius vectors and longitudes, Laplace made several errors that were detected by Airy and corrected by Bowditch.[20] In all these formulas, Bouvard calculated the numerical values of the coefficients, although the tables to which Laplace refers navigators continue to be Delambre's.[21] The entire topic is presented as the confirmation by another method of the purely analytic demonstration of the theorems on the libration of the three inner moons in Book II, Chapter 8.[22] The difference in method con-

---

[20] Bowditch (1829–39), Book VIII, no. 21, **4**, pp. 176–85.
[21] *OC*, **4**, pp. x–xi.
[22] *OC*, **1**, pp. 346–395.

sisted in the substitution of synodic for sidereal mean motions and longitudes, and of a moving for a stationary axis of rotation[23] In general, Laplace made more than he had previously done of the spectacle offered by the Jovian satellites of a gravitational system in miniature, its elements oscillating about mean values at a higher rate than those of the whole slow-motion solar system.

The two brief chapters on the satellites of Saturn and Uranus that complete Book VIII are essentially a reprinting of a paper published in the *Mémoires de l'Institut* [1801a]. The orbits of seven satellites of the most distant planets—the outermost moon of Saturn and six Uranian moons that Herschel thought he had observed—all appeared to be inclined to the plane of the ecliptic at a much greater angle than are other planetary paths. Laplace's analysis demonstrates how that anomaly can follow from the weakening of the force of gravity given the distances and ratios of the masses. He acknowledged the data to be very uncertain, however, and indeed it has since been learned that the fifth and sixth moons of Uranus do not exist.

In Book IX, the shortest in the four main volumes, Laplace developed formulas for calculating cometary perturbations from the general equations of motion set forth in Book II, Chapter 4.[24] The planetary formulas were inapplicable to orbits involving large eccentricities and inclinations, and for comets different formulas had to be applied to different parts of the same orbit. Laplace showed how to obtain numerical values for perturbations in orbital elements by means of generating functions. He would have liked to illustrate his techniques by calculating the elements of the curve to be described in the impending return of the comet last seen in 1759. Unfortunately, he was too busy and left the formulas to whoever wished to substitute the numerical data. He did complete two other examples. A chapter on the perturbation of comets that pass very close to a planet concludes that the gravity of Jupiter had drawn the perihelion of a previously invisible comet within range of sight in 1770 and then reversed the effect in 1779, after which year it had not reappeared. A second calculation shows that the same comet produced no detectable change in the length of the sidereal year in 1770 despite its proximity to the earth. Laplace felt safe in concluding that the mass of comets is so small that they can have no influence on the stability of the solar system or on the reliability of astronomical tables.

Book X, subtitled "On Different Points Concerning the System of the World," contains largely new material and marks the shifting of Laplace's interest to problems of physics even as he was completing the fourth volume. Analysis of the effects of atmospheric refraction upon

---

[23] *OC*, **4**, p. viii.
[24] *OC*, **1**, pp. 210–254.

astronomical observation is the initial topic. Laplace had already committed himself to a corpuscular emission theory of light in the remarks on aberration with which he concluded his numerical memoir on the moons of Jupiter.[25] Now his derivation of the phenomena of atmospheric refraction presupposed that model, as did the investigations of 1808 in optics proper, discussed below in chapter 24. These passages, indeed, heralded the corpuscular conception of all physical phenomena which was to guide the research program of the Arcueil school.

The first problem in Book X is to find the law governing the dependence of refrangibility on the variation of atmospheric density with altitude and temperature. An elaborate analysis of the passage of light through a refracting medium yielded a formula that Laplace reckoned to be applicable when a star had risen to an elevation of 12° above the horizon.[26] At angles greater than that, only the atmospheric pressure and temperature in the vicinity of the observer significantly affected the refraction, and these values could be read directly from the barometer and thermometer.

In order to evaluate his expressions numerically, Laplace needed to know the index of refraction of atmospheric air at a given temperature and pressure and the variation of its density respectively with pressure and with temperature. Delambre had determined the index of refraction for apparent elevations of 45° by observations of the least and greatest elevations of certain circumpolar stars at 0° and 76 centimeters of mercury. As for the pressure-volume relations of atmospheric air, physicists were all agreed on the direct proportionality of density to pressure at constant temperature. Despite many attempts at measurement, however, there was still no agreement about temperature-volume relations in gases, and Laplace engaged Gay-Lussac's assistance in examining the matter. For that purpose, Gay-Lussac calibrated a mercury against an air thermometer, took extraordinary precautions to dry the air and tubes composing the latter (for humidity was the main source of error), and found that at constant pressure of 76 centimeters of mercury, a unit volume of air at 0° expanded to a volume of 1.375 at 100°. Comparison of the two thermometers at intermediate temperatures argued for a linear expansion within that range. The final value represented the mean of twenty-five determinations, although Laplace did not say how the mean was calculated.

All this discussion, which somehow conveys a greater sense of enthusiasm than the preceding books of recapitulation and tabulation, bespeaks Laplace's growing interest in instrumentation, measurement, and

[25] [1793b]; see chapter 18.
[26] *OC*, **4**, p. 269.

minimization of observational error. No one, he observed, had yet thought how to compensate for variation in humidity in measurements of atmospheric refracton. Small though he calculated the effect to be, he gave a correction table compiled on the reasonable hypothesis that the indices of refraction of air and water vapor are proportional to their densities. In a like, almost offhand manner, he reported Gay-Lussac's ascension in a balloon to an altitude of over 6,500 meters, where the proportions of oxygen and nitrogen in the atmosphere turned out to be about the same as at ground level. As Laplace drew toward the intended conclusion of his treatise, the topics grew more recondite and more fanciful in the object: the effect of extreme atmospheric conditions on astronomical observations, the influence of differences of latitude on barometric measurements of altitude, the absorption of light by the atmospheres of the earth and sun, and the influence of the earth's rotation on the trajectories of projectiles and on free fall from great heights.[27]

Before writing Book X, Laplace evidently intended to end it with calculations, which occupy Chapter 7, contrasting consequences to be deduced from the wave theory and from the corpuscular theory of light. He there purported to show how the resistance of any ethereal medium supporting luminous oscillations would have entailed deceleration of planetary motion. The continuous impact of light corpuscles would, on the other hand, accelerate the planets, except that the effect is exactly compensated by the weakening of the sun's gravitational force through loss of mass. None of this disturbs stability, however. Since the mean motion of the earth shows no change over a two thousand year span, Laplace calculated that the sun had not lost a two-millionth part of its substance in recorded history and that the effect of the impact of light particles on the secular equation of the moon is undetectable. In a way, it would have been fitting had this chapter been the last, for Laplace applied the calculation that he had just made to the gravitational force considered as the effect of a streaming of particles through space. Thus he would have emerged full circle from his celestial mechanics, coming out just where he went in with the first calculation of the youthful probability-gravitation memoir, except that now gravity is given a velocity of 1×10^9 times the speed of light, which is to say, infinite.[28]

That chance for symmetry (if such it may be called) disappeared with the publication of two further memoirs on the theory and tables of Jupiter and Saturn [1804a] and [1804b]. Laplace immediately grafted them on to Book X, where they form the basis of Chapters 8 and 9. The

---

[27] Cf. [1803b].
[28] Cf. [1776a, 2°]; see chapter 4.

return to planetary astronomy was unconformable with the overall plan of the treatise, although it will already have been noticed that throughout its composition Laplace found ways to interpolate pieces of continuing research. He called Chapter 8 "Supplément aux théories des planètes et des satellites," which may be bibliographically confusing since these chapters did appear in the first edition, unlike the true supplements to be mentioned in a moment. At any rate, in the interval since the publication of Book VI (Volume III, 1802), Bouvard had scrutinized all the oppositions of both planets observed at Greenwich and Paris since Bradley's time in the 1750s, and Laplace himself had reviewed the theory. The result was several new inequalities, and by taking them into account the agreement between his formulas and the observations was improved. The most signal advantage of the new data was that they permitted the first precise calculation of the mass of Saturn, hitherto known only roughly through the elongations of its satellites.

"Nothing more remains for me," wrote Laplace in the concluding sentence to the preface to Volume IV, "in order to fulfill the engagement that I undertook at the beginning of this work, but to give a historical notice of the works of mathematicians and astronomers on the system of the world: that will be the object of the eleventh and last book."[29] In fact, quite a lot remained, beginning with the studies of capillary action presented to the Institute on 28 April and 29 September 1806 (86, 87) and separately printed as Supplements I and II to Book X. The years from 1806 through 1809 were evidently occupied with the further work in physics proper discussed in chapters 22–24 below, and those from 1810 through 1814 largely with probability (chapters 25 and 26). Indeed, prior to 1819, by which time Laplace was seventy years old, he published only occasionally on problems of celestial mechanics, and these papers were on minor points. His only major addition in all that fifteen-year interval was a mathematical improvement in the method of calculating planetary perturbations, presented to the Bureau des Longitudes in August 1808 (90) as a supplement to Volume III of *Mécanique céleste*.

[29] *OC*, **4**, p. xxv.

Part IV

LAPLACIAN PHYSICS AND PROBABILITY

CHAPTER 22

The Velocity of Sound

ROBERT FOX

LAPLACE apparently gained his first experience in physics in the experiments conducted jointly with Lavoisier (see chapter 14). It will be recalled that in 1777 he and Lavoisier investigated evaporation and vaporization, in 1781–1782 the expansion of glass and metals when heated, and in 1782–1783 specific heats, heats of reaction, and animal heats. Thereafter, apart from occasional further collaboration with Lavoisier in the later 1780s, Laplace took little active interest in physics between 1784 and 1801.

In 1801, however, he published a brief piece, which applies spheroidal attraction theory to the analysis of the forces exerted by an infinitely thin layer of electrical fluid spread upon such a surface.[1] And in the following year, he made one of his most enduring contributions to physics, in a paper written not by himself but by his young protégé Biot.[2] Using a knowledge of adiabatic phenomena that had only recently become available in France (even though the heating and cooling associated with the rapid compression and expansion of a gas had been quite well known among British, Swiss, and German scientists since the 1770s), Laplace had suggested to Biot how the notorious discrepancy of nearly 10 percent between the experimental value for the velocity of sound in air and the calculated value using Newton's expression, $v = \sqrt{P/\rho}$, might be removed. According to Laplace, the discrepancy arose from Newton's neglect of the changes in temperature that occur in the regions of longitudinal compression and rarefaction composing the sound wave. Hence Newton's assumption that $P \propto \rho$, which holds good only if isothermal conditions are maintained, was invalid.

Biot expressed the density of air at any point in a sound wave as $\rho' = \rho(1 + s)$, where ρ is the density of the undisturbed air and s the fractional change in density, taken as positive for compression. Provided

---

[1] [1802d]. The paper helped to establish links between gravitational theory and potential theory.

[2] Biot, "Sur la théorie du son," *Journal de physique*, **55** (1802), pp. 173–82. On this and the work described in the rest of this chapter, see Fox (1971), pp. 81–86, 161–65, and Grattan-Guinness (1990a) **1**, pp. 452–55; **2**, pp. 811–20.

isothermal conditions were maintained, it followed simply that the pressure of the air could be similarly expressed as $P' = P(1 + s)$, where P is the pressure of the undisturbed air. However, if, as Biot supposed, heating and cooling occurred, respectively in the regions of compression and rarefaction, the equation could not hold. Making the reasonable but unproven assumption that the change in temperature was proportional to s, Biot arrived at the expression $P' = P(1 + s)(1 + ks)$, where k is a constant. Hence, assuming that s is small and neglecting the terms in s, Biot could show that the velocity of sound in air is

$$v = \sqrt{\frac{P}{\rho}(1 + k)} .$$

Although Laplace's explanation of the discrepancy, as expounded by Biot, won general acceptance, replacing a variety of unsubstantiated proposals made during the eighteenth century by Lambert and (most pertinently) Lagrange among others, the experimental evidence necessary for a rigorous proof became available only over the next twenty years. By 1807, however, Biot had made the important observation that sound waves could be transmitted through a saturated vapor.[3] In this way, he confirmed that some heating and cooling must occur, for if this were not so, condensation would take place in the regions of compression and the sound would not pass. At about the same time, Poisson too contributed to the discussion, adopting a variant of Biot's approach (with due acknowledgment to Laplace but none to Biot) and arriving at a value of 0.4254 for k, if theory and observation were to agree.[4] Poisson's calculation also led to the conclusion, which continued to be cited in the literature until the midcentury, that an adiabatic decrease or increase in volume of $1/116$ would raise or lower the temperature of a gas by 1° C.

Poisson's conclusions, like Biot's, however, could not be independently verified, and the question remained in abeyance until 1816, when Laplace showed that the constant $(1 + k)$—in the form of the equation given above—was equal to γ, the ratio between the specific heat at constant pressure (c_p) and the specific heat at constant volume (c_v).[5] Since this ratio had not been reliably measured, confirmation of Laplace's result was impossible. But in 1822 experiments by Gay-Lussac

[3] Biot, "Expériences sur la production du son dans les vapeurs," *Mémoires de physique et de chimie de la Société d'Arcueil*, (1809), **2**, pp. 94–103.

[4] Poisson, "Mémoire sur la théorie du son," *Journal de l'École Polytechnique*, **7** (1808), pp. 319–92.

[5] [1816c].

and Jean-Joseph Welter, for which Laplace was clearly the inspiration, yielded values for γ that inspired confidence.[6] By observing the changes in temperature that occurred when air was suddenly allowed to enter a partially evacuated receiver (a method pioneered less successfully a decade earlier by Nicolas Clément and Charles-Bernard Desormes), Gay-Lussac and Welter arrived at the remarkably accurate figure of 1.3748. When inserted in Laplace's expression of 1816, the figure brought the theoretical value of v (337.14 meters per second) into very satisfactory agreement with the currently accepted experimental figure.

How Laplace derived the correction factor of $\sqrt{\gamma}$ in Newton's expression for v was left unclear in the paper on the subject that he published in 1816, but a reconstruction resting partly on a later paper, published in 1822, suggests the argument that may have been used. It seems that Laplace began by showing that v must be equal to $\sqrt{dP/d\rho}$.[7] In demonstrating that this quantity is in turn equal to $\sqrt{\gamma P/\rho}$, he assumed not only that the difference between c_p and c_v represents the heat required solely to bring about the increase in the volume of a gas expanding at constant pressure but also that it is the same heat that causes heating when the gas is rapidly compressed. By this argument, a decrease in the volume of a unit mass of the gas from V_0 to $(V_0 - \Delta V)$ would release an amount of heat $(\Delta V/\alpha V_0)(c_p - c_v)$ which, in adiabatic conditions, would go to raise the temperature of the gas by $\Delta V/\alpha V\{(c_p - c_v)/c_v\}$, where α is the temperature coefficient of expansion. The effect of the rise in temperature would be to increase the pressure of the compressed gas by $P_0(\Delta V/V_0)\{(c_p - c_v)/c_v\}$, in addition to the increment in pressure that would be expected for an isothermal compression. Hence the total increase in pressure is given by

$$\left(\frac{\Delta P}{\Delta V}\right)_a = \frac{P_0}{V_0} + \frac{P_0}{V_0}\left\{\frac{c_p}{c_v} - 1\right\} = \frac{c_p}{c_v}\left(\frac{\Delta P}{\Delta V}\right)_i,$$

where $(\Delta P/\Delta V)_i$ represents the pressure increment that would have been obtained under isothermal conditions. It followed that, when conditions were adiabatic, $\sqrt{dP/d\rho}$ —Laplace's expression for the velocity of sound—was equal to

$$\sqrt{\frac{c_p}{c_v} \cdot \frac{P}{\rho}}.$$

[6] Gay-Lussac and Welter, "Sur la dilatation de l'air," *Annales de chimie et de physique*, **19** (1822), pp. 436–37.
[7] [1822a].

It now remained to calculate a numerical value for c_p/c_v. At a time when no experimental data for c_v existed, this was no easy task, and in his paper of 1816 Laplace was vague about the method that he had used in arriving at a figure of 1.5. However, modern attempts at a reconstruction, although differing on matters of detail, reveal quite clearly not only the unsatisfactory nature of the argument but also the importance for Laplace of the erroneous data concerning the specific heats of gases that had been published by Delaroche and Bérard in 1813.[8] In particular, Laplace's argument rested squarely on their false observation that the specific heat of air decreases as it is compressed. Hence the similarity between Laplace's own figure of 1.5 for γ and the figure of 1.43, required in order to secure exact agreement between prediction and observation, was quite fortuitous. Nevertheless, the plausibilty of Laplace's treatment appears to have gone unquestioned, so that the measurements of γ by Gay-Lussac and Welter in 1822 provided experimental confirmation of a result on which no one, in fact, had cast serious doubt.

[8] Finn (1964), p. 15; Fox (1971), pp. 162–65; Heilbron (1993), pp. 176–78.

Short-Range Forces

ROBERT FOX

NEARLY ALL of Laplace's work in physics from 1802 was characterized by an interest in what he saw as the outstanding problems of the Newtonian tradition; in this respect his attempt to correct Newton's expression for the velocity of sound was typical. From 1805, however, his interest in Newtonian problems assumed a more mathematical character. As we shall see, much of the work to which he now turned was eventually to be severely criticized. But, whatever its shortcomings—and it did little to enhance Laplace's reputation in his later life—it contained results of enduring value, most notably in the theory of capillary action. No less importantly, it served to tighten the bond between mathematics and physics. This is not to imply that Laplace was in any sense the founder of mathematical physics: there were too many precursors in the eighteenth century for that claim to be sustained. But he did make a major contribution to the mathematization of a discipline that had hitherto been predominantly experimental.

The increasingly mathematical thrust of Laplace's work in physics is very apparent in the studies of molecular physics that he pursued, and encouraged others to pursue, for the rest of his life. But the interest was not a new one. As early as 1783, when he composed his *Théorie du mouvement et de la figure elliptique des planètes*, he elaborated a suggestion he had already advanced with Lavoisier in the *Mémoire sur la chaleur*.[1] This expressed his belief that optical refraction, capillary action, the cohesion of solids, their crystalline properties, and even chemical reactions were the results of an attractive molecular force, gravitational in nature and even identical with gravity.[2] Almost twenty years earlier, near the beginning of his career, he had remarked in the dual probability-gravitation memoir that analogy gives us every reason to suppose that gravity operates between all the particles of matter, extending down to the shortest ranges.[3] He repeated and elaborated the

---

[1] [1783a].

[2] [1784a], xii–xiii; see chapter 16. On the belief as a central theme of Laplacian physics and on the rise and decline of this style of physics, see Fox (1974).

[3] [1776a, 2°]; see chapter 5.

speculation in the first edition of *Exposition du système du monde*,[4] looking forward to the day when the law governing the force would be understood and when "we shall be able to raise the physics of terrestrial bodies to the state of perfection to which celestial physics has been brought by the discovery of universal gravitation." In this comment, which is reminiscent of the speculations on molecular forces in the Queries of Newton's *Opticks*, there lay the nucleus of a program that guided Laplace's own research in physics and that of several distinguished pupils until the 1820s.

Despite the early adumbrations of the program, it was not until 1805 that Laplace began publishing on the individual problems it raised. By then, interest in molecular forces treated in the Newtonian manner had been greatly stimulated in France by Berthollet's work on chemical affinity, in particular by his *Essai de statique chimique* (1803), in which chemical reactions were explained in terms of short-range attractive forces, supposedly of a gravitational nature, of precisely the kind postulated by Laplace in his physics. It seems likely that Laplace was strongly influenced both by the *Essai* and by Berthollet himself, whose close friend he had been since the 1780s, and that the influence was reciprocal. In any event, Laplace's first work in molecular physics was published just two years after the publication of the *Essai*, in Book X of the fourth volume of *Traité de mécanique céleste* (1805)[5] and in two supplements to the book published in 1806 and 1807.[6]

It is a measure of Laplace's closeness to the Newtonian tradition that these first studies of molecular forces were concerned with optical refraction and capillary action. Both were manifestations of action at a distance on the molecular scale that had been of special interest to eighteenth-century Newtonians such as Clairaut and Buffon as well as to Newton himself. The contributions of Clairaut appear to have been especially relevant to Laplace's work. In the 1740s, for example, Clairaut had ascribed "the roundness of drops of fluid, the elevation and depression of liquids in capillary tubes, the bending of rays of light, etc." to gravitational forces that become large at small (i.e., molecular) distances.[7] But despite the attention that both Clairaut and Buffon had paid to these and related theoretical problems, there was still no satisfactory answer by the end of the eighteenth century. In particular, although it was generally accepted that the force between the particles of ordinary matter (in the case of capillary action) and between the

---

[4] [1796], *OC*, **2**, pp. 196–98.

[5] See especially Chapter 1, pp. 231–76, *OC*, **4**, pp. 233–77.

[6] *OC*, **4**, pp. 349–417, 419–98.

[7] Clairaut, "Du système du monde dans les principes de la gravitation universelle," *MARS* (1745/1749), pp. 329–64 (338).

particles of ordinary matter and the particles of light (in the case of refraction) diminishes rapidly with distance, it had proved impossible to determine the law relating force and distance. Clairaut had tried to account for the intense short-range forces by suggesting that the law of gravitational force should contain a term inversely proportional to the fourth power of the distance, $1/r^4$.[8] Buffon, by contrast, had upheld the $1/r^2$ law, although he had observed that such a law would be modified at short range by the shape of the particles of matter.[9] Recognizing the intractability of the problem, Laplace proposed a much simpler solution that could be applied in all branches of molecular physics. In treatments that made good use of mathematical techniques developed in his earlier work on celestial mechanics, he showed that the precise form of the law was unimportant and that perfectly satisfactory theories could be given by simply making the traditional assumption that the molecular forces act only over insensible distances.

The treatment of refraction in Book X of *Mécanique céleste* centered on the specific problem of atmospheric refraction, a matter of practical as well as theoretical concern to Laplace and his colleagues at the Bureau des Longitudes. The whole discussion was conducted in terms of the corpuscular theory of light, the truth of which was assumed for the purposes of the calculation (though not explicitly endorsed as a physical reality). According to Laplace, the path of a corpuscle of light passing through the successive layers of the earth's atmosphere is determined by the varying attractive forces exerted on it by the particles of air. The measure of these forces was what Laplace, following Newton, called the refracting force (*force réfringente*), a quantity equal to ($\mu^2 - 1$), where μ is the refractive index of the air. In this analysis, ($\mu^2 - 1$) is proportional to the increase in the square of the velocity of the incident corpuscles of light, and hence, in accordance with the normal laws of dynamics, it measures the force of attraction to which they are subject. In deriving his extremely complicated differential equation for the motion of light through the atmosphere, Laplace had to assume not only the short-range character of the forces to which light corpuscles were subjected but also that the refracting force was proportional to the density of the air, ρ. In order to integrate the equation, it was necessary to make further, speculative assumptions concerning the variation of ρ with altitude (the subject of a long and properly tentative section) and to allow for the effect on ρ of the air's humidity. The result was a method of calculating the magnitude of atmospheric refraction for

[8] Ibid., pp. 337–39.
[9] Buffon, "De la nature. Seconde vue" [1765], in his *Histoire naturelle, générale et particulière*, 36 vols. (Paris, 1749–1804), **13**, pp. xii–xv.

which Laplace claimed complete reliability at any but small angles of elevation.

Despite important subsequent refinements, the core of Laplace's theory of capillary action, as expounded in the two supplements to Book X of *Mécanique céleste*, continued to guide discussions throughout the nineteenth century, and it even survives in modern textbooks.[10] However, its roots lie as firmly in the vain quest for a comprehensive physics of short-range forces as do those of his work on refraction. As in the theory of refraction, it was crucial that the forces exerted by the particles of matter on one another could be ignored at any but insensible distances (although it was equally important that these distances be finite—an assumption that distinguished Laplace's theory from others in which adhesion was seen as the cause of capillary phenomena). Hence Laplace was glad to invoke Hauksbee's observation that the height to which a liquid rises in a capillary tube is independent of the thickness of the walls of the tube.

In each of the two supplements on capillary action Laplace presented a quite distinct version of his theory. In the first, he arrived at a general differential equation of the surface of a liquid in a capillary tube by considering the force acting on an infinitely narrow canal of the liquid parallel to the axis of the tube. In the second, he treated the equilibrium of the column of liquid in a capillary tube by considering the forces acting upon successive cylindrical laminae of the liquid parallel to the sides of the tube. The two versions were in no sense inconsistent with each other, although in a number of applications the second version proved to be somewhat simpler and more fruitful.

Laplace was concerned, above all, to demonstrate the close agreement between his theory and experiment, and much of both supplements was devoted to this task. One of his most striking successes, in the first supplement, was confirmation of a proof, obtained by solving the differential equation of the liquid surface that the elevation of a liquid in a capillary tube is very nearly in inverse proportion to the tube's diameter.[11] In the same supplement and using the same version of the theory, he also showed that the insertion of a tube of radius r_1 along the axis of a hollow tube of slightly larger radius r_2 causes the liquid between the tubes to rise to a height equal to that to which it would rise in a circular capillary tube of radius $(r_2 - r_1)$;[12] in this way,

[10] See, for example, Champion and Davy (1952), pp. 172–77, and the brief study of Laplace's theory in Bikerman (1975). For a more recent account that places Laplace's investigation of both capillarity and refraction in the tradition of work pursued by Clairaut, see Heilbron (1993), pp. 150–65.

[11] (86), 23–25, *OC*, **4**, pp. 372–74.

[12] (86), 25–28, *OC*, **4**, pp. 374–78.

he confirmed a well-known observation made but not explained by Newton. Among the other classic problems treated in the supplements were the behavior of a drop of liquid in capillary tubes of various shapes (including conical tubes), the rise of liquids between parallel or nearly parallel plates, the shape of a drop of mercury resting on a flat surface, and the force drawing together parallel plates separated by a thin film of liquid.

The importance, for Laplace's theory, of the short-range character of the molecular forces cannot be overstated. The assumption allowed him repeatedly to set aside small terms involving the square of the distance, and, with the aid of this simplification, to pursue mathematical investigations that would otherwise have been impossible. It was in accordance with his belief that capillarity is a consequence of intermolecular action at a distance (albeit at a very small distance) that he was able to calculate the relative magnitude of the attractive force between the particles composing the liquid (F_1) and the force between the particles of the liquid and those of the tube (F_2) and to define the conditions that determined the shape of the liquid's surface. Neglecting variations in density near the surface of the liquid and the walls of the tube, he showed that if $F_2 > F_1/2$, the surface must be concave; otherwise it must be convex, being, in the limiting case of $F_2 = 0$, a convex hemisphere.[13]

Even as the supplements on capillary action were being written, the comparison between theory and observation was being carried still further in experiments, performed at Laplace's request, by Gay-Lussac, Haüy, and the engineer Jean-Louis Trémery.[14] These experiments gave the theory added plausibility, as Laplace himself was always ready to observe; and they certainly helped it to survive the criticism of his most important contemporary rival in the treatment of capillarity, Thomas Young. By comparison, Young's theory, which was based on the concept of surface tension rather than intermolecular attraction, was obscure and unmathematical.[15] Yet Laplace's theory (which Young saw as "unnecessarily intricate") was not without fault, and its author led the way in making modifications. In a paper that he read to the Academy of Science in September 1819, he refined the theory to take account of the effect of heat in reducing the attractive force between the particles of a liquid;[16] the net attractive force was now taken as the difference between the innate attraction (the only force considered in the

[13] (86), 44–50, *OC*, **4**, pp. 394–401.

[14] [1806a], *OC*, **4**, pp. 403–5.

[15] Young, "An Essay on the Cohesion of Fluids," *Philosophical Transactions of the Royal Society* (1805), pp. 65–87.

[16] [1819h].

supplements to *Mécanique céleste*) and a repulsive force that was supposed to be caused by the presence of heat. An even more important modification was made in 1831, when, in his *Nouvelle théorie de l'action capillaire*, Poisson remedied one of the most obvious weaknesses in Laplace's theory by taking account of the variations in density near the surfaces of the liquid and the material of the capillary tube.[17]

[17] Poisson, *Nouvelle théorie de l'action capillaire* (Paris, 1831); see esp. pp. 1–8.

CHAPTER 24

The Laplacian School

ROBERT FOX

FROM THE TIME the studies of refraction and capillary action appeared until 1815, Laplace exerted a dominating influence on French physics. The extent of his influence is equally apparent from the problems that younger men were encouraged to investigate (either directly or through prize competitions), from the nature of their answers (which with remarkable frequency served to endorse and extend the Laplacian program), and from educational syllabuses and textbooks, which seldom departed from Laplacian orthodoxy on such matters as the centrality of problems related to the imponderable fluids of heat, light, electricity, and magnetism. Yet at no time was Laplace's control total. There were always those in France who worked outside the Laplacian tradition or even in opposition to it. For example, the paper on the distribution of heat in solid bodies that Fourier read to the First Class of the Institute in December 1807 shows no sign of Laplace's influence, either in its positivistically inclined physics or in its mathematical techniques.[1] And the same is true of the revised and extended version of the 1807 paper that he submitted, successfully, in 1811 for the prize competition of the First Class.[2]

Fourier's treatment, in fact, stands in marked contrast with the discussion of the problem that Laplace incorporated as a "Note" to his paper on double refraction read to the Institute in January 1809.[3] In this note Laplace set up a model for heat transfer by reference to the molecular radiation of caloric over insensible distances. Even Poisson's papers of 1811–1813 on electrostatics owed at least as much to Coulomb as they did to Laplace, although Poisson was close to Laplace at this time and Laplace would certainly have approved of his treatment, in particular his use of the two-fluid theory of electricity.[4] Such instances

---

[1] See Grattan-Guinness with Ravetz (1972), esp. pp. 444–52, and Herivel (1975), pp. 150–91.

[2] Fourier's prize-winning entry was not published until 1824–1826, when it appeared in *MASIF*, **4** (1819–1820/1824), pp. 185–555, and **5** (1821–1822/1826), pp. 153–246.

[3] [1810a], pp. 326–42, *OC*, **12**, pp. 286–98.

[4] Poisson, "Sur la distribution de l'électricité à la surface des corps conducteurs," *MI*, **12** (1811/1812), part 1, pp. 1–92; ibid. (1811/1814), part 2, pp. 163–274. On these papers and their significance for Laplacian physics, see Hofmann (1995), pp. 112–22.

of non-Laplacian physics leave no doubt that a fruitful union of the mathematical and experimental approaches to physics would have occurred in early-nineteenth-century France quite independently of Laplace. But the fact remains that Laplace did more than any of his contemporaries to foster that union and, at least in the short term, to determine the character of the work that emerged from it.

The years of Laplace's greatest influence in physics were also those in which his personal standing was at its height, outside the scientific community as well as within it; and he seized every opportunity of furthering his scientific interests. It was a simple matter for him to direct the attention of gifted young graduates of the École Polytechnique, such as Gay-Lussac, Biot, Poisson, and Malus, to problems of his own choice, often in return for help in advancing their careers in the teaching institutions of Paris or at the Bureau des Longitudes. And, once he had become Berthollet's next-door neighbor at Arcueil and the joint patron of the informal Société d'Arcueil in 1806, he could offer his protégés the additional attractions of access to Berthollet's private laboratory and an association with the elite of Parisian science in the weekend house parties that were a feature of Arcueil life until 1813.[5] The work of Gay-Lussac, Haüy, and Trémery on capillary action (1806), of Biot on the transmission of sound in vapors (1807), and of Malus, Arago, and Biot on the polarization of light (1811–1812) was very obviously a result of direct influence of this kind.

Equally important for the course of French physics was the power that Laplace wielded in the First Class of the French Institute. Here his ability to dictate problems and solutions was no less apparent than in the more intimate atmosphere of Arcueil. It was Laplace, for instance, who persuaded the First Class to engage Biot and Arago on the experimental investigation of refraction in gases, which they described to the class in March 1806.[6] It is a measure of his influence that their results, although obviously applicable to the practical problems of astronomical refraction as well, were presented in the context of a highly theoretical discussion of short-range molecular forces and of the affinities between the particles of the eight gases examined and the corpuscles of light. Biot and Arago, in fact, adopted Laplace's analysis of refraction without question. Although they measured refractive index (μ) in their experiments, they presented their results in terms of refractive power (Laplace's *pouvoir réfringent*), that is, the quantity $(\mu^2 - 1)/\rho$; and they provided the experimental evidence—conspicu-

---

[5] Crosland (1967).

[6] Biot and Arago, "Mémoire sur les affinités des corps pour la lumière, et particulièrement sur les forces réfringentes des différens gaz," *MI*, **7** (July 1806), part 1, pp. 301–87.

ously lacking in Book X of *Mécanique céleste*—that, for any one gas, $(\mu^2 - 1)$ is proportional to ρ. It is also a mark of their allegiance to the prevailing orthodoxy of Arcueil—although in this case Berthollet was as much the inspiration as Laplace—that they speculated confidently on the analogy that they supposed to exist between chemical affinity and affinity for light, as measured by the refractive power. Such an analogy was consistent with Laplace's view that both types of affinity were gravitational in origin, so that he found it highly satisfactory to be able to show that the order in which substances appeared in the two tables of affinity were very roughly similar.

Laplace also used his position at the Institute to good effect in the system of prize competitions. There is little doubt that he was chiefly responsible for the setting of the competition for a mathematical study of double refraction that was announced in January 1808 and won by Malus, a recent recruit to the Arcueil circle, in January 1810.[7] The intention in setting this subject was clearly that Laplace's theoretical treatment of ordinary refraction, as given in *Mécanique céleste* in 1805, should be extended to embrace double refraction as well; and to this extent the competition was a success.

Double refraction had never been satisfactorily explained either in the corpuscular theory or in Huygens's wave theory. Among the corpuscularians, Newton's brief analysis of the phenomenon in terms of the two "sides" of a ray of light[8] was still endorsed in textbooks, but vaguely and without conviction; Haüy's *Traité élémentaire de physique* provides a good illustration of this.[9] Huygens's explanation, as given in the *Traité de la lumière* (1690), not only had weaknesses, particularly in its inability to explain the phenomena associated with crossed double-refracting crystals, but was also too closely allied to the wave theory to carry conviction at a time when, especially in France, the corpuscular theory was dominant.

It seems likely that the immediate stimulus for the competition on double refraction was the news, received from England in 1807, of William Hyde Wollaston's experimental confirmation of Huygens's construction for the ordinary and extraordinary rays. Huygens had used his wave theory and his notion of secondary wavelets to show that the wave surface of an extraordinary ray was an ellipsoid of revolution, whereas

---

[7] For studies of this competition, see Frankel (1974); Chappert (1977), pp. 67–76 (a useful source on all aspects of Malus's work); and Buchwald (1989), pp. 23–40. Malus's prize-winning paper was published as "Théorie de la double réfraction," in *Mémoires présentés à l'Institut National ... par divers savans ... Sciences mathématiques et physiques*, **2** (1811), pp. 305–508.

[8] *Opticks*, 4th ed. (1730), Query 26.

[9] Haüy, *Traité élémentaire de physique*, 2nd ed. (Paris, 1806), **2**, pp. 334–55.

that of an ordinary ray was a sphere. He had then deduced the properties of the ellipsoid and had established laws governing the path of the extraordinary ray at different angles of incidence. Although Huygens confirmed these laws experimentally, his method, as described in the *Traité*, was obscure; hence the need for Wollaston's systematic confirmation.[10]

Wollaston's paper appeared as an impressive confirmation of Huygens's construction and, at least by implication (for Wollaston did not endorse the wave theory), as a challenge to the corpuscularians. The challenge was one that Laplace could not resist, and his enthusiasm for a competition that was clearly intended to yield a corpuscularian counterpart to Huygens's wave theory of double refraction was only heightened by the availability of a candidate of impeccable credentials in Malus. The latter's analytical skills and commitment both to the corpuscular theory and to the doctrine of short-range forces were already apparent in his "Traité d'optique," which he read to the First Class of the Institute in April 1807,[11] and Laplace's expectations must have been high. Hence the endorsement of these principles in his prize-winning paper[12] was, we may assume, no more than the fulfillment of Laplace's expectations.

That Laplace followed the course of the competition, and Malus's work in particular, very closely is reflected in his report on the paper of December 1808 in which Malus announced his discovery of polarization (the word too was his) and related the phenomenon to the corpuscular theory.[13] Laplace used the occasion to express admiration not only for the discovery itself but also for Malus's experimental confirmation of Huygens's law of extraordinary refraction.[14] At last it was established beyond doubt that any corpuscular theory of double refraction would

[10] Wollaston,"On the Oblique Refraction of Iceland Crystal," *Philosophical Transactions of the Royal Society* (1802), pp. 381–86, read before the Royal Society on 24 June 1802. The five-year delay in the arrival in France of the report of this paper was a consequence of the patchiness of communication in time of war.

[11] Published in *Mémoires présentés à l'Institut national ... par divers savans ... Sciences mathématiques et physiques*, 2 (1811), pp. 214–302, esp. pp. 265–66.

[12] See especially Malus, "Théorie de la double réfraction," pp. 489–96.

[13] (91). Malus read his paper to the First Class of the Institute on 12 December 1808 and published it as "Sur une propriété de la lumière réfléchie," *Mémoires de physique et de chimie de la Société d'Arcueil*, 2 (1809), pp. 143–58. For Malus's treatment of polarization in terms of the forces acting on the particles of light, see especially his paper "Sur une propriété des forces répulsives qui agissent sur la lumière," *Mémoires de physique et de chimie de la Société d'Arcueil* (1809), 2, pp. 254–67 (260–67). Malus's work on polarization is well described in Buchwald (1989), pp. 41–66.

[14] *OC*, 14, p. 322. When Laplace presented his report to the First Class of the Institute, on 19 December 1808, Malus's confirmation of Huygens's law was still unpublished. See Frankel (1974), p. 233.

have to be consistent with the law, and almost immediately Laplace showed how this might be achieved, in "Mémoire sur le mouvement de la lumière dans les milieux diaphanes," which he read to the First Class of the Institute in January 1809.[15]

Laplace began by asserting, quite dogmatically, that Huygens's wave theory was inadequate for the explanation of double refraction and that the way ahead lay in devising a new explanation in terms of short-range molecular forces. His own, corpuscular theory rested on the principle of least action and on an arbitrary but plausible assumption concerning the relationship between the velocity of light inside a crystal (v), the velocity of light outside the crystal (c), and the angle (V) between the ray inside the crystal and the crystal's optic axis. Presumably by analogy with Snell's law for ordinary refraction, for which $v^2 = c^2 + a^2$, Laplace put, for the extraordinary ray, $v^2 = c^2 + a^2 \cos^2 V$. Assuming the truth of this equation, he derived expressions relating the direction of the extraordinary ray to the angle of incidence of the ray entering the crystal and the orientation of the optic axis.

Laplace's treatment was remarkably similar to Malus's. The expressions by which they both described the path of the extraordinary ray were similar in form and, by an adjustment of constants, could even be made identical. Moreover, in arriving at his expression, Malus, like Laplace, leaned heavily on the principle of least action, although he derived the dependence of v on the orientation of the extraordinary ray by assuming, as Laplace had not done, the truth of Huygens's law. The similarity between the two papers was such that the possibility of plagiarism cannot be ruled out. Malus felt that by the time Laplace wrote his paper of January 1809, he already knew the essentials of Malus's theory, which were almost certainly available to him by late 1808. And, whether or not it was intentional, Laplace's paper certainly had the effect of diminishing Malus's achievement in providing the corpuscular theory of double refraction that won the Institute's prize competition in 1810.

Taken together, the papers of Laplace and Malus could be passed off as yet another triumph for corpuscular optics: now no one could doubt that Huygens's law was consistent with the doctrine of short-range forces and the materiality of light. Yet there were weaknesses. As Young observed, the ellipsoid of revolution, which represented a wave front in Huygens's theory, was reduced in Laplace's paper to a mathematical construct without physical significance;[16] and it was by no means

---

[15] [1810a].

[16] See Young's unsigned review of Laplace's memoir, in *Quarterly Review*, **2** (1809), pp. 337–48, esp. p. 344.

obvious that there really existed molecular forces with the special directional properties required in order to explain extraordinary refraction. But in the years of Laplacian domination of French physics, such objections were readily overlooked.

Laplace's involvement in the prize competition on the specific heats of gases, which was set in January 1811, was equally apparent though less direct. The winning entry, by François Delaroche and Jacques-Étienne Bérard,[17] two young physicists close to the Arcueil circle, made a decisive contribution to Laplacian physics, and we may be sure that Laplace's wishes were prominent both in the setting of the competition and in the adjudication. The first aim was the acquisition of reliable data in a notoriously uncertain branch of experimental physics. But a quite explicit subsidiary purpose was to decide whether it was possible for some caloric to exist in a body in a combined, or latent, state (that is, without being detected by a thermometer) or whether (as William Irvine, Adair Crawford, and John Dalton had supposed) all of the caloric was present in its "sensible," or free, state and therefore as a contribution to the body's temperature. In their *Mémoire sur la chaleur* Lavoisier and Laplace had provided strong evidence against Irvine's theory,[18] so that it was predictable enough that the winners, Delaroche and Bérard, should use their measurements of the specific heats of elementary and compound gases to endorse what was clearly the Laplacian view. As a result of experiments performed entirely at Arcueil, they firmly upheld the distinction between latent and sensible caloric.

So in the competitions on both double refraction and the specific heats of gases Laplace was well served. But not all of his attempts to use the system of prize competitions were so successful. The paper on the distribution of heat in solids with which Fourier won the competition of 1811 departed significantly from Laplace's approach to the problem. In this case, Laplace, unlike Lagrange, found much to admire in Fourier's paper, but the competition set in 1809 on the theory of elastic surfaces offered him no such consolation. Before the closing date for this competition, Laplace added elastic surfaces to the list of phenomena that might be explained in terms of short-range molecular forces, so pointing the way to the kind of solution that was expected.[19] No entries

---

[17] Delaroche and Bérard, "Mémoire sur la détermination de la chaleur spécifique des différens gaz," *Annales de chimie*, **85** (1813), pp. 72–110, 113–82.
[18] [1783a].
[19] [1810a], p. 329.

of sufficient merit were received, however, and when the prize was eventually awarded, in January 1816, it went to Sophie Germain, whose paper broke pointedly both with Laplace's own guidelines and with the theory of elastic surfaces, treated in the Laplacian manner, that Poisson had presented before the First Class of the Institute in 1814.[20]

[20] Germain's paper was published in an enlarged and modified form as *Recherches sur la théorie des surfaces élastiques* (Paris, 1821). Poisson's appeared as "Mémoire sur les surfaces élastiques," *MI*, **13** (1812/1816), part 2, pp. 167–225; see esp. pp. 171–72 and 192–225. On the prize and the differences between Germain and Poisson, see Bucciarelli and Dworsky (1980), pp. 30–97, and Grattan-Guinness (1990a), **1**, pp. 461–70.

CHAPTER 25

Theory of Error

THE ACTIVITY at Arcueil was at its height from 1805 through 1809, after which interval of preoccupation with problems of physics and association with younger physicists, Laplace turned back to probability for the intensive effort that culminated in the production of *Théorie analytique des probabilités* in 1812 and the companion *Essai philosophique sur les probabilités* in 1814. The prelude to these works consists in a pair of memoirs [1810b] and [1811a] and a supplement [1810c] to the earlier paper, together with a "Notice sur les probabilités," published anonymously in the *Annuaire* of the Bureau des Longitudes [1810d] and since forgotten. Laplace presented the first of these papers before the Institute on 9 April 1810 (95). We have already seen how, prior to that, the analysis of probable error had assumed increasing importance for him in Book III of *Mécanique céleste*, particularly in the comparison of spheroidal attraction theory to geodetic data in Chapter 5, and how the problem of correcting for instrumental error in physical observations had concerned him in Book X, mainly in relation to barometric and thermometric data (see chapter 21).

The greatest novelty in the two analytical papers of 1810 and 1811 is the derivation from what is now called the central limit theorem of the least-squares method and its application to determination of the mean value in a series of observations. A workable technique had eluded him, it will be recalled, in his early investigations of error theory.[1] Legendre had published the rule in 1805 as a method for resolving inconsistencies between linear equations formed with astronomical data. His was not a probabilistic procedure, however. In 1809 Gauss published a derivation of the least-squares law from an analysis of what is now called the normal distribution, a term coined by Francis Galton.[2] It is sometimes said that this opportunity drew Laplace back to a preoccupation with the whole theory of probability in these, his advancing years, after the lapse of a quarter century since his initial immersion in its theory,

---

[1] [1774c], [1781a], and the paper (21) published in Gillispie (1979). See chapters 3 and 10. Cf. Stigler (1986), p. 361.

[2] For Laplace's derivation of the least-squares method and its relation to Gauss's, see Stigler (1986a), pp. 140–48, and for Gauss see Stigler (1981). See also Eisenhart (1964), Plackett (1972), and Sheynin (1977).

definition, and application to population problems. That is probably incorrect, however, since it was only in the addendum to the earlier memoir that he first mentioned least squares.[3] Moreover, he accompanied these memoirs with the popular "Notice sur les probabilités," mentioned above, in which Laplace enlarged on the lecture he had given before the École Normale in 1795.[4] Most of the passages from this Bureau des Longitudes piece he incorporated verbatim in the methodological and actuarial sections of the first edition of *Essai philosophique sur les probabilités* in 1814. They serve to introduce a set of tables of mortality in the Bureau's *Annuaire* and there feature a verbal statement of the central limit theorem.[5] Perhaps, therefore, it will be prudent to report his own account of the route that he had followed in finally arriving at that theorem.

The opening mathematical paper in the pair under discussion has exactly the same title, "Mémoire sur les approximations des formules qui sont fonctions de très grands nombres," as did the important sequence twenty-five years earlier that had moved probability from analysis of games of chance to population studies.[6] In Laplace's own recollection, it was the lengthy repetition of events encountered in theory of probability that had initially brought home to him the inconvenience of evaluating formulas into which the numbers of these events had to be substituted in order to achieve a numerical solution. He had then attacked the difficulty by seeking a general method for accomplishing transformations of the type that Stirling had discovered for reducing the middle term of a binomial raised to a high power to a convergent series. The method he had given transformed the integrals of linear differential or difference equations, whether partial or ordinary, into convergent series when large numbers were substituted in terms under the integral sign, the larger the numbers the more rapidly convergent the series.[7] Among the formulas he could thus transform, the most notable was that for the finite difference of the power of a variable. In probability, the conditions of the problem often required restricting consideration to the positive values even though the variable decreases through zero into the negative range.

Such was the case in the analysis of the probability that the mean inclination of any number of cometary orbits is contained within a given range. It now appears that Laplace had felt dissatisfied with everything he had so far tried on that problem. He had started it with his youthful

[3] [1810c], *OC*, **12**, p. 353.
[4] Chapter 19, and cf. Delambre, in *MI*, **12** (1811/1812), "Histoire," pp. i–ii.
[5] *Essai philosophique sur les probabilités*, pp. 110–11.
[6] [1810b], [1785a], and [1786b]; see chapter 11.
[7] [1781a] and [1785a]; see chapter 11.

analysis of the orbital inclinations of comets in order to determine whether the distribution bespeaks the same cause as the nearly coplanar arrangement of the planets.[8] Apparently, he took up the question again soon afterward, for he now says that the problem can be resolved by a method that he had given in his first comprehensive memoir on probability, namely, that the required probability could be expressed by the finite difference of the power of a uniformly decreasing variable in a formula where the exponent and the difference are the same as the number of orbits.[9] Unfortunately, a numerical solution was unobtainable in practice, and it is reasonable to surmise that this was the reason that he failed to include the problem in the printed memoir [1781a]. The obstacle stopped him for a long time, he acknowledges.

Finally, and this would appear to be the background of the 1810 memoir, he had resolved the difficulty by approaching the problem from another point of view. Assuming in general that the prior probabilities —he now says *"facilités"*—of inclination serve any law whatever, he succeeded in expressing the required probability in a convergent series. For he had come to see the problem as identical with those in which the probability is required that the mean error in a large number of observations falls within certain limits, which question he had discussed at the end of "Mémoire sur les probabilités," although without numerical examples.[10] He does not say so here, but the point of view was also akin to that taken in his calculation that the error in estimating the population of France from a given sample would fall within certain limits.[11]

Laplace could then show that if the observations are repeated an indefinite number of times, their mean result converges on a limit such that, if an equal interval on either side be made as small as one pleases, the probability that the result will be contained therein can be brought so close to certainty that the difference is less than any assignable magnitude. If positive and negative errors are equipossible, this mean term is indistinguishable from the truth. Since the methods he had so far discussed were indirect, Laplace considered it preferable to find a direct approach to evaluating the finite differences of the higher powers of the variable; and he proceeded to apply to error theory a technique that he had published the previous year in *Journal de l'École Polytechnique*.[12]

---

[8] [1776b]; see chapter 6.
[9] [1781a]; see chapter 10.
[10] [1781a], Article XIII, pp. 30–33.
[11] [1786c]; see chapter 13.
[12] [1809c], *OC*, **14**, p. 193.

The memoir contains refinements of his earlier work on generating functions and on the use of definite integrals for solving certain classes of linear partial differential and difference equations that could not be integrated in finite terms.[13] He applied to the present purpose an analysis involving the reciprocity of real and imaginary results, which he had introduced in the first memoir on approximate solutions to formulas containing very large numbers.[14]

Thus, it was by way of analyzing the distribution of cometary orbits that Laplace came to the central limit theorem, having perfected methods for evaluating the mathematical expressions developed years before. The opening articles argue the old proposition that neither the mean inclination of the cometary orbits (by now ninety-seven were known, all of which he could include in the computation), nor the proportion of direct to retrograde motions, can be supposed to result from the same cause as the arrangement of the planetary system. Article VI changes the problem of mean inclination into the problem that the mean error of a number n of observations will be contained within the limits $\pm rh/\sqrt{n}$, where r is the sum of the errors. At first, Laplace assumes the equipossibility of error in the interval h, but he goes on to the general case in which the error distribution follows any law, and obtains for the required probability the formula[15]

$$\frac{2}{\sqrt{\pi}} \sqrt{\frac{k}{2k'}} \int e^{-\frac{k}{2k'}r^2} dr. \qquad (126)$$

When $\phi(x/h)$ is the probability of the error $\pm x, k$ is

$$\int_{-h/2}^{h/2} \phi \frac{x}{h} \, dx,$$

and k' is

$$\int_{-h/2}^{h/2} \frac{x^2}{h^2} \phi\left(\frac{x}{h}\right) dx.$$

Later in 1809 (we do not know precisely when), Laplace composed a brief supplement ([1810c]) in which he returned to the choice of the

---

[13] Discussed in [1782a]. The memoir [1809e] further included an integral solution of Fourier's equation for heat diffusion. The Laplace solution was important in its own right and also because it helped Fourier find the integral solutions, since named after him, in the famous 1811 memoir with which he won the Institute's prize contest for a mathematical treatment of the difffusion of heat. See Grattan-Guinness (1990a) 2, pp. 611–14.

[14] [1785a]; the influence of Fourier was critical here; cf. chapter 29.

[15] *OC*, **12**, p. 325.

mean in a series of observations, the task that had first attracted him to error theory in the memoir on the probability of cause.[16] He invokes the procedure that he had imagined crudely there and more abstractly in the closing passages of "Mémoire sur les probabilités."[17] A curve of probability might be constructed, for which the abscissa defines the "true" instant of the observation—presumably astronomical—and the ordinate is proportional to the probability that the value is correct. The problem is to find the point on the x-axis at which the departure "à craindre" from the truth is a minimum. Now then, just as in the theory of probability, the loss "to be feared" is multiplied by its probability and the product summed, so in error theory the amount of each error, regardless of sign, is to be multiplied by its probability and the product summed. Supposing that n observations of one sort, with equal possibility of error, result in a mean value of A; that n' of another sort, following a different law of error, result in a mean of $A + q$; that n'' of yet another sort and another distribution result in a mean of $A + q'$, and so on; and that $A + X$ is the mean to be preferred among all these outcomes. From Equation (126) Laplace then derives the proposition that the required value of X will be that for which the function

$$(pX)^2 + [p'(q - X)]^2 + [p''(q' - X)]^2 + \cdots \qquad (127)$$

is a minimum, $p, p', p'', \ldots$, representing the greatest probabilities of the results given by the observations $n, n', n'', \ldots$.[18] That expression gives the sum of the squares of each result multiplied respectively by the greatest ordinate in its curve of error. It is clear from Laplace's comment that the novelty was not in the least-squares property itself. He considered its status to be merely hypothetical when the mean depended on a few observations or on an average among a number of single observations. It became valid generally only when each of the results among which it indicated the mean itself depended on a very large number of observations. Its basis had to be statistical (a word that he did not employ), and only then could it be derived from the theory of probability and employed whatever the distributions of error in instruments or observations.

Daniel Bernoulli, Euler, and Gauss are mentioned in this note, albeit rather vaguely, but not Legendre. Hard feelings about priorities in the matter of least squares had meanwhile arisen between Legendre and Gauss. Delambre, now Permanent Secretary of the scientific division of

[16] [1774], Article V; see chapter 3.
[17] [1781a], Articles XXX–XXXIII, see chapter 10.
[18] *OC*, **12**, p. 353.

the Institute, tried to make peace.[19] His contemporary account reaches the same conclusion that Laplace himself arrived at when reviewing the origin of least squares in *Théorie analytique des probabilités*.[20] Gauss had indeed had the idea first and made use of it in private calculation, but Legendre had come upon it independently and published it first. Delambre also reports Gauss's attribution of inspiration to a theorem that he had found in Laplace, namely that the value of the integral $\int_{-\infty}^{+\infty} e^{-t^2}\, dt = \sqrt{\pi}$.[21] In fact, so Legendre informed Gauss in their exchange of reproach, the theorem belonged to Euler.[22] Otherwise, Laplace escaped unscathed on the fringe of this dispute, never having claimed the least-squares rule itself, but only the generality it could assume by virtue of his derivation of it from the probability of cause. His procedure would now be called Bayesian.

Laplace followed a different procedure in the second of these papers, "Mémoire sur les intégrales définies."[23] He opened it also with a historical resume, recalling (what he had never claimed at the time) how the companion discoveries of generating functions and of approximations for formulas containing very large numbers were really complementary aspects of a single calculus.[24] He reminded his readers that the object of the former was the relation between some function of an indeterminate variable and the coefficients of its powers when the function is expanded in a series, and that generating functions had initially proved most valuable in solving difference equations in which problems of the theory of chance were formulated. The object of the latter, on the other hand, was to express variables that occur in difference equations in the form of definite integrals to be evaluated by rapidly convergent approximations. It turns out, however, that the quantity under the integral sign in such cases is nothing other than the generating function of the variable expressed by the definite integral.

The two theories thus merged in a single approach that he would henceforth call the calculus of generating functions. He also speaks of it as the exponential calculus of differential operators (*caractéristiques*).[25] It is valid both in infinitesimal and finite analysis. When a difference equation is expanded in powers of the difference taken as indeterminate but infinitesimal, and higher-order infinitesimals are held to be

[19] *MI*, **12** (1811/1812), "Histoire," pp. i–xiii.
[20] *OC*, **7**, p. 353.
[21] Cf. Equation (62), chapter 11.
[22] Plackett (1972), p. 250.
[23] [1811a].
[24] [1782a], [1785a]; see chapter 11.
[25] [1811a], *OC*, **12**, p. 360.

negligible relative to those of some lower order, the result is a differential equation. But the integral of that equation is also the integral of the difference equation wherein the infinitesimal quantities are similarly held to be negligible relative to the finite quantities. Justifying the neglect of such terms led Laplace into one of his few discussions of the foundations of the calculus and defenses of its rigor.

When first presenting the approximations for formulas containing large numbers [1785a], Laplace had shown how to evaluate several classes of the definite integrals that he employed in terms of transcendent quantities. The method depended on the passage from the real to the imaginary and resulted in series of sines and cosines. These were special cases, however, and only now was he in a position to give a direct and general method for evaluating any such expressions. That topic occupies the opening article of [1811a]. Laplace illustrated the method in three representative problems of the theory of probability. The first concerns a case of duration of play in theory of games of chance. The second is an urn problem. Two vessels contain n balls each. Of the total $2n$, half are black and half white. Each draw consists of taking a ball from each urn and placing it in the other. What is the probability that after r draws there will be x white balls in urn A? Mathematically, Laplace considered this the most interesting of his illustrations of the newly named calculus of generating functions, for his solution involved—so he claimed—the first application of partial differential equations to infinitesimal analysis in the theory of probability.[26]

Nevertheless, he reserved the fullest treatment for the third problem, the choice of a mean among the results given by different sets of observations. The main applicability being to astronomy, Laplace republished the articles containing its resolution in *Connaissance des temps* [1811b], where he intended it for readers unversed in probability.[27]

The idea was to employ the totality of a very large number of observations to correct several elements that are already approximately known. Each observation is a function of these elements, and their approximate value could be substituted in that function. Those values were to be modified by small corrections, which constituted the unknowns. The function was then expanded in a series of powers of those corrections. Squares, products, and higher powers are neglected, and the series is equated to the observed value. Those steps gave an initial equation of condition among the corrections to be applied to the elements. A second equation could be found from a second observation, a third from a third, and so on. If the observations had each been

[26] Ibid., pp. 361–62.
[27] Ibid., p. 362. For an excellent discussion, see Sheynin (1977), nos. 5.1–5.2.

precise, only one equation apiece would have been needed. But since they were subject to error, the effect of which was to be minimized, a very great number of them had to be taken in order that the errors might compensate each other in the values deduced from the total number. There was the core of the problem. How were the equations to be combined?

Here was the point at which probability entered the procedure. Any mode of combination would consist in multiplying each equation by a particular factor, or weight, and summing the products so as to form a definitive equation, or estimate, of the corrections. Employing a second factor would yield a second estimate, and so on. As many estimates would be needed as there were unknown elements. The crux consisted in choosing factors such that the mean error, positive or negative, should be a minimum. Defining mean error as the sum of the products of each error multiplied by its probability, Laplace showed algebraically that, when expected error is set equal to zero in each equation of condition, the sum of the squares of the terms representing actual errors is a minimum. As introduced by Legendre and Gauss, the method was limited (in Laplace's view) to finding the definitive equations needed for a solution. In his derivation, it also served to determine the corrections. Thus, once again, this time employing what is now called a linear regression rather than a Bayesian approach, Laplace claimed that a derivation of the method of least squares from theory of probability promoted it from the status of a rule of thumb to that of a mathematical law.

Probability: *Théorie analytique* and *Essai philosophique*

IN THE PREAMBLE to "Mémoire sur les intégrales définies," Laplace wrote, "The calculus of generating functions is the foundation of a theory that I propose to publish soon on probability."[1] Good as his word, he presented the first part of *Théorie analytique des probabilités* to the Institute on 23 March 1812 and the second part on 29 June (98, 99). There is a minor bibliographical puzzle here. The complete treatise is a quarto volume of 464 pages in the first edition and is divided into two books, Book I consisting of a "Première partie" and a "Seconde partie." The significance of the partition will be clarified in a moment. What is unclear is whether Laplace had Book I, Part I printed first and then Book I, Part II together with all of Book II; or whether, after the three-month interval, it was Book II that he saw through the press. The latter conjecture seems more logical, although the *Procès-Verbaux* of the Institute are confirmed by the "Avertissement" to the second edition, presented on 14 November 1814 (104).

In either case, the general scheme is similar to that of *Mécanique céleste*. Book I is devoted to the mathematical methods. In Book II, occupying two-thirds of the volume, they are applied to the solution of problems in probability. There is some difference in the relation of the organization to the sequence of events in Laplace's career, however. In the field of probability, his resolution of a rather larger proportion of the problems than in celestial mechanics had preceded his development of the mathematical techniques incorporated in the finished treatise. That was notably the case in the areas of games of chance and probability of cause and, to a degree, in demography. The areas of application that he explored later tended to be in the realm of mathematical statistics: error theory, decision theory, judicial probability, and credibility of witnesses.

The first edition contains a brief introduction (pp. 1–3) that was eliminated in the second (1814) and third (1820) in favor of the *Essai philosophique*. It is worth notice, nevertheless, for the interest that

---

[1] [1811a], *OC*, **12**, p. 360.

Laplace claimed for the work in bringing it before the public:

> I am particularly concerned to determine the probability of causes and results, as exhibited in events that occur in large numbers, and to investigate the laws according to which that probability approaches a limit in proportion to the repetition of events. That investigation deserves the attention of mathematicians because of the analysis required. It is primarily there that the approximation of formulas that are functions of large numbers has its most important applications. The investigation will benefit observers in identifying the mean to be chosen among the results of their observations and the probability of the errors still to be apprehended. Lastly, the investigation is one that deserves the attention of philosophers in showing how in the final analysis there is a regularity underlying the very things that seem to us to pertain entirely to chance, and in unveiling the hidden but constant causes on which that regularity depends. It is on the regularity of the mean outcomes of events taken in large numbers that various institutions depend, such as annuities, tontines, and insurance policies. Questions about those subjects, as well as about inoculation with vaccine and decisions of electoral assemblies, present no further difficulty in the light of my theory. I limit myself here to resolving the most general of them, but the importance of these concerns in civil life, the moral considerations that complicate them, and the voluminous data that they presuppose require a separate work.

Laplace never wrote that separate work, although the thought of it may well have been what led him to expand his old lecture for the École Normale into the *Essai philosophique* two years later.

In conformity with the program announced the preceding year [1811a], the general subtitle of Book I of *Théorie analytique* is "Calcul des fonctions génératrices." It consists almost entirely of a republication, with some revision, of the two cardinal mathematical investigations of the early 1780s. The "Mémoire sur les suites" [1782a], on generating functions themselves, has now become the basis of its first part, and that on the approximation by definite integrals of formulas containing very large numbers [1785a] and [1786b], the basis of its second part.[2] The introduction reiterates what Laplace had first stated in the memoir of the preceding year [1811a], to the effect that the two theories are branches of a single calculus, the one concerned with solving the difference equations in which problems of chance events are formulated, the other with evaluating the expressions that result when events are repeated many times.

---

[2] For a more detailed and mathematical summary of the latter than is given in chapter 11, see Todhunter (1865), nos. 956–68.

Laplace says in the introduction that he is now presenting these theories in a more general manner than he had done thirty years before. The chief difference in principle is that the new calculus is held to have emerged along the main line of evolution of the analytical treatment of exponential quantities. In an opening historical chapter, Laplace traces the lineage back through the work of Lagrange, Leibniz, Newton, and Wallis to Descartes's invention of numerical indices for denoting the operations of squaring, cubing, and raising magnitudes to higher integral powers, in a word, the arithmeticization of algebra, as it has been called.

The principal difference in practice between the two earlier memoirs and their revision in Book I is that Laplace omitted certain passages that had come to appear extraneous in the interval and gave greater prominence to others that now appeared strategic. The most important omission is three articles from the "Mémoire sur les suites" on the solution of second-order partial differential equations, which were important for problems of physics but not for theory of games of chance.[3] On the other hand, Laplace gave greater emphasis than in the earlier memoirs to the passage from the finite to the infinitesimal and also from real to imaginary quantities. He now develops as an argument what he had merely asserted in the immediately preceding memoir on definite integrals [1811a], namely that rigor is not impaired by the necessity of neglecting, in appropriate circumstances, infinitesimal quantities relative to finite quantities and higher-order infinitesimals relative to those of lower order. Laplace adduces, in support of the latter proposition, his solution to the problem of vibrating strings from the "Mémoire sur les suites," which affords a convincing example that discontinuous solutions of partial differential equations are possible under specified conditions.[4] He attached even greater importance to the fertility he increasingly found in the process of passing from real to imaginary quantity and discussed those procedures in the transitional section between the first and second parts. The limits of the definite integrals to be converted into convergent series are given by the roots of an equation such that when the signs of the coefficients are changed, the roots become imaginary. But it was precisely this property that led Laplace to the values of certain definite integrals that occur frequently in probability and that depend on the two transcendental quantities π and e. His early methods had been ad hoc and indirect, but since that time he had perfected direct methods for evaluating such integrals in a

[3] [1782a], Articles XVIII–XX, *OC*, **10**, pp. 54–70.
[4] *OC*, **7**, pp. 70–80.

general manner, a procedure that (he acknowledged) Euler had arrived at independently.[5]

In Book II, subtitled "Théorie générale des probabilités," Laplace turns from the calculus to probability itself. Indeed, it is fair to say that he formulated the subject for some generations to come. His treatment drew together the main types of problems from the theory of chance that had already been treated by many mathematicians, including himself, in a somewhat haphazard manner, and he proceeded to rework them in tandem with problems from the new areas of application in philosophy of science, astronomy, geodesy, instrumentation, error, population, and the procedures of judicial panels and electoral bodies. Unlike the two parts of Book I and much of *Mécanique céleste*, Book II is more than a republication of earlier memoirs with minor and incidental revision. Material from earlier work is incorporated in it, to be sure, but it is revised mathematically and fortified with new material. What is carried over without significant change from the earliest memoirs, [1774c] and [1776a], is the point of view from which the subject as a whole is treated and the spirit in which the various topics are approached.

It has been said that *Théorie analytique des probabilités* is unsystematic, rather a collection of chapters that might as well be separate than a treatise in the usual sense. Perhaps so. Its organization certainly recapitulates the evolution of the subject matter rather than some logical system within it. It is also difficult to imagine its serving either as a textbook, as the first two volumes of *Mécanique céleste* could do, or as a work of reference, in the way that the third and fourth volumes really did do. Its relation to the subject was different. Rather than drawing together the lifework of a leading contributor to a vast and classical area of science, it was the first full-scale study completely devoted to a new specialty, building out from old and often hackneyed problems into areas where quantification had been nonexistent or chimerical. Later commentators have also sometimes castigated the obscurity and lack of rigor in many passages of the analysis. Once again, it may be so. It is constitutionally and temperamentally very difficult, however, for many mathematicians to enter sympathetically into what was once the forefront of research. Important parts of *Mécanique céleste* were also in the front lines, of course—but that was the location of *Théorie analytique des probabilités* as a whole. What no one has denied is that it was a seminal work.

The first chapter gives the general principles and opens with the famous characterization of probability as a branch of knowledge re-

[5] *OC*, **7**, p. 88.

quired by the limitation of the human intelligence and serving to repair its deficiencies in part. The subject is relative, therefore, both to our knowledge and to our ignorance of the laws of a determined universe. After stating the definition of probability itself, and the rule for multiplying the probabilities of independent events, Laplace includes as the third basic principle a verbal statement of his theorem on the probability of cause, again without mentioning Bayes. Thereupon, he takes the example of the unsuspected asymmetries of a coin to consider the effect of unequal prior probabilities mistakenly taken for equal. Finally, he distinguishes between mathematical and moral expectation. The basic content of these matters was drawn from the companion memoirs [1774c] and [1776a,1°] composed thirty-nine years previously in 1773 (see chapter 3).

The actual problems discussed in the early chapters also consist in part of examples reworked from these and the other early papers on theory of chance.[6] Laplace solved them by means of generating functions and arranged them, not for their own sake, but to illustrate the typology of problems in probability at large, interspersing new subject matter where the methodology made it appropriate. Chapter 2, which is concerned with the probability of compound events composed of simple events of known probability, is much the most considerable, occupying about a quarter of Book II. In a discussion of the old problem of determining the probability that all n numbers in a lottery will turn up at least once in i draws when r slips are chosen on each draw, he adduced the case of the French national lottery, composed of ninety numbers drawn five at a time. Laplace went on to other classic problems in direct probability: of odds and evens in extracting balls from an urn, of extracting given numbers of balls of a particular color from mixtures in several urns, of order and sequence in the retrieval of numbered balls, of the division of stakes, and of the ruin or victory of one of a pair of gamblers in standard games.

Perhaps it will be useful to trace the sequence in one set of problems as an illustration of how Laplace made the connections between topics. He imagines (no. 13) an urn containing $n + 1$ balls numbered 0, 1, 2, 3,..., n. A ball is taken out and returned, and the number noted. What is the probability that after i draws, the sum of the numbers will be s? If $t_1, t_2, t_3,..., t_i$ are the numbers of balls taken in the first, second, third, etc., draws, then as long as $t_1, t_2, t_3,..., t_i$ are held fixed,

$$t_1 + t_2 + t_3 + \cdots + t_i = s. \tag{128}$$

---

[6] For a useful mathematical summary of many of them in modern terminology, see Sheynin (1976).

Only that one combination is possible. But if different numbers are taken, so that t_1 and t_2 are varied simultaneously and are capable of taking any value beginning at zero and continuing indefinitely, then Equation (128) will be given by the following number of combinations:

$$s + 1 - t_3 - t_4 - \cdots - t_i \qquad (129)$$

for t_1 can take any value above zero, and that gives

$$t_2 = s - t_3 - t_4 - \cdots - t_i,$$

up to $s - t_3 - t_4 - \cdots - t_i$, which would give $t_2 = 0$. Negative values are excluded. By like reasoning, Laplace finds that the total number of combinations that can give Equation (128) on the supposition of the indefinite variability of $t_1, t_2, t_3, \ldots, t_i$, always greater than zero, is

$$\frac{(s + i - 1)(s + i - 2)(s + i - 3) \cdots (s + 1)}{1 \cdot 2 \cdot 3 \cdots (i - 1)}. \qquad (130)$$

By the conditions of the problem, however, these variables cannot exceed n, and the probability of any particular value of t_1 from zero to n is $1/(n + 1)$.

Since the probability of t_1 equal to or greater than $(n + 1)$ is nil, it may be represented by the expression $(1 - l^{n+1})/(n + 1)$ provided that $l =$ unity. Now then, on condition that l be introduced only when t_i has reached the limit $n + 1$, and that it be equal to unity at the end of the operation, any value of t_1 can be represented by $(1 - l^{n+1})/(n + 1)$. The same is true for the other variables. Since the probability of Equation (128) is the product of the probabilities of the values $t_1, t_2, t_3, \ldots$, its expression is $[(1 - l^{n+1})/(n + 1)]^i$. The number of combinations given by that equation multiplied by their respective probabilities is then

$$\frac{(s + 1)(s + 2) \cdots (s + i - 1)}{1 \cdot 2 \cdot 3 \cdots (i - 1)} \left(\frac{1 - l^{n+1}}{n + 1} \right)^i. \qquad (131)$$

In expanding that function, l^{n+1} is to be applied only to combinations in which one variable is beginning to exceed n; l^{2n+2} only to combinations in which two of the variables begin to exceed n; and so on. Thus, if it is supposed that t_1 has grown larger than n, then by setting $t_1 = n + 1 + t_1'$, Equation (128) becomes

$$s - n - 1 = t_1' + t_2 + t_3 + \cdots, \qquad (132)$$

where t_1' increases indefinitely. If two variables, t_1 and t_2, exceed n, then setting $t_1 = n + 1 + t_1'$ and $t_2 = n + 1 + t_2'$, Equation (128) becomes

$$s - 2n - 2 = t_1' + t_2' + t_3 + \cdots. \tag{133}$$

The purpose of this manipulation is to decrease s in the function (130) by $n + 1$, relative to the system of variables $t_1', t_2, t_3, \ldots$, to decrease s by $2n + 2$, relative to the system $t_1', t_2', t_3, \ldots$, and so on. In expanding function (131) in powers of l, s is to be decreased by the exponent indicating the power of l, and when $l = 1$, the function (131) becomes

$$\frac{(s + 1)(s + 2) \cdots (s + i - 1)}{1 \cdot 2 \cdot 3 \cdots (i - 1)(n + 1)^i}$$

$$- \frac{i(s - n)(s - n + 1) \cdots (s + i - n - 2)}{1 \cdot 2 \cdot 3 \cdots (i - 1)(n + i)^i}$$

$$+ \frac{i(i - 1)}{1 \cdot 2} \frac{(s - 2n - 1)(s - 2n) \cdots (s + i - 2n - 3)}{1 \cdot 2 \cdot 3 \cdots (i - 1)(n + 1)^i} - \cdots,$$

$$\tag{134}$$

which series is continued until one factor,

$$(s - n), (s - 2n - 1), (s - 3n - 2), \ldots,$$

becomes zero or negative in value.

The formula (134)—to change the problem now—will give the probability of throwing any number s in tossing i dice each with $n + 1$ sides, the smallest number on any side being 1. If s and n are infinite numbers, formula (134) becomes the following expression:

$$\frac{1}{1 \cdot 2 \cdot 3 \cdots (i - 1)n}$$

$$\times \left[\left(\frac{s}{n}\right)^{i-1} - i\left(\frac{s}{n} - 1\right)^{i-1} + \frac{i(i - 1)}{1 \cdot 2}\left(\frac{s}{n} - 2\right)^{i-1} - \cdots \right].$$

$$\tag{135}$$

This expression, proceeds Laplace—affording his reader not so much as a new paragraph to draw breath—may be employed to determine the probability that the sum of the inclinations of orbits to the ecliptic will be contained within given limits on the assumption of equipossibility of inclination between $0°$ and a right angle. If a right angle, $\frac{1}{2}\pi$, is divided into an infinite number n of equal parts, and s contains an infinite number of these parts, then if ϕ is the sum of the inclinations of the orbits,

$$\frac{s}{n} = \frac{\phi}{\frac{1}{2}\pi}.\qquad(136)$$

Multiplying Equation (136) by ds, or $n\,d\phi/\frac{1}{2}\pi$, and integrating from $\phi - \epsilon$ to $\phi + \epsilon$ gives

$$\frac{1}{1\cdot 2\cdot 3\cdots i}\left\{\begin{array}{l}\left(\dfrac{\phi+\epsilon}{\frac{1}{2}\pi}\right)^i - i\left(\dfrac{\phi+\epsilon}{\frac{1}{2}\pi}-1\right)^i \\[2mm] +\dfrac{i(i-1)}{1\cdot 2}\left(\dfrac{\phi+\epsilon}{\frac{1}{2}\pi}-2\right)^i - \cdots \\[2mm] -\left(\dfrac{\phi-\epsilon}{\frac{1}{2}\pi}\right)^i + i\left(\dfrac{\phi-\epsilon}{\frac{1}{2}\pi}-1\right)^i \\[2mm] -\dfrac{i(i-1)}{1\cdot 2}\left(\dfrac{\phi-\epsilon}{\frac{1}{2}\pi}-2\right)^i + \cdots\end{array}\right\}.\qquad(137)$$

Formula (137) expresses the probability that the sum of the inclinations of the orbits is contained within the limits $\phi - \epsilon$ and $\phi + \epsilon$.

We shall not follow Laplace into yet another calculation (his last on this phenomenon) that the orbital arrangement of the planets results from a single cause and that the comets escape its compass, nor from that back to variation on the original problem, in which any number of balls in the urn may be designated by the same integer, nor even into his derivation by the same method that the sum of errors in a series of observation will be contained within given limits. Suffice it to indicate the sequence and the virtuosity it bespeaks.

We must, on the other hand, notice his earliest venture, this late in life, into judicial probability. Imagine a number i of points along a straight line, at each of which an ordinate is erected. The first ordinate must be at least equal to the second, the second at least equal to the third, and so on. The the sum of these i ordinates is s. The problem is to determine, among all the values that each ordinate can assume, the mean value. For that quantity in the case of the rth ordinate, Laplace obtains the expression[7]

$$\frac{s}{i}\left(\frac{1}{i} + \frac{1}{i-1} + \cdots + \frac{1}{r}\right).\qquad(138)$$

[7] *OC,* 7, p. 276.

Suppose now, however, that an event is produced by one of the i causes A, B, C,..., and that a panel of judges is to reach a verdict on which of the causes was responsible. Each member of the panel might write on a ballot the various letters in the order that appeared most probable to him. Formula (138) will now give the mean value of the probability that he assigns to the rth cause (in this application s must amount to certainty and have the value of 1). If all members of the tribunal follow that procedure, and the values for each cause are summed, the largest sum will point to the most probable cause in the view of that panel.

Laplace hastened to add that since electors, unlike judges, are not constrained to decide for or against a candidate but impute to him all degrees of merit in making their choices, the above procedure may not be applied to elections. Instead, he outlined a probabilistic scheme for a preferential ballot that would produce the most mathematically exact expression of electoral will. Unfortunately, however, electors would not in fact make their choices on the basis of merit but would rank lowest the candidate who presented the greatest threat to their own man. In practice, therefore, preferential ballots favor mediocrity, and they had been abandoned wherever tried.

The third chapter deals with limits, in the sense in which the idea figures in the frequency definitions of the discipline of probability that have developed out of it.[8] In Laplace's own terminology, his concern was with the laws of probability that result from the indefinite multiplication of events. No single passage is as clear and definite as his derivation of the central limit theorem in the memoir on approximating the values of formulas containing large numbers ([1810b]) that had brought him back to probability several years before, but the examples he adduces are much more various.

They begin with a conventional binomial problem. The probabilities of two events a and b are respectively p and $1 - p$. The probability that a will occur x times and b will occur x' times in $x + x'$ tries is given by the $(x' + 1)$th term of the binomial $[p + (1 - p)]^{x+x'}$. Laplace calculates the sum of two terms that are symmetrical on either side of the middle term of the expansion of the binomial. The formula is

$$\frac{2}{\sqrt{\pi}} \frac{\sqrt{n}}{\sqrt{2xx'}} c^{\frac{-nl^2}{2xx'}},$$

where $n = x + x'$. When $t = l\sqrt{n}/2xx'$, the sum of all such pairs is

$$\frac{2}{\sqrt{\pi}} \int c^{-t^2} dt + \frac{\sqrt{n}}{\sqrt{2xx'}} c^{-t^2}. \tag{139}$$

---

[8] Molina [1930], p. 386.

Discussing this formula and the reasoning, Laplace points out that two sorts of approximations are involved.[9] The first is relative to the limits of the a priori probability (*facilité*) of the event *a*, and the second to the probability that the ratio of the occurrences of *a* to the total number of events will be contained within certain limits. As the events are repeated, the latter probability increases so long as the limits remain the same. On the other hand, so long as the probability remains the same, the limits grow closer together. When the number of events reaches infinity, the limits converge in a point and the probability becomes a certainty. Just as he had done in his earliest general memoir on probability ([1781a]), Laplace turned to birth records to illustrate how the ratio of boys to girls gives figures from experience for prior probabilities, or *facilités*.

The latter part of the discussion contains another of the many passages scattered throughout his writings that have led modern readers to feel that Laplace must somehow have had an inkling (or perhaps a repressed belief) that random processes occur in nature itself and not merely as a function of our ignorance. The mathematical occasion here is the use, started in "Mémoire sur les intégrales définies," [1811a], of partial differential equations in solving certain limit problems (see chapter 25). The concluding example turns on a ring of urns, one containing only white and another only black balls, and the rest mixtures of very different proportions. Laplace proves that if a ball is drawn from any urn and placed in its neighbor, and if that urn is well shaken and a ball drawn from it and placed in the next further on, and so on an indefinite number of times around the circle, the ratio of white to black balls in each urn will eventually be the same as the ratio of white to black balls in all of them. But what Laplace really thought to show by such examples was the tendency of constant forces in nature to bring order into the most chaotic systems.

By comparison to the early probabilistic memoirs of the 1770s and 1780s, the fourth chapter on probability of error certainly represents the most significant development in the subject as a whole. It contains, of course, a derivation (¶ 20) of the least-squares law for taking the mean in a series of observations, which is given by essentially the same method as in the memoir on definite integrals [1811a], although in a more detailed and abstract form. There is much more to the discussion of error theory than that, however.[10] The chapter opens (¶ 18) with determinations that the sum of errors of a large number of errors— equivalent to the distribution of sums of random variables—will be

---

[9] *OC*, **7**, pp. 283–84.

[10] See Sheynin (1977), no. 6, pp. 25–34, who gives Laplace's formulations in modern notation.

contained within given limits, on the assumption of a known and equipossible law of errors. It continues (¶ 19) with the probability that the sum of the errors (again amounting to random variables), all considered as positive, and of their squares and cubes, will be contained within given limits. This is equivalent to considering the distribution of the sum of moduli. That leads to the problem of correcting values known approximately by the results of a great number of observations, which is to say by least squares, first in the case of a single element (¶ 20) and then of two or more elements (¶ 21). Mathematically, this involves a discussion of linearized equations with one unknown and with two unknowns, respectively. Laplace includes instructions on application of the analysis to the correction of astronomical data by comparison of the values given in a number of tables. He then considers the case in which the probability of positive and negative error is unequal and derives the distribution that results (¶ 22). The next-to-last section (¶ 23) deals with the statistical prediction of error and methods of allowing for it on the basis of experience. At least, that seems to be a fair statement of what Laplace had in mind in speaking of "the mean result of observations large in number and not yet made," on the basis of the mean determined for past observations of which the respective departures from the mean are known.[11] The chapter closes with a historical sketch of the methods used by astronomers to minimize error up to the formulation of least squares, in which account Laplace renders Legendre and Gauss each his due (see chapter 25).

In the fifth chapter, Laplace discussed the application of probability to the investigation of phenomena themselves and of their causes, wherein it might serve to establish the physical significance of data amid all the complexities of the world. The approach offers practical instances of his sense of the relativity of the subject to knowledge and to ignorance, to science and to nature. In the analysis of error, it is the phenomena that are considered certain, whereas here the existence and boundaries of the phenomena themselves are the object of the calculation. The main example is the daily variation of the barometer, which long and frequent observation shows to be normally at its highest at 9:00 A.M. and lowest at 4:00 P.M., after which it rises to a lower peak at 11:00 P.M. and sinks until 4:00 A.M. Laplace calculated the probability that this diurnal pattern is due to some regular cause, namely the action of the sun, and determined its mean extent. He then raised a further question that, for lack of data, he could not resolve mathematically here, but that is interesting since it came to occupy the very last calculation of his life (see chapter 28). For in theory, atmospheric tides

[11] *OC*, **7**, p. 338.

would constitute a second and independent cause contributing to the daily variations of barometric pressure. He referred his readers to his discussion of that hypothetical phenomenon, a corollary to the treatment of oceanic tides, in *Mécanique céleste*.[12] Observation of such a small effect was not yet possible, although Laplace expressed his confidence that observations would one day become sufficiently extensive and precise to permit its detection.

In short, it was calculations of this sort that Laplace had in mind when he claimed, as he here remarked again, that on the cosmic scale probability had permitted him to identify the great inequalities of Jupiter and Saturn, just as it had enabled him to detect the minuscule deviation from the vertical of a body falling toward a rotating earth. He even had hopes for its calculus in physiological investigations, imagining that application to a large number of observations might suffice to determine whether electrical or magnetic charges have detectable effects upon the nervous system, and whether animal magnetism reflects reality or suggestibility. In general—he felt confident—the same analysis could in principle be applied to medical and economic questions, and even to problems of morality, for the operations of causes many times repeated are as regular in those domains as in physics. Laplace had no examples to propose, however, and closed the chapter with a mathematical problem extraneous in subject matter but not in methodology. The problem had been imagined by Buffon in order to show the applicability of geometry to probability.[13] It consists of tossing a needle onto a grid of parallel lines, and then onto a grid ruled in rectangles, of which the optimal dimensions relative to the length of the needle constitute the problem. Laplace adapted it to a probabilistic, or in this instance a statistical, method for approximating to the value of π. It would be possible, he points out, although not mathematically inviting, to apply a similar approach to the rectification of curves and the squaring of surfaces in general.

Chapter 6, "On the Probability of Causes and Future Events," is in effect concerned with problems of statistical inference. In practice, the material represents a reworking of his early memoir on probability of cause [1774c] and of the application of inverse probability by means of approximations of definite integrals to calculations involving births of boys and girls and also to population problems at large.[14] He now had figures for Naples as well as for Paris and London. In calculating the probable error in estimates of the population of France based on the

---

[12] Book IV, *OC*, **2**, pp. 310–14.
[13] Sheynin (1976), p. 152.
[14] [1781a], [1786b], [1786c].

available samples, he made use of the partial census which, at his request, the government had instituted in 1801. The next, very brief, chapter also starts with old material, to which he gave a new turn. He recurs to his own discovery [1774c] of the effect of inequalities in the prior probabilities that are mistakenly supposed to be equal (see chapter 3). He always attached great importance to that finding, so much so that he alluded to it in the opening chapter of Book II, where definitions were laid down. The significance was that it brought out the care that needed to be taken when mathematical calculations of probability were applied to physical events. Since there are no perfect symmetries in the real world, allowance has to be made for slight deviations of parameters from assumed values in making predictions. Laplace now discussed the problem in the same connection in which he had started it, in relation to the unfairness to one of two bettors on heads or tails of unsuspected asymmetries in a coin to be tossed. That could be mitigated, he now suggested, by submitting the chance of asymmetry itself to calculation. He let the probability of throwing heads or tails be $(1 \pm \alpha)/2$, where α represents the unknown difference between the prior probabilities of throwing one or the other. The probability of throwing heads n times in a row will then be

$$\frac{(1 + \alpha)^n + (1 - \alpha)^n}{2^{n+1}}, \qquad (140)$$

and a player who bets on heads or tails consecutively will have an advantage over one who bets on an alternation.[15] Instead of that, it will be fairer to toss the two coins simultaneously and bet on the chance of their falling both heads or both tails n times running. The probability that they will fall the same way at each throw is then

$$\frac{1}{2^{n+1}}[(1 + \alpha\alpha')^n + (1 - \alpha\alpha')^n], \qquad (141)$$

which is closer to the equipossible $1/2^n$ than is probability (140).

In Chapters 8, 9, and 10, all quite brief, Laplace took up life expectancy, annuities, insurance, and moral expectation (or prudence). We do not know where he obtained his information, but it is reasonable to suppose that some of it must have been derived from occasional service on commissions appointed to review writings in this area and various actuarial schemes submitted to the government. The *Procès-Verbaux* of the Academy in its last years and of the Institute contain

---

[15] *OC*, **7**, pp. 410–11.

record of his having thus been called on from time to time. Moreover, the "Notice sur les probabilités" [1810d] was published as a rationale of the application that it was legitimate to make to tables of mortality. This piece, it will be recalled (see chapter 25), was an expansion of his École Normale lecture of 1795 (chapter 19). It was then further expanded to become the first edition of *Essai philosophique* and concludes with a summons to governments to license and regulate underwriters of insurance, annuities, and tontines, and to encourage investment in soundly managed associations. For an insurance industry and a literature did exist, although Laplace does not refer either to actual practice or to authorities. Comparison of his chapters with both would be required before a judgment could be made of what his contribution may have been.

Mathematically, his model for calculations of the "mean duration," both of life and marriage, is error theory. Given the tables of mortality covering a large population, a value for the mean length of life may be taken and the probability calculated that the mean life of a sample of stated size will fall within given limits. Calculation of life expectancy at any age follows directly. Laplace also gave a calculation for estimating the effect of smallpox on the death rate and of vaccination on life expectancy. The conclusion is that the eradication of smallpox would increase life expectancy by three years, if the growth in the population did not diminish the improvement by outrunning the food supply. Laplace did not give the data or provide numerical examples here, as he had done in his population studies. The succeeding chapter on annuities and tontines is equally abstract and gives expressions for the capital required to create annuities on one or several lives, for the investment needed to build an estate of given size, and for the advantages to be expected from participation in mutual benefit societies.

In the tenth chapter, with which Laplace concluded the first edition, he softened the asperity he had once expressed about Daniel Bernoulli's calculation of a value for moral expectation (see chapter 3) in distinguishing that notion from mathematical expectation.[16] Bernoulli had proposed that prospective benefits, in practice financial ones, may be quantified as the quotient of their amount divided by the total worth of the beneficiary. Laplace now adopted that principle as a useful guide to conduct—without attributing it this time to Bernoulli. In infinitesimal terms, where x represents the fortune and dx the increment, its benefit will be $k\,dx/x$. If y represents the moral fortune corresponding to the

---

[16] [1776a, 1°], Article XXV.

physical value, then

$$y = k \log x + \log h, \qquad (142)$$

where h is an arbitrary constant to be determined by the ratio of a value of y corresponding to a value of x.[17] But perhaps it will not be necessary to follow the calculation in order to be convinced. Laplace concluded that in the most mathematically advantageous games of chance the odds are always unfavorable over time, and that diversification is a prudent practice in the investment of wealth. On matters of this sort, he wrote more persuasively in the ordinary language of the *Essai philosophique* than in the mathematics of the *Théorie analytique*.

Laplace must have continued straight on to expand the "Notice sur les probabilités" [1810d] into *Essai philosophique*. He presented the first edition to the Institute in February 1814 (102), a year and a half after finishing *Théorie analytique*. In August of that year he also read a memoir on the probability of testimony (103). Inclusion of these two pieces, the first as the introduction and the second as a new concluding Chapter 11, together with three minor mathematical additions,[18] marks the difference between the first edition of *Théorie analytique des probabilités* and the second, completed by November 1814 (104). The two pieces have in common the tendency to move the subject further in the direction of civic relevance, the one in expounding it for laymen, the other in extending the application to concerns of life in society.

The *Essai philosophique sur les probabilités* has certainly had a longer life and almost certainly a larger number of readers than any of Laplace's other writings, including its counterpart in celestial mechanics, *Exposition du système du monde*.[19] The reason for its continuing— indeed, its growing—success has clearly been the importance that probability, statistics, and stochastic analysis have increasingly assumed in science, social science, and philosophy of science. Inevitably, Laplace's technical writings have come to have the same sort of relation to the later development of the discipline of probability that, for example, Newton's *Principia mathematica* had to the later science of mechanics. Even if there were no other reason, that would suffice to explain why most readers who wish to repair to the fountainhead of the so-called frequentist interpretation of probability, in contrast to the subjective view, have recourse to *Essai philosophique*, though germs of the latter may also be found in Laplace. But there is a complementary reason, and that is the extreme difficulty of many parts of *Théorie analytique*.

---

[17] *OC*, **7**, p. 441.

[18] *OC*, **7**, pp. 471–93.

[19] The most recent critical edition (Éditions Christian Bourgeois, 1986) has an illuminaing preface by René Thom and a fine historical postface by Bernard Bru.

Given the accessibility of *Essai philosophique* in many editions and languages, a summary scarcely seems necessary. The work itself is a summary. Instead, a reservation may be ventured, although somewhat hesitantly. If the two famous works for the layman are compared in point of intrinsic merit, it is possible to consider *Exposition du système du monde* the better book. At least, it conveys its subject with altogether greater clarity. For, once the reader is past the epistemological opening passages in *Essai philosophique*, Laplace's paraphrase of the mathematics of *Théorie analytique* is not very easy to follow. No doubt the subject matter lent itself less well to verbal summary than did astronomy or geodesy. But it may also be worth noting that the order of composition was reversed. Laplace wrote *Système du monde* as a book in its own right before he compiled *Mécanique céleste*. The former was an outline or a prospectus for the latter. In the case of *Essai philosophique*, he wrote all but the epistemological and actuarial sections after *Théorie analytique*. In its first edition, it was a précis and bears the same relation to the treatise that the initial sections in many of his memoirs do to them, that of a preface written last. Moreover, there is nothing in the epistemology that he had not already said in principle in his lecture at the École Normale in 1795 (see chapter 19) or indeed in the prolegomena of the youthful probability-gravitation memoir ([1776a], chapter 3). It is true that in the later editions of *Essai philosophique* he did enlarge upon topics likely to interest a wider public and that his observations on the credibility of witnesses and on the procedures of legislative assemblies and of judicial panels make more comprehensible the calculations of the supplementary material on the same topics added to the second and third editions of *Théorie analytique*.

Let us consider briefly the approach of the eleventh chapter of *Théorie analytique*, on the probability of testimony.[20] The inevitable urn containing numbered slips is the model analyzed. A slip is drawn, and a witness reports the number to be n. Is he telling the truth? It was Laplace's idea to apply inverse probability to such problems, taking the statement for an event and estimating the probability that it was caused by the truthfulness of the witness. There are four possibilities: (1) he is neither lying nor making a mistake; (2) he is not lying and is mistaken; (3) he is lying and not mistaken; (4) he is both lying and mistaken. With these alternatives, Laplace employs a Bayesian analysis, first for this problem and then for a series of more complicated instances involving several occurrences witnessed by more than one observer. The corre-

[20] *OC*, **7**, pp. 455–70.

sponding discussion in the fourth edition (117) of *Essai philosophique* gives numerical examples and goes on to expose what he considered the fallacy in Pascal's wager on the existence of God.[21] The probability of truthfulness in the witnesses who promise infinite felicity to believers is infinitely small. There is also a daunting estimate of the decay of reliability of historical information with the passage of time.

More interesting is the concluding article in the treatise. There the judgment rendered by a tribunal deciding between contradictory assertions is assimilated to the reports of several witnesses about the drawing of a numbered slip from an urn containing only two. The panel consists of r judges, and p is the probability that each will render a true judgment. The probability of the soundness of a unanimous verdict will then be $p^r/[p^r + (1 - p)^r]$. The value of p is given by the proportion of unanimous verdicts, denoted by i, to the total number of cases n. Since $p^r + (1 - p)^r = i/n$, or very nearly so, solving that equation will give the probability p of the veracity of each judge. Laplace then showed that

$$p = \frac{1}{2} \pm \sqrt{\frac{4i - n}{12n}} \qquad (143)$$

if the tribunal consists of three magistrates.[22]

Choosing the positive root on the assumption that each judge has a greater propensity for truth than error, Laplace calculated the case of a court half of whose verdicts are unanimous. The probability of veracity in each judge will then be 0.789, and the probability that a verdict sustained on appeal is just will be 0.981 if the finding is unanimous, and 0.789 if the vote is divided. The greater the number of judges and the more enlightened they are, the better the chance that justice will be done—for to do Laplace himself justice, he acknowledged the artificiality of these calculations in introducing the topic in *Essai philosophique*, claiming only that they might provide guidance to common sense.[23]

Laplace extended the application of inverse probability to the analysis of criminal procedures in a supplement to *Théorie analytique*, the first of four, composed in 1816 (107). For this purpose a condemnation is an event, and the probability is required that it was caused by the guilt of the accused. As always in the probability of cause, prior probabilities had to be known or assumed, and Laplace again supposed that the

---

[21] *OC*, **7**, pp. lxxix–xc.
[22] *OC*, **7**, p. 470.
[23] *OC*, **7**, p. lxxix.

probability of a truthful juror or judge lies between 1/2 and 1. In a panel of eight members of whom five suffice to convict, the probability of error came to 65/256. He felt that the English jury system with its requirement for unanimity weighted the odds too heavily against the security of society, but that the French criminal code was unjust to the accused. By one provision, if a defendant was found guilty by a majority of 7 to 5 in a court of first instance, it required a vote of 4 to 1 to overturn the verdict in a court of appeal, since a majority of only 3 to 2 there still left a plurality against him in the two courts taken together. That rule was as offensive to common sense as to common humanity.[24] In view of the severe strictures that have been passed upon Laplace's political conduct, it should be noted that he took the trouble of publishing a pamphlet expounding on mathematical grounds the urgency of reforming these savage provisions [P1816]; and in the definitive, fourth edition of *Essai philosophique* he gave his considered opinion that a majority of 9 out of 12 for conviction gave the most even balance between the interests of society in protection on the one hand and in equity on the other.[25]

Between 1817 and 1819, Laplace investigated the application of probability to sharpening the precision of geodetic data and gathered these studies for publication as the second and third supplements to the third edition (1820) of *Théorie analytique* (112, 118). In recent years scholars concerned with the history of statistics have been especially interested in these two pieces, which Laplace himself saw in the context of his theory of error. His intention was to improve on the method of least squares in the minimization of instrumental and observational error. That had also been his motivation in the opening articles of the First Supplement, where before taking up judicial probabilities, he further developed what he called the "most advantageous" method of combining equations of condition formed from observations of a single element, like those exemplified in his initial justification of the least-squares method.[26] He then applied the method to estimating the probable error in Bouvard's recent, highly refined calculations of the masses of Uranus, Saturn, and Jupiter.[27]

In the Second Supplement, Laplace turned to geodesy and compared the results of his method with the so-called method of situation of Bošković.[28] Stigler considers that this discussion contains the earliest

---

[24] *OC*, **7**, p. 529.

[25] *OC*, **7**, p. xcix. For an interesting discussion of the differences between Laplace and Fourier on error theory and statistical inference, see Callens (1997), pp. 176–89.

[26] [1811a]; see chapter 22; cf. *Théorie analytique*, Book II, No. 21, *OC*, **7**, pp. 327–35.

[27] *OC*, **7**, pp. 516–20.

[28] *OC*, **7**, pp. 531–80.

instance of a comparison of two well-elaborated methods of estimation for a general population, in which the conditions that make one of them preferable are specified. He is particularly enthusiastic about the growing statistical sophistication, as he sees it, of Laplace's later work in probability and argues that the analysis here is strikingly similar to that which led R. A. Fisher to the discovery of the concept of sufficiency in 1920.[29] In the Third Supplement, Laplace reports the result of applying his method to the extension of the Delambre-Méchain survey of the meridian from a base in Perpignan to Formentera in the Balearic Islands. The data were the discrepancies between 180° and the sums of the angles measured for each triangle. There were only twenty-six triangles in the Perpignan-Formentera chain, however, and Laplace preferred to estimate the law of error on the basis of all seven hundred triangles in that and the original survey. He could then calculate the probabilities of error of various magnitudes in the length of the meridian by the formulas already developed for his modified, or most advantageous, method of least squares. In a further article on a general method for cases involving several sources of error, Laplace obtained a paradigm equation for formulating equations of condition that relate values for observed elements to the error distributions involved. The equation served him in his later investigation of variations of the barometer as evidence for lunar atmospheric tides, the last he ever undertook (see chapter 28).

In Laplace's own life and career, perhaps it is appropriate to see the curve breaking over in 1820 with the publication of the third, and definitive, edition of *Théorie analytique des probabilités*. In 1825 he did compose a Fourth Supplement, containing a minor modification to the theory of generating functions. By then he was showing signs of age. The minutes of the Institute record that the work was presented by Laplace and by his son.

[29] Stigler (1973), pp. 441–43; cf. Sheynin (1977), pp. 41–44.

CHAPTER 27

Loss of Influence

ROBERT FOX

IT WILL BE recalled (see chapter 24) that in January 1816 Sophie Germain won the Institute's prize set in 1809 for the theory of elastic surfaces. Insofar as her paper broke sharply with the approach of the Laplacian school, it was a sign that Laplace's power was beginning to wane. But it was by no means the only, or the most important, sign. The slackening of corporate research activity at Arcueil after 1812 and the cessation of regular meetings in 1813 did not augur well, and after 1815 Laplace became an increasingly isolated figure in the scientific community of the Restoration, particularly in the realm of physics. His personal reputation also suffered from the readiness of his accommodation to yet another change of political regime. A member of the Senate, he voted for the overthrow of Napoleon and in favor of a restored Bourbon monarchy in 1814. After conveniently absenting himself from Paris during Napoleon's temporary return to power in the Hundred Days—an episode that clearly embarrassed him—he remained loyal to the Bourbons until his death, becoming a bête noire of the liberals, most notably on his refusal in 1826 to sign a declaration of the Académie Française supporting the freedom of the press.

The most serious assaults on Laplace's physics arose from the development of Fresnel's wave theory of light and its championing by Arago, a former member of the Arcueil circle, between 1815 and the early 1820s. In the face of this challenge, the Laplacian position was represented, rather typically, by Laplace's disciples. It is significant, for example, that when Fresnel won the competition of the Academy of Science for a study of the theory of diffraction in 1819,[1] his theory was measured, not against any of Laplace's writings, but against the corpuscular theory of diffraction that Biot and Pouillet had devised, probably at Laplace's instigation, in 1816.[2] And it was Biot and Poisson, not

---

[1] Fresnel, "Mémoire sur la diffraction de la lumière," *MASIF*, **5** (1821–1822/1826), pp. 339–475.

[2] The theory is expounded in Biot's *Traité de physique expérimentale et mathématique*, 4 vols. (Paris, 1816), **4**, pp. 743–75.

Laplace, who continued the open resistance to the wave theory into the 1820s, fighting vainly against the powerful unifying conception of the ether that Fresnel wielded in his own work, notably on polarization, and then, on his death in 1827, bequeathed to Cauchy and others, who continued its elaboration on into the 1850s.[3] Similarly, Laplace did not respond publicly to the growing support for the chemical atomic theory in France after 1815, even though the theory was inconsistent with Berthollet's chemistry of affinities and the whole notion of short-range chemical forces. Nor did he react to the criticism of the caloric theory that was explicit in Petit and Dulong's interpretation of their experiments on specific heats (1819) and clearly implied in Fourier's *Théorie analytique de la chaleur* (1822).[4]

New discoveries also played their part in the undermining of Laplace's position. Hans Christian Oersted's observation of the magnetic effect associated with the passage of an electric current along a wire in 1820 raised two immediate difficulties for traditional physics: One was that the force on the compass needle in the vicinity of the wire was rotational (and not central), the other that it implied (contrary to Coulomb's, and hence also to Laplacian, belief) that electricity and magnetism could interact with each other. André-Marie Ampère (an early supporter and friend of Fresnel who had never been associated with Laplace's circle) and Biot both investigated the phenomenon, and the two men soon came into conflict.[5] Ampère's explanation, which embraced not only electromagnetism but also the forces between magnets, rested on a notion of physical fluids very different from Coulomb's two fluids of magnetism, which Biot incorporated in his own explanation. Whereas Ampère treated electromagnetic and magnetic forces in terms of interactions between current-carrying conductors, Biot conceived electromagnetism as a purely magnetic phenomenon, explained by forces between the tiny magnets that he supposed to be arranged in a circular fashion around any current-carrying wire. Biot's theory, expounded in its definitive form in the third edition of his *Précis élémentaire de physique expérimentale* in 1824, was seriously flawed,

---

[3] The importance of Fresnel's work, in particular the role of polarization, for the collapse of the Laplacian orthodoxy is explored in Buchwald (1989), esp. pp. 260–90.

[4] See A. T. Petit and P. L. Dulong, "Recherches sur quelques points importans de la théorie de la chaleur," *Annales de chimie et de physique*, **10** (1819), pp. 395–413, esp. pp. 396–98 and 406–13. The criticisms of Petit and Dulong are discussed in Fox (1968–1969), esp. pp. 9–16.

[5] On the debate between Ampère and Biot and the implications of Ampère's work for the undermining of Laplacian physics, see Caneva (1980), and Hofmann (1995), chapters 7 and 8. Ampère's theory is also discussed in detail in Grattan-Guinness (1990a) **2**, pp. 917–67.

and Ampère's demonstration of its weaknesses and of the essentially Newtonian character of his own theory ensured its subsequent neglect.

Despite the direct and indirect criticism to which his style of physics was subjected, Laplace never admitted defeat. By the early 1820s few shared his apparently unswerving belief in the theories that assumed the physical reality of the imponderable fluids of heat and light. Yet between 1821 and 1823 he developed the most elaborate version of his caloric theory of gases. He expounded the theory first in papers published chiefly in the *Connaissance des temps* and then, in a definitive and modified form, in Book XII of the fifth volume of *Mécanique céleste*.[6]

In these later versions of his theory, Laplace leaned heavily on the treatment of gravitation that he had published in the first volume of *Mécanique céleste* in 1799. In particular, his expressions for the gravitational forces between spherical bodies proved to be readily applicable to the standard Newtonian model of gas structure, in which the pressure of a gas was explained by the repulsive forces that were supposed to exist between the particles composing it. The modification that the force between the particles was not only repulsive but also inversely proportional to the distance between them was easily made.

In his first paper on caloric theory in the *Connaissance des temps*,[7] which was the published version of one that he read to the Academy of Sciences in September 1821,[8] Laplace considered the equilibrium of a spherical shell taken at random in a gas. Invoking the condition that the force between the particles is effective only at short range, he showed that the pressure of the gas, P, is proportional to $\rho^2 c^2$ where ρ is its density and c the quantity of heat contained in each of its particles. The argument had a highly speculative cast and rested on the unfounded assumption that the repulsive force between any two adjacent gas particles is proportional to c^2. Even more suspect was Laplace's model of dynamic equilibrium, in which, when the temperature is constant, the particles of a gas constantly radiate and absorb caloric at an equal rate. Postulating the simplest of mechanisms for the process, Laplace pictured the radiation from any particle as resulting from the mechanical detachment of the particle's own caloric by incident radiant caloric, the density of which, $\pi(u)$, was taken as a function of the temperature, u, alone. If the fraction of incident caloric absorbed was put equal to q (a constant depending solely on the nature of the gas) and if it was assumed (quite gratuitously) that the quantity of caloric detached was

---

[6] *OC*, **5**, pp. 97–160. For a study of the theory, see Fox (1971), pp. 165–77.
[7] [1821d].
[8] (123).

proportional both to c and to the total "density" of caloric in the gas, ρc, it followed that

$$\rho c^2 = q \pi(u). \tag{144}$$

Since for Laplace $P \propto \rho^2 c^2$, Boyle's law was an immediate consequence of this equation, as was Dalton's law of partial pressures. It also followed that since the function $\pi(u)$ was independent of the nature of the gas, all gases expand to the same extent for a given increment in temperature, as Dalton (1801) and Gay-Lussac (1802) had observed. The obvious next step, of assuming $\pi(u)$ to be proportional to the (absolute) temperature as measured on the air thermometer, was not taken in the paper that he read to the Academy of Science in September 1821.[9] But it appeared very soon afterward in a paper published in the November issue of *Annales de chimie et de physique*[10] and was axiomatic in the definitive version presented in Book XII of *Mécanique céleste.*[11]

The fact that Laplace's theory was consistent with the main gas laws lent it a measure of plausibility, but one prediction in particular raised difficulties. It followed from Equation (144) that the isothermal compression of a gas to, for example, one-half of its original volume would cause c to decrease by a factor $\sqrt{2}$. Qualitatively, a reduction in the value of c was perfectly consistent with the phenomenon of adiabatic heating, but Laplace showed that a decrease by a factor $\sqrt{2}$ in isothermal compression was too great to account accurately for the error in Newton's expression for the velocity of sound (see chapter 22); in fact, it led to the impossibly high value of 2 for γ. Laplace's immediate solution was to suggest that the alternate compressions and rarefactions occur slowly enough for there to be some heat exchange with the surroundings, but by December 1821 he had abandoned this hypothesis. In a paper submitted in that month to the Bureau des Longitudes and published in the following year in the *Connaissance des temps*,[12] he argued that heat exchange does not occur but that the heat "expelled" in excess of that required to reconcile the theoretical and experimental values for the velocity of sound merely becomes latent and so ceases to contribute to the interparticle force. Hence c, in the argument outlined above, now represents not the total quantity of heat in a particle of gas but only the part of it that is "sensible" or free; the "total heat" of a particle is Q or $(c + i)$, i being its latent or combined heat.

[9] [1821d].
[10] [1821a].
[11] *OC*, **5**, p. 125.
[12] [1822a].

It was now a simple matter for Laplace to restate his fundamental expression for the velocity of sound ($\sqrt{dP/d\rho}$) in a form that involved c. Assuming $P \propto \rho^2 c^2$, it followed that the velocity of sound is equal to

$$\sqrt{\frac{2P}{\rho}\left(1 + \frac{\rho}{c}\frac{dc}{d\rho}\right)};$$

and this in turn implied that

$$1 + \frac{\rho}{c}\frac{dc}{d\rho}$$

is equal to $\gamma/2$.

In a supplement to his paper of December 1821, dating almost certainly from early in 1822, Laplace also showed how he could express γ in an equation involving Q, ρ, and P.[13] To do this, he assumed that Q was a function of any two of P (that is, $\rho^2 c^2$), ρ, and the absolute temperature, so that in adiabatic conditions ($\Delta Q = 0$)

$$dP\left(\frac{\partial Q}{\partial P}\right)_\rho + d\rho\left(\frac{\partial Q}{\partial \rho}\right)_P = 0, \qquad (145)$$

and

$$\gamma = -\frac{\rho}{P}\left\{\frac{(\partial Q/\partial \rho)_P}{(\partial Q/\partial P)_\rho}\right\}. \qquad (146)$$

In these ways, Laplace showed how the investigation of c and Q, both of them quantities that were intimately related to his speculations on the state of caloric in bodies and not susceptible to a direct test, could proceed simply through the measurement of γ.

Laplace's argument gave heightened theoretical significance to the experiments to determine γ that Gay-Lussac and Welter conducted in 1822. Particularly fruitful for his purpose was the observation that γ remained very nearly constant over a wide range of temperature and pressure. When the condition $\gamma = $ constant was introduced into Equation (146), it followed, by integration, that Q must be a function of $P^{1/\gamma}/\rho$ or, inserting the absolute temperature u, of $uP^{(1/\gamma-1)}$. By postulating the simplest possible relationship between Q and $P^{(1/\gamma-1)}$ —proportionality—Laplace could then show that

$$Q = KuP^{(\frac{1}{\gamma}-1)}, \qquad (147)$$

where K is an unknown constant, determined by the nature of the gas.

[13] [1822b].

In *Mécanique céleste*,[14] Equation (147) appeared as

$$Q = F + KuP^{(\frac{1}{\gamma} - 1)}, \tag{148}$$

but for Laplace's purposes the two expressions were equivalent. One important result in particular followed with either Equation (147) or Equation (148). This was that the ratio between the volume-specific heats of any gas at two different pressures P_0 and P_1 but at the same temperature is equal to $(P_1/P_0)^{1/\gamma}$. In reality, the volume-specific heats in such circumstances should be in the ratio (P_0/P_1), but close agreement with Delaroche and Bérard's erroneous results for the variations of specific heat with pressure endorsed Laplace's conclusion and the assumptions from which it was derived.

By his own lights, Laplace had secured a major triumph in giving at least one branch of the caloric theory the quantitative character that it had always conspicuously lacked. Yet most contemporaries must have seen his triumph as illusory. There was no independent evidence to confirm his assertions concerning the state of caloric in gases which were clearly determined by the requirement that the deductions made from them should agree with the gas laws.[15] And logically, despite his elaborate expressions involving the hypothetical entities c and i, much of Laplace's argument rested on far simpler premises than he intimated. In 1823 this point was made implicitly but unmistakably in two papers by his most loyal pupil, Poisson, who reviewed and extended several aspects of the Laplacian theory of heat. Most of Poisson's results had already been obtained by Laplace, but new ground was also broken. In the first paper Poisson derived the now familiar expressions for adiabatic changes in volume, $TV^{\gamma-1} = $ constant, and $PV = $ constant;[16] and in the second he arrived at the false conclusion that the principal specific heats of a gas, c_p and c_v, are equal to $BP^{(1/\gamma-1)}$ and $(1/\gamma)BP^{(1/\gamma-1)}$, where B is a constant.[17] No reader could miss the point that all this was achieved without any mention of Laplace's elaborate mechanisms. Poisson merely assumed, as any supporter of the caloric theory would have done, that the heat content of a gas was a function of its pressure and density.

Possibly the most telling evidence of Laplace's diminished status in physics toward the end of his life is the almost total indifference with which his work on caloric was received. It aroused neither overt opposi-

---

[14] *Mécanique céleste*, **5**, p. 128; *OC*, **5**, p. 143.

[15] For discussion of this point, see Fox (1971), pp. 173–74.

[16] Poisson, "Sur la vitesse du son," *Connaissance des temps* (1826/1823), pp. 257–77.

[17] Poisson, "Sur la chaleur des gaz et des vapeurs," *Annales de chimie et de physique*, 2nd series, **23** (1823), pp. 337–52.

tion nor support except, somewhat ambiguously, from Poisson, and it stimulated no further research. Yet in 1824 Laplace appeared as confident as ever that a physics based on imponderable fluids and short-range molecular forces could be achieved. In that year, in the "Avertissement" to the fifth edition of *Exposition du système du monde*, he wrote that he intended to make molecular forces the subject of a special supplement. But the intention was never fulfilled.[18] The physics of the sixth edition (which appeared in 1827, the year of Laplace's death, and again, as a quarto, in 1835) was virtually identical to that of the fourth edition (1813).

So in his last years there were few who endorsed Laplace's approach to physics. The great days of Arcueil were now a distant memory, and the once loyal disciples were no longer involved in the problems that their master had identified. In 1822 Biot virtually retired from the scientific community, following his conflict with Arago over the wave theory of light and his defeat by Fourier in the election for one of the two posts of permanent secretary of the Academy of Science. And he remained aloof, though not inactive, for some years, leaving Poisson to carry on Laplacian physics into the late 1820s and 1830s. But even Poisson was far from being an uncritical admirer. In his *Nouvelle théorie de l'action capillaire* (1831) he criticized and corrected a number of shortcomings in Laplace's theory of capillarity;[19] and in *Théorie mathématique de la chaleur* (1835), despite a prefatory discussion of the properties of caloric, he totally ignored—as he had done in 1823—the model that Laplace had perfected in the early 1820s.

[18] See Bibliography, Section B, p. 282.
[19] See chapter 23.

The Last Analysis

ON THE PUBLICATION of Volume IV of *Mécanique céleste* in 1805, Laplace had undertaken to complete the original plan with an eleventh and final book giving an account of the work of predecessors and contemporaries in the science of astronomy. Instead, he had become immersed in physics and in probability, and by the time in the 1820s when he got down to the histories he had promised, his own work was beginning to be history. Meanwhile, notably in 1819 and the early 1820s, he had published other investigations on particular topics of celestial mechanics and decided to collect these pieces and append them to the historical summaries of the areas they concerned. Thus the intended Book XI of *Mécanique céleste* became Books XI through XVI. Laplace had them printed as they were completed beginning in March 1823 and assembled them as the fifth volume at the end of 1825.[1]

The historical notices are more detailed than the concluding chapters of *Exposition du système du monde* and are written in a matter-of-fact rather than an inspirational vein. It would be interesting to know at what point in the great investigations of his own career he had studied the works of his predecessors, for he evidently knew them very well indeed. His histories are still worth consulting for his own sense of what he himself had contributed to the several topics.

It is doubtful that many of the new investigations assembled for Volume V made much difference to the further development of celestial mechanics. On the whole, their day was past. According to the titles, Book XI is about the figure and rotation of the earth;[2] Book XII is about the attraction and repulsion of spheres and the motion of elastic fluids, though it really contains the comprehensive development of Laplace's caloric theory of heat and gases discussed in Section 27;[3] Book XIII is about the oscillation of fluids surrounding the planets;[4] Book XI is about the rotation of heavenly bodies about their centers of gravity;[5] Book XV is about the theory of planetary and cometary

---

[1] (125, 126, 129–31, 133).
[2] Cf. [1818b], [1819a], [1819b], [1820e], [1820f], [1820g].
[3] Cf. [1820h], [1821d], [1822a], [1822b], [1822d].
[4] Cf. [1815c], [1819d], [1819e], [1820d], [1821a], [1823a].
[5] Cf. [1809d], [1824a].

motions;[6] and Book XVI is about satellite theory.[7] Those comprehensive designations are a bit misleading, however. Most of Volume V consists of minor emendations to the data and improvements on fine points of the analysis of particular phenomena treated under those headings in the four main volumes.

Several novelties and peculiarities are worth signaling, however. Inevitably, Laplace had been in touch with the work of Cuvier, whose position and influence in the biological sciences during the Napoleonic period and afterward paralleled his own eminence in mathematical quarters. The first substantive chapter of Book XI consists of calculations purporting to show how the depth and configuration of the seas can be reconciled with the geological evidence for catastrophic inundations and extreme climatic changes in the history of the earth. Chapter 4 also concerns the theory of the earth, in respect to its cooling. As far back as 1809, Laplace had learned of Fourier's investigations of the diffusion of heat and had referred in print to his colleague's pathbreaking but still unpublished paper of 1807.[8] Despite the difference in their approaches (which he did not mention), Laplace opened his chapter on the heat of the earth by expressing his pleasure that Fourier's two fundamental equations

$$\frac{\partial^2 V}{\partial x^2} + \frac{\partial^2 V}{\partial y^2} + \frac{\partial^2 V}{\partial z^2} = k\frac{\partial V}{\partial t}, \tag{149}$$

which expresses the diffusion of heat inside the earth, or any comparable body, where V is the heat of any point, and

$$-\frac{\partial V}{\partial r} = fV - fl, \tag{150}$$

which expresses the transmission of heat through the surface, were simply modifications respectively of his general equation (94) on the attraction of spheroids (the potential function), and of his own Equation 2 in Book III, no. 10 of *Mécanique céleste*.[9] Laplace did not allude to the ambivalence in the background of his further relations with Fourier and with his own disciple, Poisson, who was bested in the rivalry between the two.[10] He went on to analyze the temperature gradient beneath the surface and the rate of cooling of the earth in expressions analogous to

[6] Cf. [1813a], [1819c], [1821c].
[7] Cf. [1809c], [1812a], [1820a], [1820b], [1820c], [1820i], [1821b].
[8] [1810a], *OC*, **12**, p. 295; see Grattan-Guinness (1972), pp. 444–52.
[9] *OC*, **2**, p. 30; **5**, pp. 82–83.
[10] Grattan-Guinness (1972), pp. 462–63.

those that he had developed years before [1785b] for spheroidal attraction theory. His purpose was to estimate whether the shrinkage of the earth on cooling was sufficient to decrease the angular velocity and thus to alter the length of the day detectably. While acknowledging that his parameters were hypothetical, he felt safe in concluding that the effect, if it existed at all, did not amount to 0.01 seconds since the time of Hipparchus.

In Book XIII, Laplace thought to redeem a promise, or an assurance, other than historical. It will be recalled (see chapter 7) that the atmospheric tides had interested him since the earliest memoirs on the ebb and flow of the seas [1779b]. He recurred to the subject briefly and inconclusively in *Mécanique céleste*, Book IV, and more confidently in Book II of *Théorie analytique des probabilités*. Discussing the significance of data in Chapter 5, where the most important example came from barometric readings, Laplace there predicted that one day records would be sufficiently full and accurate to permit detecting the gravitational influence of sun and moon among the other, in effect much larger, causes that determine atmospheric pressure. Now, in 1823, taking his point of departure from his previous analysis of mean sea level, he set out to apply his own version of the least-squares analysis to the detection of significance in the variations of the barometer that could be correlated with the relative positions of earth, sun, and moon [1823a]. The Third Supplement (118) to *Théorie analytique des probabilités*, published in 1820, again described his least-squares method—now somewhat modified—as the "most advantageous." As in the original distinction from Legendre's approach [1811a], Laplace meant by "most advantageous" that he combined the equations of condition for the unknowns so as to determine the most probable values (see chapter 25). In an excellent discussion, Stigler describes the modification as weighted least squares.[11] Laplace now did have access to a series of barometric measurements recorded at the Observatory in Paris—quite probably at his instigation—three times a day, at 9:00 A.M., noon, and 3:00 P.M. from 1 October 1815 through 1 October 1823. He published his findings in *Connaissance des temps* [1823a] and reprinted them with little change in Volume V of *Mécanique céleste*.[12]

Laplace's idea was to determine whether the operation of gravitational influence can be detected over those eight years by comparing variations in barometric pressures in the four days surrounding syzygies (when the sun, moon, and earth are in line) and quadratures (when they make a right angle). From the conditions, he formed the linear

---

[11] Stigler (1975).
[12] *OC*, 5, pp. 184–88, 262–68.

equations

$$x \cos(2iq) + y \sin(2iq) = E_i,$$

$$y \cos(2iq) - x \sin(2iq) = F_i, \qquad (151)$$

in which x and y are the unknowns to be estimated; i indicates the day of each set of data and has the value -1, 0, $+1$, or $+2$; q represents the synodic motion of the moon; and E_i and F_i are computed from the data by the following formulas:

$$E_i = A''_i - A_i + B_i - B''_i,$$

$$F_i = \{2A'_i - (A_i + A''_i) - 2B'_i + (B_i + B''_i)\}. \qquad (152)$$

In these expressions, A_i denotes the eight-year mean of the 9:00 A.M. measurements for the ith day after syzygy; B_j for the ith day after quadrature; A'_i and B'_i the means for the noon values; and A''_i and B''_i for the 3:00 P.M. values. Thus, E_i is the mean barometric change between 9:00 A.M. and 3:00 P.M. for the ith day after syzygy, minus the same change for the ith day after quadrature; while F_i is proportional to the difference between the mean rates of change for those days. These expressions had to be combined by the modification that he had brought to his "most advantageous" method in the Third Supplement to *Théorie analytique des probabilités*.[13]. He multiplied each of the four equations for E by a factor of three and by the corresponding coefficient of x, and multiplied each of the equations for F by the corresponding coefficient of y, obtaining

$$x(8 + \Sigma \cos 4iq) + y\Sigma(\sin 4iq) = 3\Sigma E_i \cos 2iq - \Sigma F_i \sin 2iq,$$

$$y(8 - \Sigma \cos 4iq) + x\Sigma(\sin 4iq) = 3\Sigma E_i \sin 2iq + \Sigma F_i \cos 2iq. \quad (153)$$

Substituting the data from the observations and calculating x and y (0.10743 and -0.017591 respectively), Laplace found the range of the lunar atmospheric tide to be 0.05443 millimeters of mercury and the time of the maximum tide in syzygy to be 18 minutes and 36 seconds after 3:00 P.M. There then remained the problem of determining the probability that these observations really did exhibit the existence of a lunar atmospheric tide. For it was not enough to compare the variations at syzygy and quadrature with those assumed to follow from irregular or random causes, as he had just done. Unless the probability of error in the conclusion is contained within very narrow limits, it might be that the data exhibit only the overall effects of irregular causes, a fallacy to

---

[13] *OC*, **7**, pp. 608–16.

which the science of meteorology was prone. In what amounted to the application of a central limit theorem (the phrase was never his), Laplace calculated the probability that chance alone would produce a variation in the means no greater than that indicated by the eight-year accumulation of observations. The value was 0.843, an unconvincing figure in the tidal quest. All that could be said was that there would be "some implausibility" in attributing the variation to chance alone.[14] To increase to near certainty the probability that tides are responsible, the very small effects detected would need to rest on something like nine times the 1,584 thrice-daily readings, or approximately forty thousand observations.

In the next four years Bouvard continued the program of recording barometric readings at the Paris Observatory, which data he compared with a comparable series assembled on similar principles by Ramond at Clermont-Ferrand, in the Puy-de-Dôme. Bouvard then recalculated the whole corpus, an enormous labor that led Laplace to take up his theory of lunar atmospheric tides yet again. In this, his last paper [1827b], composed in his seventy-eighth year, he modified once more his method of calculating probable error. Whether or not he had in the interval become aware of the importance of independence as between E_i and F_i in Equation (152)—Stigler thinks he may well have revised his approach for just that reason[15]—his new calculation did combine the equations of condition in such a way that terms depending on the same measurements served to determine only one unknown. Moreover, in explaining it, Laplace now put a distance between his approach and that of least squares, for which several mathematicians have given proofs that he found "not at all satisfactory."[16] The art consists in choosing the factors by which the equations of condition formed from the data are to be multiplied in combining them into a system of final equations.

Laplace now compares his procedure with that followed in planetary astronomy. In order to correct the elliptical elements, observed longitudes are equated to theoretical longitudes, each modified by the relevant correction. A large number of equations of condition are thus formed. Each is multiplied by the coefficient of the initial correction. Adding all of them gives the first final equation. The same procedure is followed for each successive correction, until there are as many final equations as there are corrections. But the longitude does not depend on a single observation. It is derived from two observations by different instruments, one giving right ascension and the other declination. The

[14] *OC*, **5**, p. 268.
[15] Stigler (1975), p. 503.
[16] *OC*, **5**, p. 491.

law of errors may not be the same for both or have the same effect on the longitude. It was in determining the most advantageous factors (given all these complications) that Laplace considered this method to be superior to least squares, and he referred to the general expression that he had given for it in the Third Supplement to *Théorie analytique des probabilités*:

$$l^{(i)}x + p^{(i)}y + q^{(i)}z + \cdots = a^{(i)} + m^{(i)}\gamma^{(i)} + n^{(i)}\lambda^{(i)} + r^{(i)}\delta^{(i)} + \cdots ,$$

$$(154)$$

where *l*, *p*, *q*, *a*, *m*, *n*, and *r* are coefficients given by the conditions; γ, λ, and δ are errors in the observations arising from different circumstances, and *x*, *y*, and *z* are unknowns to be estimated.[17]

Obviously, his equations of condition (152) for the variations in the lunar tide had been formed in that mold, and now, four years later, he combined them in a manner conformable to the astronomical illustration just given of the "most advantageous" method. Having thus revised the method, Laplace let Bouvard perform the calculation, which yielded values of 0.031758 for *x* and 0.01534 for *y*, and a difference of 0.01763 millimeters of mercury for the range of the atmospheric tide. Again, the probability had to be calculated that the results show the existence of a regular cause, in this case the gravitational pull of the moon, rather than mere chance. If the value for *x* were the effect of chance alone, the probability that it would fall within the limits ± 0.031758 would be

$$\frac{1}{\sqrt{\pi}} \int g e^{-g^2 l^2} \cdot dl,$$

where *g* is given by observation, and the integral is evaluated between those same limits. That works out to be only 0.3617, and it would again need to be very close to unity to be convincing evidence of the existence of a lunar tide. The value for *y* gave even a lower probability, and the detectability of a lunar tide in Paris had, therefore, still to be considered moot despite the additional data.[18]

Corollary information proved more amenable to the search for causes. Ramond in his observations in Clermont-Ferrand had discovered, and Bouvard in Paris had confirmed, that the daily variation of the barometer between 9:00 A.M. and 3:00 P.M. varied with the seasons, the mean increase being 0.557 millimeters of mercury in the three months from November through January and 0.940 in the following quarter. In the

[17] *OC*, **7**, p. 612.
[18] *OC*, **5**, p. 500.

remaining two quarters, the values were intermediate between those extremes. Did these differences result from cause or chance? Again, only probability could decide. On making the calculation, Laplace found that the two values just cited do argue the existence of a regular cause with a high degree of probability, but that the intermediate values and the annual mean of 0.762 can reasonably be attributed to the effects of chance. Bouvard had also noticed that the mean variation is positive in every month of the year, and in a corollary calculation Laplace found the pattern probabilistically predictable.[19] The manuscript for this paper, published posthumously [1827b] with another on elliptical motion and the calculation of planetary distances, was found among Laplace's papers after his death and was incorporated in further printings of *Mécanique céleste*, Volume 5, as a supplement.

[19] For comparisons of Laplace's capabilities with the resources of modern statistical science in relation to this problem, see Stigler (1975). pp. 509–15, and Sheynin (1977), pp. 56–58.

Part V

THE LAPLACE TRANSFORM

CHAPTER 29

Laplace's Integral Solutions to Partial Differential Equations

IVOR GRATTAN-GUINNESS

LAPLACE'S NAME is most widely used today by mathematicians when referring to the "Laplace transform" method of solving differential, difference, and integral equations. Thus it is meet to outline here the contexts in which it arises in Laplace's writings and its later development into a systematized theory.[1]

To begin at the end: The modern definition of the Laplace transform $\bar{f}$ of f is

$$\bar{f}(s) = \int_0^\infty e^{-su} f(u)\, du, \qquad \mathrm{Re}(s) > 0. \tag{155}$$

The essence of Laplace transform theory is to convert the given f-problem via Equation (155) to a problem in $\bar{f}$, solve that, and then convert back to the f-solution. The theory itself includes addition, convolution, and shifting theorems, results on transforms of derivatives and integrals, and especially the inverse theorem by means of which we have a general rule to get back to f from $\bar{f}$:

$$f(x) = \frac{1}{2\pi\sqrt{-1}} \int_C \bar{f}(s)\, e^{sx}\, ds, \tag{156}$$

where C is a certain kind of contour in the s-plane. Note that the problem context mentioned at the beginning, and the various theorems stated after Equation (155), are needed to characterize the theory; integral forms similar to Equations (155) and (156) are not sufficient on their own to earn the name of "Laplace transform."

Let us now trace some of this history. Historians of the calculus are accustomed to finding seeds of later ideas in Euler, and such is the case

---

[1] I am indebted to Jock MacKenzie and Stephen Stigler for advice on this chapter.

here.[2] In a paper published in 1744 he examined

$$z = \int X(x)\, e^{ax}\, dx \qquad (157)$$

as a possible form of solution of differential equations,[3] and a more extensive discussion was given in 1753.[4] However, his preference was for functional or, to a lesser extent, power-series solutions of differential equations (and sometimes combinations of both), although iterative indefinite integrals of a function would also sometimes be used.[5] Euler's ideas to date on solutions of differential equations were outlined in detail in *Institutiones calculi integralis* (1768–1770); integral solutions of the form of Equation (157) were given some space, as was the similar form[6]

$$\int X(x) x^{\lambda}\, dx. \qquad (158)$$

The influence of Euler on Lagrange was very profound; and of particular interest for the subsequent influence on Laplace is Lagrange's 1773 paper on finding the mean of a set of observations (of which Laplace was aware, as we saw in chapter 3). Lagrange considered a few special discrete and continuous distributions and evaluated the probability of errors falling within given limits. This involved him in "Laplace-transform"–looking integrals such as

$$\int X(x)\, e^{-\alpha x} a^{x}\, dx, \qquad (159)$$

where X is a rational function, and their conversion into infinite series.[7] Although we can interpret the results in terms of modern Laplace transform theory, Lagrange, like Euler, did not see his results in quite that way.

The first significant signs of Laplace's interest in such expressions occur in his "Mémoire sur les suites" [1782a]. There his solution

---

[2] On the Laplace transform, see Deakin (1981–1982).

[3] See esp. article 6 of L. Euler. "De constructione aequationum" (1744), in *Opera omnia*, 1st series, **22**, pp. 150–61.

[4] See articles 6 ff. of L. Euler. "Methodus aequationes differentiales" (1753), in *Opera omnia*, 1st series, **22**, pp. 181–213.

[5] See, for example. esp. article 28 of L. Euler, "Recherches sur l'intégration de l'équation" (1766), in *Opera omnia*, 1st series, **23**, pp. 42–73. On these and other references, see Petrova (1975).

[6] *Institutiones calculi integralis*, **2** (1769), chapters 3–4 and 5 respectively; in *Opera omnia*, 1st series, **12**.

[7] See esp. articles 37–42 of Lagrange, "Mémoire sur l'utilité de la méthode" (1773), *OL*, **2**, pp. 171–234.

method by successive approximations (see chapter 11) involved formulas such as

$$u = \sum_{r=1}^{\infty} a_r(s, s_1) \int^{(r)} \phi(s)(ds)^r + \sum_{r=1}^{\infty} b_r(s, s_1) \int^{(r)} \psi(s_1)(ds_1)^r, \quad (160)$$

where ϕ and ψ are arbitrary. This kind of result was already given in his first presentation [1777a] of the method; but later [1782a] he came close to a Laplace transform of the form (158) for integral values of μ, for he used iterative integration by parts to obtain

$$\int \phi(z) z^{\mu} \, dz = C + \sum_{r=0}^{\mu} (-1)^r \mu Pr \phi_{r+1}(z) z^{\mu-r}, \quad (161)$$

where ϕ_r is the rth indefinite integral of ϕ. However, the purpose was not to explore the properties of Equation (161) for themselves but to obtain this integral solution of a general linear second-order partial differential equation:

$$u = (2x)^{-m/2} \left\{ \int_0^{x+at} \lrcorner \left(\frac{x + at - z}{2x} \right) \phi(z) \, dz \right.$$

$$\left. + \int_0^{x-at} \lrcorner \left(\frac{x - at - z}{2x} \right) \psi(z) \, dz \right\} \quad (162)$$

where $\lrcorner$ is the solution of a certain second-order ordinary differential equation.[8]

More promising material for our purpose occurs in Laplace's paper [1785a] on approximating to functions of very large numbers. He took the linear difference equation

$$S(s) = \sum_r A_r(s) \, \Delta^r y(s) \quad (163)$$

(where Laplace's s is real) and applied two transforms to it:

$$y(s) = \int x^s \phi(x) \, dx, \quad (164)$$

akin to Equation (161) and now sometimes called the "Mellin transform," which proved particularly useful when $S \neq 0$ and the "almost-Laplace" transform

$$y(s) = \int e^{-sx} \phi(x) \, dx, \quad (165)$$

[8] *OC,* **10**, p. 54; **9**, pp. 24 ff.; **10**, p. 66; **10**, p. 68.

which is helpful when $S \equiv 0$.[9] In the course of using these two trans-forms he derived a few basic properties; for example, from Equation (165) he had a theorem on (forward) differences:[10]

$$\Delta^r y(s) = \int e^{-sx} (e^x - 1)^r \phi(x) \, dx \qquad (166)$$

while Equation (164) provided a similar result for differences,

$$\Delta^r y(s) = \int x^s (x - 1)^r \phi(x) \, dx, \qquad (167)$$

and also one for derivatives,[11]

$$\frac{d^k y(s)}{dx^k} = \int x^s (\log x)^k \phi(x) \, dx. \qquad (168)$$

"In many circumstances," Laplace commented prophetically, "these forms [165] ... are more useful than the preceding ones (164)."[12] It is worth noting also that earlier parts of this paper feature integrals the integrands of which involve e^{-t^2} for the Laplace transform of the error function is of some importance in certain applications.[13] Later in the paper Laplace urged general solutions to differential equations using the form of Equations (164) or (165).[14]

It is this work that is normally cited as the origin of the term "Laplace transform"; but we must also look at the effect on Laplace of Fourier's 1807 monograph on heat diffusion. Fourier had found the "diffusion equation," in forms such as

$$\frac{\partial^2 y}{\partial x^2} = \frac{\partial y}{\partial t}, \qquad (169)$$

to represent the physical phenomenon, and the "Fourier series" solution form

$$y = \sum_{r=0}^{\infty} (a_r \cos rx + b_r \sin rx) \, e^{-r^2 t} \qquad (170)$$

to solve it.

[9] OC, **10**, pp. 212, 236–48.
[10] Ibid., p. 236.
[11] Ibid., pp. 242–47; see also pp. 278–91.
[12] Ibid., p. 148.
[13] This is well conveyed in, for example, Murnaghan (1962).
[14] OC, **10**, p. 253.

A strong controversy ensued at the Institut de France about this work, partly because solutions of the form of Equation (170) had been considered and rejected in the eighteenth century.[15] A criticism particularly relevant to our current discussion is that the initial condition function f is not explicitly encased in the solution (170)—as it is in integral and functional solutions—but appears only in the integrals that define its coefficient; to the mathematical mind of the time, the explicit involvement of f helped to justify the generality of any solution. Lagrange remained opposed to Equation (170), as did Poisson and Biot; but Laplace accepted Fourier's results. In the memoir [1810a] he constructed a Newtonian intermolecular force model to obtain the heat transfer term in Equation (169).[16] More significantly, he published a miscellany paper on analytical methods [1809e] that not only related to Fourier's work but also developed techniques presented in "Mémoire sur les suites" [1782a] and the paper [1785a] on very large numbers.

As a special case of advancing the results of "Mémoire sur les suites" [1782a], Laplace considered Fourier's diffusion equation (169) and solved a problem that is conspicuous by its absence from Fourier's 1807 monograph: to solve Equation (169) for an infinite range of values of x, where the periodicity of the trigonometric functions rules out Equation (170). Poisson, already aware in 1806 of the trend of Fourier's work, had offered this power-series solutions of Equation (169):[17]

$$y = \sum_{r=0}^{\infty} \frac{x^{2r}}{(2r)!} f^{(r)}(t) + \sum_{r=0}^{\infty} \frac{x^{2r+1}}{(2r+1)!} g^{(r)}(t). \qquad (171)$$

In the miscellany paper [1809e] Laplace used the same type of solution of Equation (169). Applying the initial condition

$$y = \phi(x), \quad \text{when } t = 0, \ -\infty \leq x \leq \infty, \qquad (172)$$

he obtained

$$y = \sum_{r=0}^{\infty} \frac{t^r}{r!} \phi^{(2r)}(x), \qquad (173)$$

---

[15] For further details on these matters, see Grattan-Guinness with Ravetz (1972).

[16] *OC*, **12**, p. 293.

[17] S.-D. Poisson, "Mémoire sur les solutions...," in *Journal de l'École polytechnique*, cahier 13, **6** (1806), pp. 60–116, esp. pp. 109–11.

and then submitted it to an ingenious manipulation. The result

$$\int_{-\infty}^{\infty} z^{2r} e^{-z^2} dz = \frac{(2r)!}{4^r r!} \sqrt{\pi} \tag{174}$$

is proved in the paper on very large numbers [1785a], where "t_{2r}" is misprinted from the original for "t^{2r}".[18] I have used z for t in Equation (174), although Laplace did not explicitly recall his proof here, and the obvious

$$\int_{-\infty}^{\infty} z^{2r+1} e^{-z^2} dz = 0 \tag{175}$$

converted Equation (173) to

$$y = \frac{1}{\sqrt{\pi}} \int_{-\infty}^{\infty} \sum_{r=0}^{\infty} \frac{(2z\sqrt{t})^r}{r!} \phi^{(r)}(x) e^{-r^2} dz. \tag{176}$$

The integrand of Equation (176) contains a Taylor expansion, so that we have finally the integral solution[19]

$$y = \frac{1}{\sqrt{\pi}} \int_{-\infty}^{\infty} \phi(x + 2z\sqrt{t}) e^{-z^2} dz. \tag{177}$$

This solution form preserved the tradition of containing the initial condition function ϕ explicitly. Fourier himself now realized that an integral solution would work for an infinite range of values of x, and by ingenious if unrigorous manipulations of infinitesimals he very quickly found the "Fourier transform" and its inverse. They took forms such as

$$\underline{f}(q) = \sqrt{\frac{2}{\pi}} \int_0^{\infty} f(u) \cos qu\, du \tag{178}$$

and

$$f(x) = \sqrt{\frac{2}{\pi}} \int_0^{\infty} \underline{f}(q) \cos qx\, dq. \tag{179}$$

[18] OC, 10, p. 269.
[19] OC, 14, pp. 184–93.

and led to double-integral solutions of the diffusion equation (169) in which, as in Equation (177), the initial condition function f is encased:[20]

$$y = \frac{2}{\pi} \int_0^\infty \int_0^\infty f(u) \cos qu \cos qx\, e^{-q^2 t}\, du\, dq. \qquad (180)$$

Thus two new integral (and the Fourier series) solutions of linear partial differential equations were produced in a very short time. During the next decade an intense development of these methods occurred in a variety of physical contexts. Fourier integrals were the most popular, but Poisson used Laplace's form (177) whenever he could. Because Laplace himself was not prominently involved, we shall not pursue the details here;[21] but several of his results of this period are worthy of notice. He used Fourier integrals in the paper [1810b], which continued the purpose of that on very large numbers [1785a].[22] In the memoir on definite integrals [1811a] he returned to another old interest, finding the mean of a set of observations, and derived more Fourier integrals from integrals such as

$$\int_0^\infty e^{-ax} x^{-\omega}\, e^{\sqrt{-1}\, rx}\, dx \qquad (181)$$

which may be compared with the results in the earlier work [1785a].[23] Elsewhere in this paper [1811a] he may possibly have revealed another influence from Fourier, for he showed that the set of functions[24]

$$\left\{ \frac{(-2)^i}{1 \cdot 3 \cdots (2i-1)\sqrt{\pi}} \int_{-\infty}^\infty e^{-s^2}(\mu + s\sqrt{-1})^{2i}\, ds, \quad i = 1, 2, \ldots \right\}$$

$$(182)$$

was orthogonal over $(-\infty, \infty)$ with respect to the weighting function $e^{-\mu^2}$ —an analysis that corresponds closely to Fourier's 1807 demonstration of the orthogonality of the (misnamed) "Bessel functions" $\{J_0(a_i)\}$.[25] The $\{a_i\}$ are the roots of a certain transcendental equation with respect to the weighting function x—and an orthogonality expansion similar to

---

[20] Fourier's most detailed treatment is in articles 342–85 of his *Théorie analytique de la chaleur* (1822), in *Oeuvres*, **1**, pp. 387–448.

[21] Some hints are given in n. 7, chapters 21 and 22; and in chapters 6–10 of Burkhardt (1908).

[22] *OC*, **12**, pp. 334–44.

[23] *OC*, **12**, p. 363; *OC*, **10**, p. 264.

[24] *OC*, **12**, p. 382.

[25] See Grattan-Guinness with Ravetz (1972), chapter 16, n. 7.

Fourier's.[26] This work constitutes Laplace's anticipation of a form of the "Hermite polynomials" of half a century later.[27] The expansion is now misnamed "the Gram-Charlier expansion." Finally in this association of Laplace with Fourier, we recall from chapter 28 Laplace's use of Fourier analysis in estimating the age of the earth.

Laplace's *Théorie analytique des probabilités* also deserves mention; for, as in his earlier work on mathematical probability (chapters 3 and 10) he included, especially in Book I, Part 1, treatments of the generating function in the form

$$\sum_r p_r t^r \tag{183}$$

rather than as what we now call the moment generating function of the distribution function of the discrete random variable t:

$$\sum_t e^{\lambda t} f(t). \tag{184}$$

He also used a continuous analogue of Equation (183),

$$y(x) = \int t^{-x} T(t)\, dt, \tag{185}$$

which harks back to his Equation (164), and at one point he adapted his earlier Equation (168) to define the then rather novel fractional derivative of $y(x)$:

$$\frac{d^i y(x)}{dx^i} = \int t^{-x} \left(\log \frac{1}{t} \right)^i T(t)\, dt, \tag{186}$$

where i is *not* necessarily an integer.[28] He also used a continuous version of Equation (184),

$$y(x) = \int e^{-xt} f(t)\, dt, \tag{187}$$

where an inverse transform is attempted[29] and also in treating again the "Hermite polynomials" (in Book II, Chapter 3). Various other of his earlier results were given an airing: in Book I, Part 2, some Fourier integrals were evaluated, and the use of Equations (164) and (165) in

[26] Ibid., p. 384.
[27] See Molina (1930).
[28] *OC*, **7**, p. 86.
[29] *OC*, **7**, p. 136.

solving difference equations and in developing asymptotic theory was again dealt with. Laplace also made some use of characteristic functions for discrete distributions (for example, in Book I, Part 1 and Book II, Chapter 4—and already in the memoir [1810b] for a uniform distribution).[30]

However, not surprisingly, the full relationship between mathematical probability and harmonic analysis does not seem to have been grasped at this time. For example, we now know that the distribution function of a sum of independent random variables is the convolution of the component distribution functions $f(x)$ and $g(x)$; that is, if f and g are transformed into $F(p)$ and $G(p)$, then the product $F(p)G(p)$ is the transform of the "convolution" function

$$\int_0^x f(q)g(x - q)\, dq \tag{188}$$

to use the latter name ("composition" and *Faltung* have also been used). Some special cases had already arisen, in Lagrange (1773) and Laplace [1782a] cited above, and elsewhere. For example, in the important manuscript of 1777 published by Gillispie (1979), Laplace sought distribution functions $g(u)$ compatible with a given distribution of errors in a collection of independent observations. He found an expression for the probability that the sum of the errors of n such observations equalled a given measure nx; it involved an integral

$$\int_0^s f(u)g(s - u)\, du \tag{189}$$

where f was known and s was related to nx. He then argued that $g(u)$ could take the form of a polynomial of the nth degree with known or determinable constant coefficients.[31] In 1814 Poisson highlighted the use of convolution integrals in connection with Fourier-integral solutions of the differential equations in deep-water dynamics, but the emergence of these integrals was gradual and slow.[32] Again, the existence of the integral in Equation (185) is better secured by the use of complex variables in defining the characteristic function of f:

$$X(\lambda) = \int_{-\infty}^{\infty} e^{\sqrt{-1}\,\lambda t} f(t)\, dt; \tag{190}$$

---

[30] *OC*, **12**, p. 309.

[31] (21), Gillispie (1979), pp. 232–37; see chapter 10.

[32] See esp. S.-D. Poisson, "Mémoire sur la théorie des ondes," in *MASIF*, **1** (1816), pp. 71–186; and for discussion, Grattan-Guinness (1990a), pp. 667–74.

but this move had to wait for a time, although Cauchy was very adept at handling complex variable forms of the Fourier integral (178).[33]

Thus we see a number of opportunities that later generations were to grasp. It is appropriate here only to outline the later development of the Laplace transform, which has been our principal theme. The first systematic study of its basic properties was carried out in an 1820s manuscript by Abel, first published in 1839.[34] He started from generating functions, and noted the multivariate transform

$$\phi(x, y, z, \ldots) = \int e^{xu+yv+zp+\cdots} f(u, v, p, \ldots) \, du \, dv \, dp \cdots . \quad (191)$$

Meanwhile, in 1833 Robert Murphy had taken the transform in the form

$$g(x) = \int t^{x-1} \phi(t) \, dt, \quad (192)$$

where ϕ is a rational function, and made explicit the formula for the inverse:[35]

$$\phi(t) = \frac{1}{t}\left(\text{coefficient of } \frac{1}{x} \text{ in } \frac{g(x)}{t^x}\right). \quad (193)$$

It was the difficulty of finding a general formula for the inverse, as well as the difficulty of solving integral equations of any kind, that prevented the Laplace transform from revealing its power for so long.[36] Fourier's quick success in obtaining his inverse transform (Equation 179) was crucial to the much more rapid development of his methods. It should be pointed out that when the phrase "Laplace transformation" is found in nineteenth-century mathematical literature, the reference is normally either to Laplace's method of reducing partial differential equations or to his method of cascades (compare chapters 11 and 6 respectively). When progress did come, it was through the aid of Fourier analysis. For example, in 1859 Riemann converted the transform,

$$g(s) = \int_0^\infty h(x) x^{-s} \, d(\log x), \quad s = a + b\sqrt{-1}, \quad (194)$$

---

[33] See esp. A.-L. Cauchy, "Sur les intégrales des équations linéaires" (1823), in *Oeuvres*, 2nd series, **1**, pp. 275–357; and its continuation by operational means in "Sur l'analogie des puissances et des différences" (1827), ibid., 2nd series, **7**, pp. 198–254. The form of (188) received some attention in the context of probability in S.-D. Poisson, *Recherches sur la probabilité des jugements* (1837), chapter 4.

[34] N. H. Abel, "Sur les fonctions génératrices et leurs déterminantes," in *Oeuvres complètes*, ed. B. Holmboe (1839), **2**, pp. 77–88. He slightly misstates Equation (189); *Oeuvres complètes*, ed. L. Sylow and S. Lie (1881), **2**, pp. 67–81.

[35] R. Murphy (1833), pp. 353–408; see p. 362.

[36] Literature on this history includes Bateman (1911), and Hans Hahn (1911).

to a sum of Fourier sine and cosine integrals, and he inverted both, by means of the appropriate versions of Equation (179), to end up with[37]

$$h(y) = \frac{1}{2\pi\sqrt{-1}} \int_{a-\infty\sqrt{-1}}^{a+\infty\sqrt{-1}} g(s)y^s \, ds. \qquad (195)$$

Dini studied such inversions, in contour form, in 1880, and again Fourier analysis provided the means.[38] Laplace transform theory had to await such events as Poincaré's 1885 analysis of asymptotic solutions to differential equations,[39] and especially the development of Heaviside's operational calculus in the 1890s.[40] Modernly recognizable proofs of the inversion formula were produced early in the twentieth century,[41] and they helped substantially in the exegesis of Heaviside's ideas in the 1920s with the "operational calculus" and its applications. J. R. Carson's work was especially significant in systematizing the operational calculus, based on the Laplace transform, for its use in electrical circuit theory.[42] Thus textbooks on both theory and applications began to appear in the 1930s and early 1940s.[43] Then, with the theory well launched, some of its development was classified during World War II, for it was used in problems such as wave guide design for radar systems.

Perhaps Laplace, ever sensitive to the needs of government, would have appreciated that.

---

[37] B. Riemann. "Über die Anzahl der Primzahlen unter einer gegebenen Grösse" (1859), in *Gesammelte mathematische Werke*, 2nd ed. (Leipzig, 1892), pp. 144–55; see p. 149.

[38] See esp. articles 62–88, of Ulisse Dini, *Serie di Fourier* (Pisa, 1880), in his *Opere*, **4**.

[39] Henri Poincaré, "Sur les équations linéaires" (1885), in *Oeuvres*, **1**, pp. 226–89. Amusingly, Poincaré misnames it "Bessel transformation" throughout the paper and puts a corrective note at the end. The *Oeuvres* edition silently makes the correction, although the original page numbers are incorrectly given. On the long and tortuous history of the transform, and the curious business of its name, see Deakin (1981–82).

[40] Oliver Heaviside, *Electromagnetic Theory*, 3 vols. (London, 1893–1912). See Lützen (1979) for an account.

[41] See esp. H. M. MacDonald (1902–1903), H. Mellin (1910), and T. J. I'A. Bromwich (1916).

[42] See esp. Carson (1926),(1922), and (1930).

[43] See esp. Doetsch (1937), Widder (1946), and for applications Carslaw and Jaeger (1941). For bibliography, see Gardner and Barnes (1942), pp. 359–82.

Conclusion

LAPLACE HIMSELF might be shocked, but to call his place in history apostolic does not force the term unduly. Canonical is the word for his image of an infinite intelligence that recalls the past and predicts the future state of all things from knowledge of the position and motion of every particle at any moment. Belief in a fully determined universe was nonetheless a faith for being naturalistic rather than religious. Laplace, even like Einstein after him, conceived the vision in his youth before setting out to prove it. A grand faith, and one not without works after all, determinism formed the expectations held of the exact sciences in the century and more following Laplace's death. It carried forward into Einstein's great vision of an underlying order in nature. Such an outlook on the world is nonetheless magnificent for proving unattainable and is not to be demeaned by facile skepticism about the capacity of the human mind to attain rationality in any guise at all, scientific or other.

Scientific critics, on the other hand, are entitled to take Laplace's assurance of stability in the solar system as an instance of overconfidence—though not, as is sometimes implied, of naïveté. Already in the 1850s LeVerrier found that the calculations of values for eccentricity in _Mécanique céleste_ hold good only if the mass of the planet in question forms a considerable part of the total mass of the planetary system, and that they were invalid for bodies as small as the earth. More recently, indeed very recently, it has been calculated in the light of chaos theory that the motions of the planets become unpredictable after some 100 million years.[1]

Even without benefit of later knowledge, the historian can scarcely fail to notice how Laplace's entire astronomical program begged the question in a fundamental manner. The rigor he pleads for is in the object of his investigations, the system of the world. The mathematical procedures by which he purports to demonstrate it were not rigorous, even by the standards of his time. Laplace was a virtuoso in formulating rapidly convergent series, in obtaining mathematical expressions embracing terms to represent a multitude of physical phenomena, in justifying the neglect of inconvenient quantities in order to reach solutions, and in giving the widest possible extension to his conclusions.

---

[1] Laskar (1995) gives an illuminating brief account of the problem of stability of the solar system from Laplace to the present.

He was an indefatigable calculator manipulating expressions that occupy many lines in arguments that run for very many pages. The outcomes depend, virtually without exception, on approximate solutions to his problems and not on strict proofs.

Poisson and Fourier, respectively Laplace's most faithful disciple and an occasional rival, expressed appreciations, the one at his funeral, the other as permanent secretary of the Academy of Science. Poisson recalled how throughout the careers of Laplace and Lagrange their memoirs on the same subject had succeeded each other:

> There was, moreover, a difference between their geniuses that would be noticed by whoever read their works. Whether the question was the libration of the moon or some problem of number theory, Lagrange most often seemed to see in the questions he treated only the mathematics of which they were the occasion, and from there came the high value he put on the elegance of the formulations and the generality of the methods. For Laplace, on the contrary, mathematical analysis was an instrument that he bent to his purposes for the most varied applications, but always subordinating the method itself to the content of each question. Perhaps posterity will judge the one to have been a great mathematician and the other a great scientist who sought knowledge of nature through the instrument of the most advanced mathematics.[2]

Fourier put it this way in his éloge:

> It cannot be said that it was given to him to found an entirely new science, ... but Laplace was born to perfect everything, to deepen everything, to push back all the boundaries, to solve what was thought to be insoluble. He would have completed the science of the heavens if that science could be completed.
>
> One meets with the same characteristics in his research on the theory of probability, a completely modern science, of immense scope, where misunderstandings of its object have given rise to the most false interpretations, but the applications of which will one day embrace the whole field of human knowledge.[3]

The judgment of an anonymous commentator (almost certainly Augustus de Morgan) in *The Dublin Review* is blunter:

> The genius of Laplace was a perfect sledge hammer in bursting purely mathematical obstacles; but, like that useful instrument, it gave neither finish nor beauty to the results. In truth, in truism if the reader please,

---

[2] "Discourse de M. Poisson aux funérailles de ... Laplace," prefixed to the 6th edition of *Exposition du système du monde* (1835).

[3] *MASIF*, **10** (1831), pp. lxxxi–cii (Bibl. J, p. 307).

Laplace was neither Lagrange nor Euler, as every student is made to feel. The second is power and symmetry, the third power and simplicity; the first is power without either symmetry or simplicity. But, nevertheless, Laplace never attempted the investigation of a subject without leaving upon it the marks of difficulties conquered: sometimes clumsily, sometimes indirectly, always without minuteness of design or arrangement of detail; but still, his end is obtained and the difficulty is conquered.[4]

Fourier's observations suggest one respect in which Laplace was unique among the handful of scientists of his stature in modern history. Alone among them, he never thought to change the world picture within which he came of age intellectually. He took upon himself the role of vindicator of the science he received rather than of innovator. Not that he made no discoveries or improvements, in physics as well as mathematics: the theory of capillary action, calculation of the speed of sound, the refractive indices of different gases, the cyclical character of the long inequality of Jupiter and Saturn, generating functions, theory of inverse probability, the potential function, the Laplace transform. Any one of these, and others among his lesser contributions, would have made the reputation of another scientist. But Laplace was nothing revolutionary, whether in his science or his politics. He had no notion of changing the Newtonian system of the world. He sought rather to perfect it, to demonstrate its stability as Newton had failed to do, to exhibit its cogency in the microcosm of physics no less than the macrocosm of astronomy, and all by the means proper to his generation and milieu, that is to say, by mathematical analysis.

Consistently enough, the range of his knowledge of the science of his time, and earlier times, was nothing short of encyclopedic, and in the modern meaning of the term. The point comes out most obviously to the reader of *Exposition du systéme du monde* and also in Book X of *Mécanique céleste*. In his analytical memoirs the sheer extent of his information is rather dispersed among the formulas, but it is no less impressive. Throughout, the detail, the geometric and numerical detail, of the apparent motions of the sun, of the moon, of each of the planets in turn, of the Jovian satellites, of the rise and fall of tides, of physical and chemical phenomena in general—Laplace had at his fingertips everything observed and tabulated about these matters from earliest times until his own day. He appears to have read the entire literature of the exact sciences, to have kept up to date with ongoing research throughout his lifetime, and to have had extraordinary faculties of retention or recall or both. Erudition and theoretical conservatism are

---

[4] "Theory of Probabilities," *The Dublin Review* **3** (1837), p. 348 (April, pp. 338–54; July, pp. 237–48).

not ordinarily considered to be defining characteristics of scientific creativity. On the contrary. The combination, nevertheless, animated by the athletic virtuosity of a mathematical talent exercised with great power of concentration, is what distinguishes Laplace among great scientists of all time.

The question of originality has also a more narrow bearing. We have had frequent, indeed almost regular, reason to notice that the starting point of an investigation was a memoir or finding by Lagrange dealing with the subject in question. That is true of Laplace's youthful paper on difference equations, of the determination of a mean value in a series of observations, of the anomalous acceleration of the moon, of the long-term invariance of the mean motion of the planets and the variation of their orbital eccentricities, of the calculus of operations, of the assignment of limits to error distributions, of planetary perturbations and the great inequality of Jupiter and Saturn. With respect to Lagrange, Laplace always made full and appreciative acknowledgment, and their relations appear to have been cordial. He was less meticulous with others. There were episodes of friction with Legendre (who, however, on the occasion of Laplace's death proposed that his place in the Academy of Science remain vacant for six months as a mark of respect). The author of the *Dublin Review* article cited above complained that Laplace failed systematically to mention Taylor, de Moivre, Daniel Bernoulli, Maclaurin, and others while deploying techniques they had developed as if they were part of the armory of analytical weapons he was himself forging. Even allowing for the laxer standards of the eighteenth century in the matter of citation, other readers of Laplace may feel the complaint to be not unfounded.

Certain ironies attach to Laplace's role in the history of probability, though of the magnitude of the impetus he gave the subject there can be no doubt. The division between the frequentist and the subjective interpretations is a later distinction of which no one had an inkling in his time. Indeed, the roots of both positions may be found in Laplace's views. He was a frequentist in his phenomenonology and a subjectivist in his epistemology. On the one hand, he considered, and his calculations depend on this, that the probability of events that occur in large numbers converges on a limit in proportion to the number of events. On the other hand, he held that probability is relative to knowledge, not to nature, and that it amounts to nothing but good sense reduced to calculation.

By the end of his career Laplace had come to appreciate, what proved to be the case in the next generation, that the major contributions of probablity in the foreseeable future would lie in its applicability to political and social sciences and practices. The argument in *Essai philosophique sur les probabilités* culminates in considerations pertaining to demography, credibility of testimony, procedures of judicial panels and electoral bodies, actuarial determination of insurance risk and annuities, medical statistics, and physiological and psychological research. Condorcet, to whom Laplace was never generous, was the one who had pointed to such a prospect on reviewing his seminal paper on inverse probability (see chapter 3). Laplace himself had come to the subject in a merely mathematical manner from his investigations of difference equations. In his eyes, initially at least, and throughout in his heart of hearts, the importance of inverse probability was that it made possible analysis of causality, reduction of error, and proof of the regularity of phenomena in a determined universe.

Matters fell out otherwise. In the event, Laplace's legacy pried open the first chink in the armor of deterministic physics. In the 1830s and 1840s the Belgian astronomer turned sociologist Adophe Quetelet set out to form a social physics on the basis of the regularities, which he saw as positivistic laws, that statistically minded bureaucrats were busily discovering in such human phenomena as birth rates, death rates, life expectancy, and the incidence of marriage, divorce, disease, crime, and suicide. His profile of the protagonist in these dramas, the "average man" (*l'homme moyen*), derived from application to the data of Laplace's theory of error. Quetelet's enthusiasm for the bell-shaped curve knew no limits. He made it a measure of the normal distribution, not merely of errors of observation, but of variation in the occurrence of societal and, by extension, of natural events. The notion of a science of society appealed to reform-minded scientists in the liberal atmosphere of the midcentury. In 1850 John Herschel published a full and careful article on Quetelet's work in the *Edinburgh Review*. James Clark Maxwell was even then engaged with the kinetic theory of gases. Scholars now agree that reading Herschel's essay suggested to Maxwell the strategy of calculating the mean velocity of molecules by recourse to a distribution function identical in form with the Laplacian law of errors: "The velocities," he supposed, "are distributed among the particles according to the same law as the errors are distributed among the observations in the theory of 'the method of least squares.'" That analysis marked a turning point in physics. It presupposed the operation of an order of chance in nature and thereby inaugurated the science of statistical mechanics.[5]

---

[5] John Herschel, "Quetelet on Probabilities," *Edinburgh Review*, **92**, pp. 1–57, a review of Quetelet, *Theory of Probability Applied to the Moral and Social Sciences*, the 1846 translation of *Lettres ... sur la théorie des probabilités appliquée aux sciences morales et politiques* (Brussels, 1846); Maxwell, "Illustrations of the Dynamical Theory of Gases," *Scientific Papers*, ed. W. D. Niven (2 vols., 1890), **1**, pp. 377–409. See Gillispie (1963); Everitt (1974); Brush (1977), **1**, pp. 184–87; Porter (1986), chapters 4, and 5.

With regard to classical physics more generally, Laplace's influence is not to be measured by the eventual failure of the model of inter-particulate forces of attraction and repulsion. The young physicists who came out of the École Polytechnique at the turn of the century were both united with and divided among themselves in various ways on problems other than particles versus waves. Important though the issue was in optics, it is not a question of winners and losers even there. The legacy to later physics drew on point masses and central forces in the contributions to optics by Biot, to acoustics by Sophie Germain and Poisson, to electrostatics and magnetism by Poisson in his founding of potential theory. It drew on waves, interference, and an elastic ether in Fresnel's optics and on a punctiform ether for elasticity in Cauchy. It drew on electrodynamic corpuscles and forces now tangential, and again radial, in Ampère. It insisted on exclusion of all models in Fourier, for whom the physics simply was the mathematics. In practice the fre-quently cited hyphenation Newtonian-Laplacian breaks down. Neither Ampère nor Fourier was anything Laplacian. Both nailed the flag of Newton to their masts.

The mathematical techniques throughout were very varied. The tradi-tional approach of rational mechanics obtained differential equations to be solved by various methods of approximation. That was characteristic of Poisson, Germain, and Fresnel, whatever their differences in physics. There was also empirical curve-fitting—Biot did a lot of that. Those of an engineering temperament employed a geometric style. Malus was a pupil of Monge, not Laplace, and his first work on optics was an exercise of differential geometry. Cauchy introduced rigorous methods into mechanics. Navier applied Fourier analysis to the design of suspen-sion bridges. Only one thing can be said of the physics that mattered in France from 1800 until the death of Laplace: For good or for ill, all of it was mathematical. For that he was not alone responsible, of course. Nevertheless, his presence, his example, his standing in the Institute counted enormously, more powerfully than any other personal factor.

Comparison to Einstein holds only with respect to their commitment to a deterministic order of things. Otherwise theirs were contrasting roles and personalities. Einstein, a revolutionary in science if there ever was one, and equally a rebel against the political and social conventions of his time, cut a figure of cosmic benevolence in the world at large, and this in spite of the failure of his family life and the vacuum of his personal relationships. Laplace has ever been a less sympathetic person-age, perhaps even unsympathetic. In common with other mathematical eminences of a certain type, he did project what persons with less exacting temperaments may perceive as arrogance. Still, the adverse judgments may be unfair. He had his opponents, to be sure, but he did form close personal attachments, with Biot, with Poisson, with Bouvard, with Cuvier, and he led a warm and tranquil family life.

Perhaps, too, the image of scientific lawgiver is overdrawn. It should be remembered that, though he did indeed draw the terms of the three famous competitions set by the Institute for mathematical treatments of physical phenomena, the second and third of the prizes, for heat diffusion and for refraction of light, went to people supposed to be opponents of the so-called Laplacian orthodoxy, to Fourier and to Fresnel. The first award crowned the memoir by Malus that ended by vindicating the Huygens construction for double refraction. There Laplace devised a more economical analysis even while the memoir was under adjudication. It is hard to imagine that he meant to steal his protégé's thunder. Insensitivity is the more probable explanation, along with inability to restrain himself when he hit upon a better solution.

Certainly Laplace, in this the opposite of Einstein, had no feeling that the world is too much with us. The dedication to Napoleon was not reprinted in the second and third editions of *Théorie analytique des probabilités*, and neither that nor its predecessor in Volume 2 of *Mécanique céleste* appears in the *Oeuvres complètes*. These apostrophes to power have incurred Laplace much odium since 1815 and have been taken by his detractors to epitomize a willingness to serve every set of masters in the state quite without regard to principle. It may have been so; his voice was rarely if ever raised in opposition to any action of any government in power. Fairness, however, requires the observation that his political conduct was no different from that of the scientific community as a whole. His eminence there exposed him to closer and more jealous scrutiny than has been directed at his colleagues, and his personality and influence may also have aroused greater hostility than was provoked by others. Fairness also requires recalling that the government of the restored monarchy showed no scruple in associating his reputation with its own, anticlimactic attempt at prestige.

In 1816 he was elected to the Académie Française, and in 1817 Louis XVIII elevated him in the peerage to the dignity of marquis. The reason for that was obvious, however, whereas the relation between Napoleon and the scientific community presents a problem that calls for further study and deeper insight. It was more than a straightforward matter of patronage, important though that was.[6] Some special affinity was involved, comparable perhaps to the interdependence between artist and despot discerned by Jakob Burckhardt in *The Civilization of the Renaissance in Italy*. With all due allowances for differences in century and locus of talent, when this new cultural pact, which recruited

[6] See Crosland (1967); Fischer (1988); Dhombres and Dhombres (1989).

scientists as courtiers, finds its analyst, he too may discover a clue to motivations in the illegitimacy of both parties with regard to traditional sources respectively of authority and of knowledge. The key to institutionalization, on the other hand, was the systematic need that authority, in the form of the modern state, and knowledge, in the form of modern science, were just then beginning to develop for each other in practical fact.

Fourier in his éloge remarked that Laplace retained his extraordinary memory to a very advanced age. He always ate and drank very lightly and showed no sign of enfeeblement before his last two years. Magendie was his doctor and Bouvard was with him constantly at the end. There is a bust by Houdon, and Guérin did his official portrait as president of the Senate in 1803. Most likenesses show a thin, pointed face with narrow lips and prominent nose. Later portraits suggest a slight tendency to dewlaps in the final years.

Laplace was buried in Père Lachaise. It was the wish of his son, who died in 1874 at the age of eighty-four, that his body should be transferred to his native Calvados. Accordingly, the family had a tomb erected in the form of a small Greek temple on their estate in St. Julien de Mailloc, a village in the canton of Orbec. In 1888 Laplace was there reinterred in company with the remains of his wife, his daughter, and his son.[7] At the same time the monument placed over his original grave was moved to Beaumont-en-Auge, where it stands in a high place in the cemetery and commands a view of the countryside. There could be no more advantageous setting in which to act on the suggestion proposed in the opening sentence of *Exposition du système du monde*. "If on a clear night, and in a place where the whole horizon is in view, you follow the spectacle of the heavens, you will see it changing at every moment. The stars rise or set. Some begin to show themselves in the east, others disappear in the west. Several, such as the Pole Star and the Great Bear, never touch the horizon in our climate … ."

[7] Edgar C. Smith, Letter to the editor, *Nature* 119 (2 April 1924), pp. 493–94.

Abbreviations

AC	*Annales de chimie*
AP	*Archives parlementaires*
BSPM	*Bulletin de la Société Philomathique de Paris*
CT	*Connaissance des temps*
CX	*Correspondance de l'École Polytechnique*
HARS	*Histoire de l'Académie Royale des Sciences de Paris* (the preface to the corresponding volume of *MARS*)
JP	*Journal de physique*
JX	*Journal de l'École Polytechnique*
MARS	*Mémoires de l'Académie Royale des Sciences de Paris*
MASIF	*Mémoires de l'Académie Royale des Sciences de l'Institut de France* (**1** [1816/1818])
MI	*Mémoires de l'Institut National des Sciences et Arts; sciences mathématiques et physiques*, 14 vols. (thermidor an VI [1798]–1818), for the Directory and the Napoleonic period
OC	*Oeuvres complètes de Laplace* [H]
OL	*Oeuvres de Lagrange*, 14 vols. (1867–1892)
PVAS	*Registre des Procès-Verbaux des séances de l'Académie Royale des Sciences de Paris*
PVIF	Académie des Sciences, *Procès-Verbaux des séances ... depuis la fondation de l'Institut jusqu'au mois d'août 1835*, 10 vols. (Hendaye, 1910–1922)
RSC	Royal Society of London, *Catalogue of Scientific Papers (1800–1863)*, III (London, 1869), pp. 845–48
SE	*Mémoires de mathématique et de physique, présentés ... par divers sçavans* (often cited as *Savants étrangers*), 11 vols. (1750–1786)

Bibliography

This bibliography comprises four sections: (A) Laplace's occasional writings; (B–I) Laplace's published works; (J–N) secondary literature by subject; and (O) secondary works by author and date.

Section A
Laplace's Occasional Writings

A. Laplace conducted an extensive correspondence with leading scientists throughout his lifetime. The record of his thoughts must have been important to him for he kept copies of his own letters. It was long thought that the bulk of the correspondence, along with all the other papers in the possession of the family, had been consumed in a fire that swept through the château of Mailloc in Normandy in 1925. The proprietor was then Laplace's great-great-grandson, the comte de Colbert-Laplace. In a letter to Karl Pearson, the latter told of his intention to publish the Laplace-Lagrange correspondence and of his grandmother's recollections of her grandfather (see Pearson [1929]). Also, F. N. David (1965) has stated that much personal and scientific material, including all correspondence with English scientists, was destroyed during the British bombardment of Caen in 1944, but no details are given and no authority is cited.

All is not lost, however. Roger Hahn has discovered that in fact many manuscripts survived the fire in the family property. He arranged for their purchase by the Bancroft Library of the University of California at Berkeley, where they may now be consulted. Hahn (1994) is an analytical calendar of these papers and of all other letters that he has been able to locate in many libraries in Europe and America.

A number of Laplace letters are available in published sources, beginning with *OC*, **14** (H). There are fourteen letters to Lagrange and twelve from Lagrange in *Oeuvres de Lagrange*, **14** (Paris, 1892). Letters to and from Laplace also figure in the *Oeuvres de Lavoisier, Correspondance* (ed. successively by René Fric, Michelle Goupil, and Patrice Bret (6 vols. to date) (Paris, 1955–). René Taton (1953) has published a few letters and other pieces illustrating Laplace's relations with Lacroix from 1789 until 1815. Yves Laissus (1961) has published letters to Alexis Bouvard (20 Feb. 1797) and J.-B. Delambre (29 Jan. 1798).

In 1886 Charles Henry published six letters from Laplace to Condorcet and d'Alembert, of which the originals are in the Condorcet papers at the Institut de France, and four together with a fragment on the orbits of comets from the papers of the Abbé A.-G. Pingré are in the Bibliothèque Sainte-Geneviève. These documents are reprinted in *OC*, **14**, pp. 340–71. See below (I, 1, 34).

Two letters by Laplace as Minister of the Interior, dated 17 frimaire and 26 frimaire an VIII (8 Dec. and 17 Dec. 1799), the latter accompanied by a report, appear in the *Moniteur universel* (an VIII), no. 78 (18 frimaire), 307–8, and no. 87 (27 frimaire), 343–45.

Autographs of Laplace other than letters are not indicated in Hahn (1994). Perhaps the most interesting are reports on his examination of artillery cadets in 1784, 1785, and 1786, and his recommendations for reform of the system in July and August 1789, at the Archives de la Guerre, X^D 249 (Vincennes).

Sections B through I
Laplace's Published Works

Before beginning, it may be well to exorcise a phantom. The *Catalogue général des livres imprimés* of the Bibliothèque Nationale lists among the writings of Laplace *Essai sur la théorie des nombres. Second supplément* (1825). The volume corresponding to that call number (V. 7051) has "Laplace" penciled on the flyleaf, but it is in fact Legendre's 2nd supp. to his *Essai sur la théorie des nombres* (1808). Another copy is bound with the 2nd edition of that work (1825).

In what follows we will give first Laplace's major treatises in the order in which they were published during Laplace's lifetime, followed by details of the translations, the two collected editions, and the individual memoirs. The place of publication is Paris, unless indicated otherwise.

B. **Exposition du système du monde**, 2 vols. (an IV [1796]); 2nd edition (an VII [1799]); 3rd edition (1808); 4th edition (1813); 5th edition (1824). An edition printed by de Vroom in Brussels in 1826 and 1827 appears to be a reprint of the 5th edition, even though the latter printing is called *sixième édition* on the title page. It seems probable that the true 6th edition, for which Laplace was reading proof at the end of his life, was delayed until 1835, when it appeared in Paris, prefaced by the éloge delivered by Fourier on 15 June 1829 before the Institut de France [J]. This edition occupies vol. 6 of [G] and [H] below. It was reprinted in Paris by Fayard in 1984. For the differences between it and the 5th edition, see the discussion under [H].

C. *Traité de mécanique céleste*: vols. 1 and 2 (an VII [1799]); vol. 3 (an XI [1802]); vol. 4 (an XIII [1805]); vol. 5 (1823–1825). A 2nd edition was published in 4 vols. (1829–1839). This work occupies vols. 1–5 of [G] and [H] below.

D. *Théorie analytique des probabilités* (1812): 2nd edition (1814); 3rd edition (1820). With 3 supplements published in 1816, 1818, and 1820, respectively, this work, with a 4th supplement (1825), occupies vol. 7 of [G] and [H] below. A facsimile of the 1st edition was published by Éditions Culture et Civilisation (Brussels, 1967).

E. *Essai philosophique sur les probabilitès* (1814), originally published as the "Introduction" to the 2nd edition of [D]; 2nd edition (1814); 3rd edition

(1816); 4th edition (1819); 5th edition (1825). It is included with [D] in vol. **7** of both [G] and [H] below. A facsimile of the 1st edition was published by Éditions Culture et Civilisation (Brussels, 1967). A reprinting of the 5th edition (Paris, 1986) has a preface by René Thom and a postface by Bernard Bru. There is a new English translation of the 5th edition with notes by Andrew I. Dale (New York, 1995).

F. **Translations of *Mécanique céleste***: J. C. Burckhardt published a German translation of vols. **1** and **2** as *Mechanik des Himmels*, 2 vols. (Berlin, 1800–1802). There are two early English translations of Book I, respectively by John Toplis (London-Nottingham, 1814) and by Thomas Young, the latter under the title *Elementary Illustrations of the Celestial Mechanics of Laplace* (London, 1821). Henry Harte published a translation of Books I and II, 2 vols. (Dublin, 1822–1827). These efforts were entirely superseded by the splendid work of Nathaniel Bowditch, *Mécanique céleste by the Marquis de Laplace, Translated with a Commentary*, 4 vols. (Boston, 1829–1839). Bowditch's commentary in the footnotes is an indispensable vade mecum for the study of Laplace, explaining and filling out the demonstrations, and containing a great body of historical as well as mathematical and astronomical elucidation. Bowditch made the translation between 1815 and 1817. He held off publishing it in the expectation that Laplace intended to issue a revised edition of vol. **1** in order to incorporate the material from the supplement to vol. **3** (90), which developed an improvement Poisson had made in the proof of the invariability of planetary mean motions by taking account of the square of the perturbing masses. Bowditch expected that Laplace would also revise Book III, vol. **2**, in order to remedy certain defects in the calculation of the action of spheroids pointed out by James Ivory (see Todhunter [1873], **2**, pp. 221–24). Bowditch did not translate vol. **5**.

G. *Oeuvres de Laplace*, 7 vols. (1843–1847), reprints [B], [C], and [D] above, with [E] included as the introduction to [D]. Its publication, initiated by Laplace's widow, was eventually subsidized by the State. In 1848 the National Assembly voted to underwrite a further edition, but that never went forward.

H. *Oeuvres complètes de Laplace*, 14 vols. (1878–1912), was financed by a bequest from Laplace's son, General the Marquis de Laplace, who died on 7 October 1874. His will entrusted the task to the Académie des Sciences. There is correspondence in the Laplace dossier in its archives concerning the arrangements between his niece, the comtesse de Colbert-Laplace, the permanent secretaries, and the publisher, Gauthier-Villars. General Laplace expressly directed that the edition was to contain neither commentary nor extraneous elements. He based this injunction on what he took to be the wishes of his father, who had often said in the last months of his life that no corrections should be made in the works of savants after their death; modifying their writings in any way could only distort the record of their initial thoughts and be prejudicial to the history of science.

In only one respect (apart from correction of typographical errors) was the new edition to depart from [G] in its printing of the major treatises in the first seven volumes. Before his death, Laplace had begun correcting the proofs for a 6th edition of *Exposition du systéme du monde* [B]. His intention had been to return to the 4th edition and restore its chapters 12 ("De la stabilité et de l'équilibre des mers"), 17 ("Réflexions sur la loi de la pesanteur universelle"), and 18 ("De l'attraction moléculaire"), which he had omitted from the 5th edition. In the foreword to the projected 6th edition Laplace said that he intended a separate work bringing together "the principal results of the application of analysis to the phenomena due to molecular actions distinct from universal attraction, which had just been much extended." Since he did not have the time for that, his son considered it consistent with his principles to restore these chapters from the 4th edition.

For the rest, Volumes **8–12** contain the individual memoirs published by the Académie des Sciences through 1793 and by the Institut de France after 1795. Volume **13** contains writings published in *Connaissance des temps* from 1798 until Laplace's death, and Volume **14** memoirs reprinted from *Journal de l'École Polytechnique* (most notably his course on mathematics given at the École Normale in 1795); articles from *Journal de physique, Annales de physique et chimie, Journal des mines*, and so on; scattered items of correspondence; and fragments concerning annuities, rents, and matters of public interest. There is a very inadequate "Table analytique."

Despite their instructions, the editors did modernize much of Laplace's notation, for which reason it is preferable to have recourse to the original printing whenever possible, particularly in the case of the earlier memoirs. Since that is possible only in large research libraries, it has seemed practical to give quotations from the original sources and to make page references to the *Oeuvres complètes* whenever the work in question is contained in them. Unfortunately, however, the edition is not exhaustive; the editors simply went through the journals mentioned above and reprinted what they found seriatim. Thus, they missed Laplace's earliest papers [1771a, 1771b, 1774a] published in Leipzig and Turin, as well as his first major treatise, *Théorie du mouvement et de la figure elliptique des planètes* (1784), and other, lesser writings. It is quite possible that further publications that escaped their net may also have eluded us.

Individual Memoirs

I. What follows is (Part 1) a chronology of Laplace's work in the order of the composition of particular writings, and (Part 2) a bibliographical listing of his memoirs in the order of publication. As will become evident, the two sequences are not everywhere the same. Prolonged immersion in these confusing details has convinced us that any redundancy in this double listing will be more than compensated by the facility it creates for keeping the problem of tracing the development of Laplace's research distinct from the problem of tracing the history of the influence of his publications. The cross-references make it possible to relate the two sequences at any juncture.

The first list has been established from the records of the Académie des Sciences. The *Procès-Verbaux*, or minutes of the semiweekly meetings of that body, were transcribed into a register that is conserved in the Archives de l'Académie des Sciences in the Institut de France. Before Laplace was elected to membership, the papers that he submitted were referred to commissions for evaluation, according to the normal practice. The record of these reports is maintained in a separate register (which, however, is not complete); and many of the reports themselves remain in the archives, classified by date. The sequence of these early communications, (1)–(13) in our numbering, was investigated by Stephen M. Stigler (1978). The manuscripts of the papers themselves were normally returned to the contributor and have not been conserved in the archives. In rare instances the text was transcribed in the *Procès-Verbaux*. After Laplace's election to the Academy, the only record we have consists in most instances of the original title of his communications and the date on which he read or simply submitted them. In certain instances he formally requested recognition of priority—*pour prendre date*. Unless otherwise indicated, the dates and titles below (through 1793) are from the *Procès-Verbaux*, which carry through to the suppression of the Academy on 8 August of that year by the Revolutionary Convention, and thereafter from the published *Procès-Verbaux* of the Institut de France.

Since the foundation of the Institut de France in 1795, the Académie des Sciences has formed one of its constituent bodies, as its Classe des Sciences Physiques et Mathématiques from 1795 to 1814, and under the name Acadèmie des Sciences again since the Restoration. For the period 1795–1835 its *Procès-Verbaux* have been published together with the reports of committees, which are even more valuable. In the Academy before 1793 and in the Institute after 1795, Laplace served on many commissions concerned with evaluating the work of others or with special projects. We have included in this listing only the reports for which he was primarily responsible as author or spokesman. His other involvements in the affairs of the Institute were manifold, particularly in relation to the prize programs. They may be followed by means of the indexes to each volume of the *Procès-Verbaux*.

As for the second listing, that of publications, a word is needed about the organization and dating of the Academy's memoirs. Under the Old Regime, an annual volume was published under the general title *Histoire et Mémoires de l'Académie Royale des Sciences de Paris*. The two sections are separately paginated and for this reason are cited separately (see the table of abbreviations below). The *Histoire* consists of announcements—prizes, distinguished visitors, works received, and so on—together with an abstract of the memoirs, prepared by the permanent secretary. The *Mémoires*, which constitute the bulk of each volume, are the scientific papers themselves published by the members. Confusion in dating easily arises, because publication was always two to four years in arrears. Thus the volume for the year 1780 appeared in 1784 and contained memoirs submitted in 1783—or at any time between the nominal and publication dates. Our method is to indicate that volume as *MARS* (1780/1784), with the latter date that of publication, and to specify memoirs in the bibliography by the date of publication. The *Connaissance des temps* presents the reverse problem. The volume of this almanac "for 1818," for example, contained the ephemerides for that year but appeared in 1815. Thus, it is cited *CT* (1818/1815), with the latter date still that of publication.

Journals proliferated after 1795, and Laplace then fell into the practice of publishing short papers and abstracts of his long memoirs in the *Connaissance des temps* and other periodicals mentioned in the list of abbreviations. More often than not, he would publish the same piece, sometimes with minor modifications, in several journals. The Laplace entry in the Royal Society of London, *Catalogue of Scientific Papers, 1800–1863*, III (1869), cites all the journals in which each piece appeared, as well as the translations that appeared in Germany and in Britain. We have limited ourselves to the most readily accessible journal; and when more than one is indicated, we have cited the number of the memoir (indicated by *RSC*) as listed in the catalog. The editors took the ostensible date of the various journals at face value and thus cannot always be relied on for their dating.

In addition Laplace published a few reports concerned with the administration of the Institute and the École Polytechnique. Also, the opinions that he delivered in his political capacity as a senator in the Napoleonic regime and a member of the Chambre des Pairs after the Restoration were printed in the proceedings of these bodies. Although we are not noticing his rare interventions in debates, it has seemed practical to include these episodic political and administrative pieces in the chronology of publications and to distinguish them from the scientific writings by a "P" placed before the date.

SECTION I PART 1
IN ORDER OF COMPOSITION

(1) 28 Mar. 1770. "Recherches sur les maxima et minima des lignes courbes." Referees: Borda and Condorcet. Report, 28 Apr. 1770. printed in Bigourdan (1931). Published as [1774a] below.

(2) 18 July 1770. "Sur quelques usages du calcul intégral appliqué aux différences finies." Referees: Borda and Bossut. Report, 1 Sept. 1770, is in the archives of the Academy. The last paragraph is printed in Bigourdan (1931). An early draft of [1771b].

(3) 28 Nov. 1770. "Sur une méthode pour determiner la variation de l'écliptique du mouvement des noeuds et de l'inclinaison de l'orbite des planètes." Referees: Condorcet and Bossut. Report, 12 Dec. 1770, is in the archives of the Academy.

(4) Date unrecorded, but probably Dec. 1770. "Sur la détermination de la variation de l'inclinaison et les mouvements des noeuds de toutes les planètes et principalement la variation de l'obliquité de l'écliptique." Referees: Dionis du Séjour, Bezout, and Condorcet. Report, 9 Jan. 1771, is in the archives. An entry in *PV* for that same date refers to a "Suite" to a memoir with an almost identical title to (4) and names d'Alembert and Condorcet as referees.

This is probably an error, since (4) was itself a sequel to (3); and it is more likely that d'Alembert and Condorcet were referees for (5).

(5) 19 Jan. 1771. "Sur le calcul intégral." No referees recorded in *PV*, and the archives contain no report; but see (4) above. The memoir was translated into Latin and was published in *Nova acta eruditorum* [1771a].

(6) 13 Feb. 1771. "Sur le calcul intégral des suites récurrentes et la détermination de l'orbite lunaire." Referees: Condorcet and Bossut. Report, 20 Mar. 1771, is in the archives. The register of reports further specifies the topics as "1° Sur l'intégration de l'équation linéaire d'un ordre quelconque. 2° Sur une generalization de la méthode qu'il a déjà employé pour les series récurrentes. 3° Sur une application des formules de son mémoire sur l'obliquité de l'écliptique à des équations de l'orbite lunaire."

(7) 4 May 1771. "Sur les perturbations du mouvement des planètes causées par l'action de leurs satellites." Referees: Bezout and Bossut. Report, 15 May 1771, is in the archives.

(8) 17 May 1771. "Sur le calcul intégral appliqué aux différences finies à plusieurs variables." Referees: Borda and Bossut. There is a report in the archives, but it is dated the same day as the memoir and is by Condorcet. It describes the memoir as an extension of a previous work, which must be (2), on which Borda and Bossut had reported. Evidently a revised draft of [1771b].

(9) 27 Nov. 1771. "Une théorie générale du mouvement des planètes." Referees: d'Alembert, Bezout, and Bossut. The report, 29 Jan. 1772, is in the archives.

(10) 5 Feb. 1772. "Sur les suites récurrentes appliquées à la théorie des hasards." Referees: Dionis du Séjour and Le Roi. Report, 26 Feb. 1772, is in the archives. Probably a draft of [1774b]. The clerk failed to record the term "récurro-récurrent," although the referees allude to it.

(11) 2 May 1772. "Recherches pour le calcul intégral." Referees: Le Roi and Condorcet. The report, 6 May 1772, is in the archives and identifies the topic as "sur les solutions des équations différentielles non comprises dans l'intégral général." The subject of singular solutions occupies the first part of [1775a]. This paper, (11), was probably combined with (12) and was superseded by (14).

(12) Date unrecorded in *PV*. "Nouvelles recherches sur les intégrales particulières." Referees: Le Roi and Condorcet, Report, 30 May 1772, is in the archives.

(13) 10 Mar. 1773. "Recherches sur l'intégration des équations différentielles aux différences finies et sur leur application à l'analyse des hasards." Reading continued 17 Mar. Referees: Le Roi, Borda, and Dionis du Séjour. Report, 31 Mar. 1773, is in the archives. Revised version presented "pour retenir date," 7 Dec. 1773. Published as [1776a, 1°]. This was the last memoir that Laplace submitted prior to his election to the Academy on 31 Mar., the very date of the report. The referees were dazzled and wound up their account with the following judgment: "Tel est le mémoire. ... Nous en avons dit beaucoup de choses avantageuses dans le courant de notre rapport, et nous sommes persuadés que le petit nombre de savants qui le liront en porteront le même

jugement, et nous croions même qu-ils ajouteront à nos éloges. Enfin nous ne craignons pas d'avancer que cet ouvrage donne dès à présent à son auteur un rang très distingué parmi les géomètres."

(14) 14 July 1773. "Recherches sur les solutions particulières des équations différentielles." Reading continued 21 July. Published with (15) and (17) as [1775a].

(15) 27 Apr. 1774. "Une suite du mémoire sur les équations séculaires des planètes." The reference is probably to an early draft of (17) and hence of the second part of [1775a]. The memoir of which it is said to be the sequel was probably [1776a, 2°].

(16) 31 Aug. 1774. "Sur le calcul intégral." Probably the draft of the first part of [1776c].

(17) 17 Dec. 1774. "Sur les inégalités séculaires des planètes." Presented "*pour retenir date.*" This memoir must have become the second part of [1775a] and may have been a revision of (15).

(17a) 6 Sept. 1775. Report (with d'Alembert, Borda, Bezout, and Vandermonde) on Dionis du Séjour, *Essai sur les phénomènes relatifs aux disparitions périodiques de l'anneau de Saturne* (Paris, 1776); *OC*, **14**, pp. 333–39. It is unclear why the editors of *OC* saw fit to include this one among Laplace's reports from the registers of the old Academy—and nothing else.

(18) 28 Feb. 1776. Began reading "Un mémoire sur les nombres." The only trace that exists of what was probably this piece is a remark in a letter from Lagrange thanking Laplace for sending him several memoirs and observing about one of them, "Votre démonstration du théorème de Fermat sur les nombres premiers de la forme $8n + 3$ est ingénieuse" (Lagrange to Laplace, 30 Dec. 1776, *OL*, **14**, p. 67). See also Laplace's reply, 3 Feb. 1778, acknowledging receipt of a further communication from Lagrange on Fermat's theorem (ibid., p. 74). The only other recorded involvement of Laplace with number theory was his service as referee, together with Bezout, on two committees reviewing works by the abbé Genty on prime numbers, 23 Aug. 1780 and 18 July 1781 (*PV*, **99**, fols. 219–20; and **100**, fol. 155); and on a third with Lagrange and Lacroix, 26 Mar. 1802 (*PVIF*, **2**, p. 485). See (78).

(19) 4 Dec. 1776. There is no record in the *PV*, but a marginal note in "Recherches sur le calcul intégral aux différences partielles" [1777a] states that the memoir was submitted on this date, having been read in 1773.

(20) 22 Jan. 1777. "Recherches sur la loi de la pesanteur à la surface d'un sphéroïde homogène en équilibre." Laplace deposited the text on 25 Jan. and observed at the outset that this paper was a continuation of what he had begun in [1776d]. It is transcribed in full in *PV*, **96**, fol. 17–25, and is virtually identical with the opening section of [1778a]; see *OC*, **9**, pp. 71–87. The printed version omits an undertaking that Laplace placed at the end of the manuscript memoir. He there proposed in a further memoir to investigate the figure of Saturn and the law of gravity resulting from the action of its rings. That intention he fulfilled in [1789a].

(21) 8 Mar. 1777. "Recherches sur le milieu qu'il faut choisir entre les résultats de plusieurs observations." This memoir is transcribed in *PV*, **96**, fol. 122–42. Published by Gillispie (1979).

(22) 9 Apr. 1777. "Sur la nature du fluide qui reste dans le récipient de la machine pneumatique." No trace remains of this piece.

(23) 7 May 1777. "Un mémoire sur les oscillations des fluides qui recouvrent les planètes." On 31 May, Laplace read an addition to this memoir, which he is recorded as having withdrawn. After revision, it was combined with (20) in [1778a].

(24) 15 Nov. 1777. A marginal note in [1778a] records this date for submission of the memoir, combining (20) and (23). The first installment of the investigation of tidal phenomena, "Recherches sur plusieurs points du système du monde."

(25) 18 Feb. 1778. "Un mémoire sur le calcul intégral." Probably a draft for all or part of [1780a].

(26) 13 May 1778. "Recherches sur les ondes, pour servir à son mémoire imprimé dans le volume de 1775." The reference is to [1778a], and this investigation of wave motion was printed as the final article (XXXVII) of the second sequel, [1779b]; see *OC*, **9**, pp. 301–10.

(27) 7 Oct. 1778. A marginal note in [1779a] records the submission of the memoir. No mention in *PV*. This was the first sequel on tides and on the motion of the earth.

(28) 25 Dec. 1778. A marginal note in [1779b] specifies the date of submission. See also (26), which was appended. This was the second sequel on tides and on the motion of the earth.

(29) 16 June 1779. A marginal note gives the date for submission of "Mémoire sur l'usage du calcul aux différences partielles dans la théorie des suites" [1780b].

(30) 7 July 1779. "Une addition à son mémoire actuellement sous presse sur la précession des équinoxes." The reference is to [1780c]. Laplace apparently read the addition before reading the memoir. See (31).

(31) 18 Aug. 1779. "Un mémoire sur la précession des équinoxes." The draft of [1780c].

(32) 1 Sept. 1779. "Un écrit où il répond à quelques objections faites contre son mémoire sur la précession des équinoxes, imprimé en 1776" [*sic*].

(33) 31 May 1780. "Mémoire sur le calcul aux suites appliqué aux probabilités." Draft of [1781a], submitted for publication on 31 July according to a marginal note.

(34) 21 Mar. 1781. Read "Un mémoire sur la détermination des orbites des comètes." A draft of the analytical part (Articles I–VII) of [1784b]. Laplace evidently revised the calculations (36) after learning of Herschel's "comet" (actually Uranus). It is also likely that the draft read to the Academy did not contain the instructions for application in Article VIII, an early version of which Laplace communicated to the abbé Pingré no later than November 1782. It was published with fragments of correspondence by Henry (1886); *OC*, **14**, pp. 355–68) and is virtually identical with Article VIII of the published memoir ([1784b]; *OC*, **10**, pp. 127–41).

(35) 2 May 1781. "Une application de sa méthode à la comète qui paroît actuellement."

(**36**) 13 June 1781. "Un mémoire sur une méthode de calculer l'orbite des comètes." The dossier for this session includes a note containing the calculations in Laplace's hand.

(**37**) 28 July 1781. "Des eléments de la comète de M. Herschel, déterminés par un nouveau calcul." The numerical data are transcribed in *PV*, 100, fols. 160–61.

(**38**) 21 Dec. 1781. An entry records on behalf of Lavoisier and Laplace the date on which they deposited the description of a new "pyromètre" by means of which the elongation of solid bodies under the influence of heat could be measured to an accuracy of "0.01 lignes," which is to say about 0.001 inches. Accompanying the account of the instrument, which had been constructed, was a series of experiments on the dilation of glass and metals. These must certainly have been among the earlier experiments carried out in the garden of the Arsenal in 1781 and 1782 that were published after Lavoisier's execution, "De l'action du calorique sur les corps solides, principalement sur le verre et sur les métaux, et du rallongement ou du raccourcissement dont ils sont susceptibles" [1793c], in *Oeuvres de Lavoisier*, **2** (1862), pp. 739–59. This instrument may well have been the one for which Biot reconstructed the design in his *Traité de physique* (1816), aided by the recollections of Laplace and Madame Lavoisier (*Oeuvres de Lavoisier*, **2**, p. 760).

(**39**) 2 Mar. 1782. An entry records on behalf of Lavoisier and Laplace a series of experiments already begun that show that substances emit negative electricity in passing from the liquid to the gaseous state and positive electricity in the reverse process. These experiments were clearly the basis of [1784c].

(**40**) 22 Jan. 1783. "Un mémoire sur la planète d'Herschel." The gist of this memoir, together with the results of the calculations in (42), was evidently incorporated in [1784a], part 1, nos. 14–17, pp. 28–59; but it does not appear that Laplace ever published (35), (36), (37), and (42) per se. According to an annotation on a handlist of Laplace's memoirs contained in the dossier at the Academy of Science concerning publication of *OC*, there was a memoir on "Eléments de la nouvelle planète Ouranus," in the *Mémoires de l'Académie Impériale et Royale des Sciences et Belles-Lettres de Bruxelles* in 1788. In fact, there is nothing by Laplace in that collection, but **5** (1788), pp. 22–48, does contain a memoir by F. von Zach (or "de Zach," as he is called there) entitled "Mémoire sur la nouvelle planète Ouranus," presented on 20 May 1785. Laplace and von Zach were in frequent correspondence, and Laplace supplied him with the elements of the orbit, pp. 43–44.

(**41**) 15 Mar. 1783. Report (with Bezout and d'Alembert) on Legendre's memoir on "Attraction des sphéroïdes homogènes," Transcribed in *PV*, **102**, fols. 85–87. A review of the state of the problem. The Legendre memoir was published in *SE*, **10** (1785), pp. 411–34.

(**42**) 21 May 1783. "Une note d'où il résulte que d'après ses calculs, et ceux de M. Méchain, la planète Herschel est la même chose qu'une étoile observée par Mayer et qui ne se retrouve plus." Data given in *PV*, **102**, fols. 119–20.

(**43**) 24 May 1783. "Un mémoire sur l'attraction des sphéroïdes elliptiques." Presented in addition on 31 May. This piece almost certainly constituted Part II of [1784a].

(44) 18 June 1783. "Un mémoire, fait conjointement avec M. Lavoisier, sur une nouvelle méthode de mesurer la chaleur." Laplace is recorded as having read the memoir [1783a].

(45) 25 June 1783. Lavoisier and Laplace announced that they had repeated the combustion of combustible air (hydrogen) combined with dephlogisticated air (oxygen) in the presence of several observers and obtained pure water. This demonstration, which occupied the public meeting of the Academy for St. Martin's Day, took place exactly one week after Laplace had read their joint *Mémoire sur la chaleur* [1783a] (*PV*, **102**, fol. 104 [18 June 1783], fol. 144 [25 June 1783]). Lavoisier then published this and other experiments, "Mémoire dans lequel on a pour objet de prouver que l'eau n'est point une substance simple," in *MARS* (1781/1784), 468–94; also *Oeuvres de Lavoisier*, **2** (1862), pp. 334–59.

(46) 23 July 1783. Laplace read experiments done in England on the freezing of mercury (*PV*, **102**, fol. 159). It is not clear whose experiments these were, although Laplace's interest in them almost certainly pertained to the work on heat that he continued with Lavoisier through 1783 into 1784, which is reported in "Mémoire contenant les expériences faites sur la chaleur, pendant l'hiver de 1783 à 1784, par P. S. de Laplace & A. L. Lavoisier." Published after Lavoisier's death—see (38)—and in *Oeuvres de Lavoisier*, **2** (1862), pp. 724–38.

(47) 3 Dec. 1783. Laplace requested the appointment of a commission to review his treatise on the motion and figure of the planets [1784a]. The referees, Dionis du Séjour, Borda, and Cousin, reported on 31 Jan. 1784, recommending publication under the *privilège* of the Academy, as was necessary for a work issued independently of the *MARS*. The report is given in *PV*, **103**, fols. 72–76.

(48) 11 Aug. 1784. "Un mémoire sur l'équilibre des fluides sphéroïdes." The draft of [1785b].

(49) 25 June 1785. A memoir "Sur lés probabilités," intended as sequel to his "mémoire de 1782." The reference is to [1785a], the memoir on approximate solutions to problems involving functions containing terms raised to very high powers, and the present memoir was the draft of its sequel [1786b]. Laplace continued the reading on 28 June.

(50) 19 Nov. 1785. "Un mémoire sur les inégalités séculaires des planètes." Condorcet recorded the memoir on this date, and Laplace read it on 23 Nov. This is the draft of [1787a].

(51) 30 Nov. 1785. "Un mémoire sur la population de la France." The draft of [1786c].

(52) 6 May 1786. Laplace presented a theorem on the motions of Jupiter and Saturn "*pour retenir date.*" On 10 May he read a draft of the first memoir on the theory of Jupiter and Saturn [1788a].

(53) 15 July 1786. "Un 2e mémoire sur la théorie de Jupiter et de Saturne." The draft of [1788b].

(54) 21 July 1787. On behalf of a committee consisting also of Cousin and Legendre, Laplace read a *compte-rendu* of R.-J. Haüy, *Exposition raisonée de*

la théorie de l'électricité et du magnétisme, d'après les principes d'Aepinus (Paris, 1787). The account is very appreciative, and the report is transcribed in *PV*, **106**, fols. 290–93.

(55) 19 Dec. 1787. "Un mémoire sur l'équation séculaire de la lune." A first draft of [1788c].

(56) 2 Apr. 1788. "Un mémoire sur l'équation séculaire de la lune." Probably a revision rather than an extension of (55), for the memoir [1788c] is a brief one.

(57) 26 Apr. 1789. A commission composed of Lagrange, Lalande, and Méchain presented a *compte-rendu* of tables for Jupiter and Saturn prepared by Delambre, which compare the predictions from Laplace's theory of the two planets to the record of observations. The report is transcribed in *PV*, **108**, fols. 92–99.

(58) 18 July 1789. "Un mémoire sur l'inclinaison de l'écliptique." This is probably the draft of Articles II–VII of [1793a].

(59) 17 Apr. 1790. "Un mémoire sur la théorie des satellites de Jupiter." The draft of [1791a].

(60) 15 Dec. 1790. Laplace began "Un mémoire sur le flux et le reflux de la mer." Delayed in publication [1797a].

(61) 21 Jan. 1796 (1 pluviôse an IV). Laplace presented "Un mémoire sur les mouvements des corps célestes autour de leurs centres de gravité." *PVIF*, **1**, p. 6. Draft of [1798a]. Readied for publication on 5 Jan. 1797 (16 nivôse an V).

(62) 7 Oct. 1796. (16 vendémiaire an V). Report (with Lagrange) on two memoirs by Flaugergues. "De l'aberration de la lumière" and "Du phénomène de l'apparence de l'étoile sur le disque de la lune dans les occultations." *PVIF*, **1**, pp. 114–15.

(63) 20 Apr. 1797 (1 floréal an V). Read a "Mémoire sur les équations séculaires du mouvement des noeuds et de l'apogée de l'orbite lunaire et sur l'aberration des étoiles." *PVIF*, **1**, 203. Draft of [1798b].

(64) 10 Jan. 1798 (21 nivôse an VI). Read the preamble of a "Mémoire sur les équations séculaires du mouvement de la lune, de son apogée et de ses noeuds." *PVIF*, **1**, p. 330. Text of [1799a].

(65) 21 Dec. 1798 (I nivôse an VII). Laplace, Lacepède, and Fourcroy submitted a report on the questions to be presented to the Institut d'Égypte. The report was combined with those from the second and third classes and was printed in *Histoire de la classe des sciences morales et politiques de l'Institut de France*, **3** (prairial an IX [May–June 1801]), pp. 5–19.

(66) 5 Apr. 1799 (16 germinal an VII). A report (with Lagrange) on several memoirs of Parceval on "le calcul intégral aux différences partielles." *PVIF*, **1**, 546–47.

(67) 7 Sept. 1799 (21 fructidor an VII). Presented copies of *Exposition du systéme du monde* and *Mécanique céleste*. *PVIF*, **1**, p. 619.

(68) 2 Nov. 1799 (11 brumaire an VIII). A report (with Coulomb and Lefèvre-Gineau—it is not clear that Laplace was the author) on a memoir by Libes on the role of caloric in elasticity. *PVIF*, **2**, 21–22.

(69) 12 Nov. 1799 (21 brumaire an VIII). A report (with Napoleon Bonaparte and Lacroix) on a memoir of Biot on "les équations aux différences mêlées." *PVIF*, **2**, pp. 30–32. See also the joint report with Prony (read by the latter on 6 frimaire an VIII [27 Nov. 1799], *PVIF*, **2**, pp. 45–48), on Biot's memoir "Considerations sur les intégrales des équations aux differénces finies," *PVIF*, **2**, pp. 45–48, published in full, *MI*, **3** (prairial an IX [May–June 1801]), "Histoire," pp. 12–21.

(70) 2 Mar. 1800 (11 ventôse an VIII). Read a "Mémoire sur le mouvement des orbites des satellites de Saturne et d'Uranus." *PVIF*, **2**, p. 118. Abstracted [1800b]. Draft of [1801a].

(71) 15 June 1800 (26 prairial an VIII). Read a "Mémoire sur la théorie de la lune." *PVIF*, **2**, p. 177. Draft of [1801b].

(72) 20 June. 1800 (1 messidor an VIII). Announced Bouvard's application of a new equation contained in the previous memoir (71) to observations by Maskelyne, yielding a flattening of $1/314$. *PVIF*, **2**, p. 179.

(73) 2 Dec. 1800 (11 frimaire an IX). Proposed continuing in the *Mémoires de l'Institut* the notes on the French population contained in the final volumes of the Academy under the Old Regime. *PVIF*, **2**, p. 274.

(74) 10 June 1801 (21 prairial an IX). Read a "Mémoire sur la théorie de la lune." *PVIF*, **2**, p. 359. Not separately published. May have been incorporated in *Mécanique céleste*, Book VII (Volume **3**, 1802).

(75) 12 Nov. 1801 (21 brumaire an X). Report (with Delambre) on a memoir "Sur la théorie de Mars" by Lefrançois-Lalande. *PVIF*, **2**, pp. 426–29.

(76) 26 Jan. 1802 (6 pluviôse an X). Read a "Mémoire sur une inégalité à longue période, qu'il vient de découvrir dans le mouvement de la lune" *PVIF*, **2**, p. 457. Not separately published at this time. May have been incorporated in *Mécanique céleste*, Book VII (Volume **3**, 1802). Laplace returned to this topic in [1811c] and [1812a]. See also (94).

(77) 12 Mar. 1802 (21 ventôse an X). Read a "Mémoire sur la théorie lunaire." *PVIF*, **2**, p. 476. Not separately published. May have been incorporated in *Mécanique céleste*, Book VII (Volume **3**, 1802).

(78) 27 Mar. 1802 (6 germinal an X). Report (with Lagrange and Lacroix) on a memoir of Genty on number theory. *PVIF*, **2**, p. 485.

(79) 17 Nov. 1802 (26 brumaire an XI). Announces measures that the government has adopted to resume making exact estimates of the size of the population by taking samples in various regions and calculating the factor by which the annual number of births is to be multiplied. *PVIF*, **2**, p. 595.

(80) 29 Dec. 1802 (8 nivôse an XI). Presented Volume **3** of *Traité de mécanique céleste*. *PVIF*, **2**, p. 606.

(81) 2 May 1803 (12 floréal an XI). Read a set of observations on the tides, upon which a committee consisting of himself, Levêque, and Rochon was appointed. *PVIF*, **2**, p. 659. See [1803a].

(82) 12 Sept. 1803 (25 fructidor an XI). Read a "Mémoire sur les tables de Jupiter et la masse de Saturne." *PVIF*, **2**, p. 703. Draft of [1804a].

(83) 27 May 1805 (7 prairial an XIII). Presented Volume **4** of *Mécanique céleste* (the minute mistakenly says Volume **14**). *PVIF*, **3**, p. 216.

(84) 14 Oct. 1805 (22 vendémiaire an XIV). Read a "Mémoire sur la diminution de l'obliquité de l'écliptique." *PVIF*, **3**, p. 262. There is no record of publication.

(85) 23 Dec. 1805 (2 nivôse an XIV). Read a "Mémoire sur les tubes capillaires." *PVIF*, **3**, p. 293. Published in part [1806a].

(86) 28 Apr. 1806. Presented "Théorie de l'action capillaire." *PVIF*, **3**, p. 344. Originally issued (1806) as a separate booklet under the above title, this piece was incorporated in later printings of *Mécanique céleste*, Volume **4**, as the first Supplement to Book X. *OC*, **4**, pp. 349–417. Abstracted [1806c].

(87) 29 Sept. 1806. Read a "Suite à sa théorie de l'action capillaire." *PVIF*, **3**, p. 431. Abstracted [1807a]. A printed copy, separately issued although sometimes bound with (86), was presented to the Institut on 6 July 1807. *PVIF*, **3**, p. 353. This piece was incorporated in later printings of *Mécanique céleste*, Volume **4**, as the second Supplement to Book X. *OC*, **4**, pp. 419–98.

(88) 24 Nov. 1806. Read a memoir on "L'adhésion des corps à la surface des fluides." *PVIF*, **3**, p. 451. Printed in part as [1806d].

(89) 21 Mar. 1808. Presented a copy of the 3rd edition of *Exposition du système du monde*. *PVIF*, **4**, p. 36.

(90) 17 Aug. 1808. Presented *Supplément au Traité de mécanique céleste* to the Bureau des Longitudes. This appendix concerns the theory of planetary perturbations developed in Books II and VI and was incorporated in later printings of Volume **3** (*OC*, **3**, pp. 325–50). Laplace also presented a copy to the Institut on 26 Sept. 1808. *PVIF*, **4**, p. 106. The last section (5) corrects an error in the sign in the expressions for the fifth power of the eccentricities and inclinations of the orbits in the theory of the inequalities of Jupiter and Saturn, *OC*, **3**, pp. 349–50.

(91) 19 Dec. 1808. Report (with Haüy, Chaptal, and Berthollet) on the memoir of Malus, "Sur divers phénomènes de la double réfraction de la lumière." Laplace's report is printed in *PVIF*, **4**, pp. 145–47; and in *OC*, **14**, pp. 321–26. *RSC* **22**.

(92) 30 Jan. 1809. Read a memoir on "La loi de la réfraction extraordinaire de la lumière dans les milieux transparents." *PVIF*, **4**, p. 159. Abstracted [1809b]. Published as [1810a].

(93) 10 Apr. 1809. Reported (with Lacroix) on a memoir of Poisson on "La rotation de la terre." *PVIF*, **4**, pp. 190–92.

(94) 18 Sept. 1809. Read a memoir on "La libration de la lune." *PVIF*, **4**, p. 253. This memoir was probably combined with (76) in [1811c] and [1812a].

(95) 9 Apr. 1810. Read a report on "Les probabilités." *PVIF*, **4**, p. 341. A draft of [1810b].

(96) 21 May 1810. Reported (with Biot and Arago) on a memoir by Daubuisson on "La mesure des hauteurs par le baromètre." *PVIF*, **4**, pp. 350–52.

(97) 29 Apr. 1811. Read a memoir on "Les intégrales définies." *PVIF*, **4**, p. 475. A draft of [1811a].

(98) 23 Mar. 1812. Presented the "première partie" of *Théorie analytique des probabilités*. *PVIF*, **5**, p. 34.

(99) 29 June 1812. Presented the "seconde partie" of *Théorie analytique des probabilités*. *PVIF*, **5**, p. 69. The first edition was thus issued in at least two installments.

(100) 24 May 1813. Presented the 4th edition of *Exposition du systéme du monde*. *PVIF*, **5**, p. 214.

(101) 2 Aug. 1813. Read a memoir on "Les éléments des variations des orbites planétaires." *PVIF*, **5**, p. 235. There is no record of publication.

(102) 14 Feb. 1814. Presented *Essai philosophique sur les probabilités* (1814). *PVIF*, **5**, p. 3l6. Incorporated with (104) as its introduction. The version printed in *OC*, **7**, pp. v–cliii, is (117), the 4th edition (1819).

(103) 8 Aug. 1814. Read a memoir on "La probabilité des témoignages." *PVIF*, **5**, p. 386. Incorporated in (104), the 2nd edition of *Théorie analytique des probabilités*, Book II, Chapter 11; *OC*, **7**, pp. 455–85.

(104) 14 Nov. 1814. Presented the 2nd edition of *Théorie analytique des probabilités*, incorporating (102) and (103), together with three minor additions. *PVIF*, **5**, p. 422. *OC*, **7**, pp. 471–93.

(105) 10 July 1815. Read a memoir on "Les marées." *PVIF*, **5**, p. 527. A draft of [1815c].

(106) 18 Sept. 1815. Read a memoir on "Les probabilités, dans lequel il détermine la limite de l'erreur qui peut rester après qu'on a déterminé les valeurs les plus probables des inconnues." *PVIF*, **5**, p. 554. The draft of [1815a].

(107) 26 Aug. 1816. Presented a supplement to *Théorie analytique des probabilités*. *PVIF*, **6**, p. 73. Incorporated in the 3rd edition (1820) as Supplement 1. *OC*, **7**, pp. 497–530. See (118).

(108) 28 Oct. 1816. Read a "Note sur la pendule." *PVIF*, **6**, p. 108. Probably the draft of [1816a].

(109) 25 Nov. 1816. Read "Note sur l'action réciproque des pendules et sur la vitesse du son dans les diverses substances," *PVIF*, **6**, p. 113. A draft of [1816b].

(110) 23 Dec. 1816. Read a "Note sur la vitesse du son." *PVIF*, **6**, p. 131. Related to [1816c] and [1816d]. See the marginal note in the latter.

(111) 4 Aug. 1817. Read a memoir on "L'application du calcul des probabilités aux opérations géodésiques." *PVIF*, **6**, p. 208. A draft of [1817a].

(112) 2 Feb. 1818. Presented a second Supplement to *Théorie analytique des probabilités*. *PVIF*, **6**, p. 263. The Supplement consists of (111) together with additions and was incorporated in the 3rd edition, *Théorie analytique des probabilités* (1820). *OC*, **7**, pp. 531–80. See (118).

(113) 18 May 1818. Read a "Mémoire sur la rotation de la terre." *PVIF*, **6**, p. 316. The draft of [1819a].

(114) 3 Aug. 1818. Read a memoir "Sur la figure de la terre et la loi de la pesanteur à sa surface." *PVIF*, **6**, p. 350. Abstracted in [1818b] and published in [1819b].

(115) 26 May 1819. Read a memoir "Sur la figure de la terre" before the Bureau des Longitudes; see the marginal note in *CT* (1822/1820), p. 284. The draft of [1820e].

(116) 13 Sept. 1819. Read a memoir "Considérations sur les phénomènes capillaires." *PVIF*, **6**, p. 487. Presumably the draft of [1819h].

(117) 25 Oct. 1819. Presented the 4th edition of *Essai philosophique sur les probabilités*. *PVIF*, **6**, p. 504. Replaced the 1st edition (102) as introduction to *Théorie analytique des probabilités* (104) in the 3rd edition (1820). *OC*, **7**, pp. v–cliii.

(118) 20 Dec. 1819. Read a "Mémoire sur l'application du calcul des probabilités aux opérations géodésiques." *PVIF*, **6**, p. 515. The draft of what became the 3rd Supplement to the 3rd edition of *Théorie analytique des probabilités* (1820). *OC*, **7**, pp. 581–616. Abstracted in [1819] and printed as [1819g].

(119) 19 Jan. 1820. Read a memoir "Sur les inégalités lunaires dues à l'aplatissement de la terre" before the Bureau des Longitudes. See the marginal note in [1820a].

(120) 28 Feb. 1820. Presented the 3rd edition of *Théorie analytique des probabilités*, containing the revised Introduction (117) and three supplements (107), (112), and (118).

(121) 29 Mar. 1820. Read "Sur le perfectionnement de la théorie et des tables lunaires" before the Bureau des Longitudes. See the marginal note in [1820b].

(122) 12 Apr. 1820. Read "Sur l'inégalité lunaire à longue période, dépendante de la différence des deux hemisphères terrestres" before the Bureau des Longitudes. See the marginal note in [1820c].

(123) 10 Sept. 1821. Read a "Mémoire sur l'attraction des corps sphériques et sur la répulsion des fluides élastiques." *PVIF*, **7**, p. 222.

(124) 12 Dec. 1821. Presented a memoir on elastic fluids and the speed of sound to the Bureau des Longitudes. See the marginal note in [1822a].

(125) 17 Mar. 1823. Presented Book XI of *Mécanique céleste*. *PVIF*, **7**, p. 457. *OC*, **5**, pp. 6–96.

(126) 21 Apr. 1823. Presented Book XII of *Mécanique céleste*. *PVIF*, **7**, p. 480. *OC*, **5**, pp. 99–160.

(127) 8 Sept. 1823. Presented a memoir "Sur le flux et le reflux de la mer." *PVIF*, **7**, p. 538. Published as [1823a].

(128) 5 Jan. 1824. Presented the 5th edition of *Système du monde*. *PVIF*, **8**, p. 3.

(129) 9 Feb. 1824. Presented Book XIII of *Mécanique céleste*. *PVIF*, **8**, p. 24. *OC*, **5**, pp. 164–269.

(130) 26 July 1824. Presented Book XIV of *Mécanique céleste*. *PVIF*, **8**, p. 117. *OC*, **5**, pp. 273–323.

(131) 13 Dec. 1824. Presented Book XV of *Mécanique céleste*. *PVIF*, **8**, p. 162. *OC*, **5**, pp. 327–87.

(132) 7 Feb. 1825. Laplace, together with his son, presented a fourth supplement to *Théorie analytique des probabilités*. *PVIF*, **8**, p. 182. *OC*, **7**, pp. 617–45.

(133) 16 Aug. 1825. Presented Volume **5** of *Mécanique céleste*, Book XVI being the final book. *PVIF*, **8**, p. 261. *OC*, **5**, pp. 389–465.

(134) 23 July 1827. The Academy received a Supplement to *Mécanique céleste*, Volume **5**; the manuscript had been found among Laplace's papers after his death. *PVIF*, **8**, p. 571. *OC*, **5**, pp. 469–505.

SECTION I, PART 2

IN ORDER OF PUBLICATION

[1771a] "Disquisitiones de calculo integrale," in *Nova acta eruditorum, Anno 1771* (Leipzig, 1771), pp. 539–59; not in *OC*. A draft was presented to the Academy on 19 Jan. 1771; see (5).

[1771b] "Recherches sur le calcul intégral aux différences infiniment petites, et aux différences finies," in *Mélanges de philosophie et de mathématiques de la Société royale de Turin, pour les années 1766–1769* (*Miscellanea Taurinensia*, IV), date of publication not given but probably 1771, pp. 273–345. A typographical error numbers pp. 273–88 as 173–88. Laplace read an early draft to the Academy on 18 July 1770 and a revised version on 17 May 1771; see (2) and (8). Not in *OC*.

[1774a] "Disquisitiones de maximis et minimis, fluentium indefinitarum," in *Nova acta eruditorum, Anno 1772* (Leipzig, 1774), pp. 193–213. Not in *OC*. First draft read to the Academy on 28 Mar. 1770 (1).

[1774b] "Mémoire sur les suites récurro-récurrentes et sur leurs usages dans la théorie des hasards," in *SE*, **6** (1774), pp. 353–71; *OC*, **8**, pp. 5–24. A draft was presented on 5 Feb. 1772. Referees: Dionis du Séjour and Le Roi, who reported on 26 Feb. 1772 (10).

[1774c] "Mémoire sur la probabilité des causes par les événements," in *SE*, **6** (1774), pp. 621–56; *OC*, **8**, pp. 27–65. *PV* contains no mention of this memoir. It seems probable, however, that it was composed between March and December 1773, concurrently with the revision of [1776a, 1°].

[1775a] "Mémoire sur les solutions particulières des équations différentielles et sur les inégalités séculaires des planètes," in *MARS* (1772, part 1/1775), pp. 343–77; *OC*, **8**, pp. 325–66. Laplace read the first part (*OC*, **8**, 325–54) on 14 and 21 July 1773 (14), having started the topic in a paper of 2 May 1772 (11). The second part (ibid., pp. 354–66) was registered on 17 Dec. 1774 (17).

[1775b] "Addition au mémoire sur les solutions particulières," in *MARS* (1772, part 1/1775), pp. 651–56; *OC*, **8**, pp. 361–66.

[1776a] "Recherches, 1°, sur l'intégration des équations différentielles aux différences finies, et sur leur usage dans la théorie des hasards. 2°, sur le principe de la gravitation universelle, et sur les inégalités séculaires des planètes qui en dépendent," in *SE* (1773/1776), pp. 37–232; *OC*, printed as two memoirs, **8**, pp. 69–197, 198–275. A note printed in the margin (*SE* [1773/1776], p. 37) gives 10 Feb. 1773 as the date on which 1° was read. The register of *PV* gives 10 Mar., with reading continued on 17 Mar. See (13). We do not know when Laplace readied 2° for publication, although it must have been before 27 Apr. 1774, when he presented a further memoir called a "Suite"; see (15) and (17). As printed, 2° probably contains elements from (3), (4), (6), (7), and (9), but most of it goes beyond anything suggested by these titles.

[1776b] "Mémoire sur l'inclinaison moyenne des orbites des comètes, sur la figure de la terre, et sur les fonctions," in *SE* (1773/1776), pp. 503–40; *OC*, **8**, pp. 279–321. There is no record of when Laplace presented these topics.

[1776c] "Recherches sur le calcul intégral et sur le système du monde," in *MARS* (1772, part 2/1776), pp. 267–376; *OC*, **8**, pp. 369–477. Laplace read the first part on 31 Aug. 1774 (16).

[1776d] "Additions aux recherches sur le calcul intégral et sur le système du monde," in *MARS* (1772, part 2/1776), pp. 533–54; *OC*, **8**, pp. 478–501.

[1777a] "Recherches sur le calcul intégral aux différences partielles," in *MARS* (1773/1777), pp. 341–402; *OC*, **9**, pp. 5–68. A marginal note says that this memoir was read in 1773 and was submitted for publication on 4 Dec. 1776 (19).

[1778a] "Recherches sur plusieurs points du systéme du monde," in *MARS* (1775/1778), pp. 75–182; *OC*, **9**, pp. 71–183. A combination of (20) and (23), the latter revised and submitted on 15 Nov. 1777 (24).

[1779a] "Recherches sur plusieurs points du système du monde" (Suite), in *MARS* (1776/1779), pp. 177–267; *OC*, **9**, pp. 187–280. A marginal note specifies that this continuation of [1778a], with which the articles are numbered consecutively, was submitted on 7 Oct. 1778.

[1779b] "Recherches sur plusieurs points du système du monde" (Suite), in *MARS* (1776/1779), pp. 525–52; *OC*, **9**, pp. 283–310. Submitted on 25 Dec. 1778. See (26) and (28).

[1780a] "Mémoire sur l'intégration des équations différentielles par approximation," in *MARS* (1777/1780), pp. 373–97; *OC*, **9**, pp. 357–79. There is no record of anything to which this is more likely to have corresponded than (25), read on 18 Feb. 1778.

[1780b] "Mémoire sur l'usage du calcul aux différences partielles dans la théorie des suites," in *MARS* (1777/1780), pp. 99–122; *OC*, **9**, pp. 313–35. A marginal note dates the submission 16 June 1779 (29).

[1780c] "Mémoire sur la précession des équinoxes," in *MARS* (1777/1780), pp. 329–345; *OC*, **9**, pp. 339–54. A marginal note is confirmed by the entry in *PV* (31) that Laplace read this memoir formally on 18 Aug. 1779, even though he had already read an "addition" to it on 7 July (30). Apparently it elicited some criticism and discussion (32).

[1781a] "Mémoire sur les probabilités," in *MARS* (1778/1781), pp. 227–32; *OC*, **9**, pp. 383–485. On 31 May 1780 Laplace read a "Mémoire sur le calcul aux suites appliqué aux probabilités" (33) and submitted the memoir for publication on 19 July, according to a marginal note.

[1782a] "Mémoire sur les suites," in *MARS* (1779/1782), pp. 207–309; *OC*, **10**, pp. 1–89. Not mentioned in *PV*.

[1783a] *Mémoire sur la chaleur*, written with Lavoisier. This separate printing issued from the Imprimerie Royale. The memoir as published by the Academy a year later is in *MARS* (1780/1784), pp. 355–408; *OC*, **10**, pp. 149–200; also in *Oeuvres de Lavoisier*, ed. J. B. Dumas, **2** (Paris, 1862), pp. 283–333.

[1784a] *Théorie du mouvement et de la figure elliptique des planètes* (Paris, 1784). Not in *OC*. Part I of this treatise probably represents a revision and expansion of (9), together with material on comets from (35), (36), (37), (40), and (42), and other up-to-date matter. Part II almost surely consists of a memoir "Sur l'attraction des sphéroïdes elliptiques" (43) that Laplace read on 24 and 31 May 1783 and no doubt draws also on the Legendre memoir that he discussed in (41). Laplace requested a commission of review on 3

Dec. 1783 (47). A finished copy was presented to the Academy on 24 Feb. 1784 (*PV*, **103**, fol. 37).

[1784b] "Mémoire sur la détermination des orbites des comètes," in *MARS* (1780/1784), pp. 13–72; *OC*, **10**, pp. 93–146. The same title as (34), probably enlarged.

[1784c] "Mémoire sur l'électricité qu'absorbent les corps qui se réduisent en vapeurs," written with Lavoisier, in *MARS* (1781/1784), pp. 292–94; *OC*, **10**, pp. 203–5: also in *Oeuvres de Lavoisier*, **2**, pp. 374–76.

[1785a] "Mémoire sur les approximations des formules qui sont fonctions de très grands nombres," in *MARS* (1782/1785), pp. 1–88; *OC*, **10**, pp. 209–91. Not mentioned in *PV*.

[1785b] "Théorie des attractions des sphéroïdes et de la figure des planètes," in *MARS* (1782/1785), pp. 13–196; *OC*, **10**, pp. 341–419. Read on 11 Aug. 1784 (48).

[1786a] "Mémoire sur la figure de la terre," in *MARS* (1783/1786), pp. 17–46; *OC*, **11**, pp. 3–32. Not mentioned in *PV*.

[1786b] "Mémoire sur les approximations des formules qui sont fonctions de très grands nombres" (Suite), in *MARS* (1783/1786), pp. 423–67; *OC*, **10**, pp. 295–338. Laplace read the draft on 25 and 28 June 1785 (49).

[1786c] "Sur les naissances, les mariages et les morts à Paris, depuis 1771 jusqu'en 1784, et dans toute l'étendue de la France, pendant les années 1781 et 1782," in *MARS* (1783/1786), pp. 693–702; *OC*, **11**, pp. 35–46. Read on 30 Nov. 1785 (51).

[1787a] "Mémoire sur les inégalités séculaires des planètes et des satellites," in *MARS* (1784/1787), pp. 1–50; *OC*, **11**, pp. 49–92. Recorded on 19 Nov. 1785 and read on 23 Nov. (50).

[1788a] "Théorie de Jupiter et de Saturne," in *MARS* (1785/1788), pp. 33–160; *OC*, **11**, pp. 95–207. Laplace read the draft on 10 May 1786 (52).

[1788b] "Théorie de Jupiter et de Saturne" (Suite), in *MARS* (1786/1788), pp. 201–34; *OC*, **11**, pp. 211–39. Read on 15 July 1786 (53). This memoir is Part III of [1788a].

[1788c] "Sur l'équation séculaire de la lune," in *MARS* (1786/1788), pp. 235–64; *OC*, **11**, pp. 243–71. The draft was read on 19 Dec. 1787 (55) and was revised on 2 Apr. 1788 (56).

[1789a] "Mémoire sur la théorie de l'anneau de Saturne," in *MARS* (1787/1789), pp. 249–67; *OC*, **11**, pp. 275–92.

[1789b] "Mémoire sur les variations séculaires des orbites des planètes," in *MARS* (1787/1789), pp. 267–79; *OC*, **11**, pp. 297–306.

[1790a] "Sur la théorie des satellites de Jupiter," in *CT* (1792/1790), pp. 273–86. Not in *OC*.

[1791a] "Théorie des satellites de Jupiter," in *MARS* (1788/1791), pp. 249–364; *OC*, **11**, pp. 309–411. Read on 17 Apr. 1790 (59).

[1793a] "Sur quelques points du système du monde," in *MARS* (1789/1793), pp. 1–87; *OC*, **11**, pp. 477–558. On 18 July 1789 Laplace read a paper on the inclination of the plane of the ecliptic (58), which may probably have been the draft of Articles II–VII; *OC*, **11**, pp. 481–93.

[1793b] "Théorie des satellites de Jupiter" (Suite), in *MARS* (1789/1793), pp. 237–96; *OC*, **11**, pp. 415–73. Not mentioned in *PV*.

[1793c] "De l'action du calorique sur les corps solides, principalement sur le verre et sur les métaux," in *Oeuvres de Lavoisier*, **2** (1862), pp. 739–59, written by Laplace and Lavoisier. For the original printing, undated but ca. 1803, see *Dictionary of Scientific Biography* **8**, p. 87. The year assigned here is that in which Lavoisier composed the report of his experiments with Laplace in the winter of 1781–1782 (38) and also those reported in [1793d]. Not in *OC*.

[1793d] "Mémoire contenent les expériences faites sur la chaleur pendant l'hiver de 1783 à 1784," in *Oeuvres de Lavoisier*, **2** (1862), pp. 724–38, written by Laplace and Lavoisier. See [1793c]. Not in *OC*.

[P1796] *Discours prononcé aux deux conseils ... au nom de l'Institut national des sciences et des arts*. 17 Sept. 1796 (1er jour complémentaire an IV). Report on the first year of the Institut. Not in *OC*.

[1797a] "Mémoire sur le flux et le reflux de la mer," in *MARS* (1790/1797), pp. 45–181; *OC*, **12**, pp. 3–126. Laplace began the reading on 15 Dec. 1790 (60).

[1797b] "Sur le mouvement de l'apogée de la lune et sur celui de ses noeuds," in *BSPM*, **1** (1797), pp. 22–23. Not in *OC*. *RSC* 1.

[1797c] "Sur les équations séculaires du mouvement de la lune," in *BSPM*, **1** (1797), pp. 99–101. Not in *OC*. *RSC* 2.

[1798a] "Mémoire sur les mouvements des corps célestes autour de leurs centres de gravité," in *MI*, **1** (an IV [1795–1796]/thermidor an VI [July–Aug. 1798]), pp. 301–76; *OC*, **12**, pp. 129–87. Presented on 21 Jan. 1796 (6l).

[1798b] "Sur les équations séculaires des mouvements de l'apogée et des noeuds de l'orbite lunaire," in *CT* (an VIII [1799–1800]/pluviôse an VI [Jan.–Feb. 1798]), pp. 362–70; *OC*, **13**, pp. 3–14. Read to the Institute on 20 Apr. 1797 (63). The calculations were deferred to [1799a].

[1798c] "Mémoire sur la détermination d'un plan qui reste toujours paralléle à lui-même, dans le mouvement d'un système de corps agissant d'une manière quelconque les uns sur les autres et libres de toute action étrangère," in *JX*, **2**, 5e cahier (prairial an VI [May–June 1798]), pp. 155–59; *OC*, **14**, pp. 3–7.

[1798d] "Sur les plus grandes marées de l'an IX," in *CT* (an IX [1800–1801]/fructidor an VI [Aug.-Sept. 1798]), pp. 213–18; *OC*, **13**, pp. 15–19.

[1799a] "Mémoire sur les équations séculaires des mouvements de la lune, de son apogée et de ses noeuds," in *MI*, **2** (an VII [1798–1799]/fructidor an VII [Aug.-Sept. 1799]), pp. 126–82; *OC*, **12**, pp. 191–234. Read on 10 Jan. 1798 (64). The calculations for [1798b].

[1799b] "Sur la mécanique," in *JX*, **2**, 6e cahier (thermidor an VII [July–Aug. 1799]), pp. 343–44; *OC*, **14**, pp. 8–9.

[1799c] "Sur quelques équations des tables lunaires," in *CT* (an X [1801–1802]/fructidor an VII [Aug.-Sept. 1799]), pp. 36l–65; *OC*, **13**, pp. 20–24.

[1799d] *Allgemeine geographische Ephemeriden*, **4**, no. 1 (July 1799), pp. 1–6. Not in *OC*. Laplace's calculation, requested by the editor, F. X. von Zach, in

support of the statement in *Exposition du système du monde* (1796, **2**, p. 305) that the force of gravity exerted by a luminous body 250 times larger than the sun would prevent light rays escaping from its surface. The proof is translated in Hawking and Ellis (1973), appendix A, pp. 365–68. I owe this reference to the kindness of John Stachel.

[1800a] "Sur l'orbite du dernier satellite de Saturne," in *BSPM*, **2** (1800), p. 109; not in *OC*.

[1800b] "Sur les mouvements des orbites des satellites de Saturne et d'Uranus," in *CT* (an XI [1802–1803]/messidor an VIII [June–July 1800]), pp. 485–89; not in *OC*. Read on 1 Mar. 1800 (70); an abstract of [1801a].

[1800c] "Sur la théorie de la lune," in *CT* (an XI [1802–1803]/1800), pp. 504–6; not in *OC*. An abstract of [1801b] on a periodic inequality in the nutation of the lunar orbit.

[1800d] "Leçons de mathématiques professées à l'École normale en 1795," in *Séances de l'École normale* (an VIII [1799–1800]), I, pp. 16–32, 268–80, 381–93; II, pp. 3–23, 130–34; II, pp. 116–29, 302–18; III, pp. 24–39; IV, pp. 32–70, 223–63; V, pp. 201–19; VI, pp. 32–73; reprinted in *JX*, **2**, 7e and 8e cahiers (June 1812), pp. 1–172; *OC*, **14**, pp. 10–177, *RSC* 8. Reprinted with annotation in Dhombres (1992).

[P1800] Rapport sur la situation de l'École polytechnique, 24 Dec. 1800 (3 nivôse an IX). Submitted to the minister of the interior on behalf of the Conseil de Perfectionnement. Not in *OC*.

[1801a] "Mémoire sur les mouvements des orbites des satellites de Saturne et d'Uranus," in *MI*, **3** (prairial an IX [May–June 1801]), pp. 107–27; *OC*, **12**, pp. 237–53. Read on 1 Mar. 1800 (70).

[1801b] "Mémoire sur la théorie de la lune," in *MI*, **3** (prairial an IX [May–June 1801]), pp. 198–206; *OC*, **12**, pp. 257–63. Read on 15 June 1800 (71). *RSC* 11.

[1801c] "Sur la théorie de la lune," in *CT* (an XII [1803–1804]/fructidor an IX [Aug.–Sept. 1801]), pp. 493–501; not in *OC*. On an inequality in the lunar parallax; an abstract of [1801b].

[1801d] "Sur un problème de physique, relatif à l'électricité," in *BSPM*, **3**, no. 51 (prairial an IX [May–June 1801]), pp. 21–23. Not in *OC*. A concluding editor's note reads; "Nous devons au C. Laplace cette application à l'électricité, des formules relatives à la théorie de la figure de la terre."

[1803a] "Sur les marées," in *BSPM*, **3**, no. 74 (floréal an XI [Apr.–May 1803]), p. 106; not in *OC*. *RSC* 13. See (81).

[1803b] "Mémoire sur le mouvement d'un corps qui tombe d'une grande hauteur," in *BSPM*, **3**, no. 75 (prairial an XI [May–June 1803]), pp. 109–15; *OC*, **14**, pp. 267–77.

[1803a] Two notes on the relation of pressure to temperature among the molecules of an enclosed gas, contributed to Claude-Louis Berthollet, *Essai de statique chimique*, 2 vols. (1803), **1**, pp. 245–47. n. 5; **1**, pp. 522–23, n. 18; *OC*, **14**, pp. 329–32.

[1804a] "Sur les tables de Jupiter et sur la masse de Saturne," in *CT* (an XIV [1805–1806]/nivôse an XII [Dec. 1803–Jan. 1804]), pp. 435–40; *OC*, **13**,

pp. 25–29. Laplace read a draft at the Institute on 12 Sept. 1803 (82). *RSC* 16.

[1804b] "Sur la théorie de Jupiter et de Saturne," in *CT* (an XV [1806–1807]/frimaire an XIII [Nov.–Dec. 1804]), pp. 296–307; *OC*, **13**, pp. 30–40. *RSC* 20.

[1805] "Rapport sur le projet ... portant rétablissement du calendrier grégorien." Sénat conservateur, 9 Sept. 1805 (22 fructidor an XIII), in *AP*, **8**, pp. 722–23; not in *OC*.

[1806a] "Sur la théorie des tubes capillaires," in *JP*, **62** (Jan. 1806), pp. 120–28; *OC*, **14**, pp. 217–27. An abstract of (85), read on 23 Dec. 1805. *RSC* 17.

[1806b] "Sur l'attraction et la répulsion apparente des petits corps qui nagent à la surface des fluides," in *JP*, **63** (Sept. 1806), pp. 248–52; *OC*, **14**, pp. 228–32. Read in part on 29 Sept. 1806 (87). *RSC* 18.

[1806c] "Sur l'action capillaire," in *JP*, **63** (Dec. 1806), pp. 474–77; also *CX*, **1** (1804–1808), pp. 246–56; *OC*, **14**, pp. 233–46. An abstract of (86), read on 28 Apr. 1806. *RSC* 15.

[1806d] "De l'adhésion des corps à la surface des fluides," in *JP*, **63** (Nov. 1806), pp. 413–18; *OC*, **14**, pp. 247–53. An abstract of (88), read on 24 Nov. 1806. *RSC* 19.

[1807a] "Supplément à la théorie de l'action capillaire," in *JP*, **65** (July 1807), pp. 88–95; not in *OC*. An abstract of (87), read on 29 Sept. 1806.

[1809a] "Mémoire sur la double réfraction de la lumière dans les cristaux diaphanes," in *BSPM*, **1**, no. 18 (Mar. 1809), pp. 303–10; *OC*, **14**, pp. 278–87. An abstract of (92), read on 30 Jan. 1809. *RSC* 23.

[1809b] "Sur la loi de la réfraction extraordinaire de la lumière dans les cristaux diaphanes," in *JP*, **68** (Jan. 1809), pp. 107–11; *OC*, **14**, pp. 254–58. An abstract of the passages of [1810a] (92) that deal with the application of the principle of least action to double refraction, in consequence of forces of attraction and repulsion acting at undetectable distances. *RSC* 23.

[1809c] "Sur l'anneau de Saturne," in *CT* (1811/July 1809), pp. 450–53; *OC*, **13**, pp. 41–43. *RSC* 24.

[1809d] "Mémoire sur la diminution de l'obliquité de l'écliptique qui résulte des observations anciennes," in *CT* (1811/July 1809), pp. 429–50; *OC*, **13**, pp. 44–70. *RSC* 30.

[1809e] "Mémoire sur divers points d'analyse," in *JX*, **8**, 15ᵉ cahier (Dec. 1809), pp. 229–65; *OC*, **14**, pp. 178–214.

[1810a] "Mémoire sur le mouvement de la lumière dans les milieux diaphanes." in *MI*, **10** (1809/1810), pp. 300–42; *OC*, **12**, pp. 267–98. Read at the Institut on 30 Jan. 1809 (92). The date is mistakenly given as 1808 in *OC*, **12**, p. 267n. *RSC* 27.

[1810b] "Mémoire sur les approximations des formules qui sont fonctions de très grands nombres, et sur leur application aux probabilités," in *MI*, **10** (1809/1810), pp. 353–415; *OC*, **12**, pp. 301–53. Read on 9 Apr. 1810 (95). Delambre summarized this paper in *MI*, **11** (1810/1811), "Histoire," pp. iii–v.

[1810c] "Supplément au mémoire sur les approximations des formules qui sont fonctions de très grands nombres," in *MI*, **10** (1809/1810), pp. 559–65; *OC*, **12**, pp. 349–53.

[1810d] "Notice sur les probabilités," in *Annuaire publié par le Bureau des longitudes* (1811/1810), pp. 98–125. Not in *OC*.

[1810e] "Sur la dépression du mercure dans une tube de baromètre, due à sa capillarité," in *CT* (1812/July 1810), pp. 315–20; *OC*, **13**, pp. 71–77. Bouvard published corrections to the table of data, which are printed in *OC*, **13**, pp. 334–41. *RSC* 31.

[1811a] "Mémoire sur les intégrales définies, et leur application aux probabilités, et spécialement à la recherche du milieu qu'il faut choisir entre les résultats des observations," in *MI*, **11** (1810/1811), pp. 279–347; *OC*, **12**, pp. 357–412. Read on 29 Apr. 1811 (97). There is a prefatory summary by Delambre, in *MI*, **12** (1811/1812), "Histoire," pp. i–xiii, reviewing the state of the question of least squares.

[1811b] "Du milieu qu'il faut choisir entre les résultats d'un grand nombre d'observations," in *CT* (1813/July 1811), pp. 213–23; *OC*, **13**, p. 78. *RSC* 32.

[1811c] "Sur l'inégalité à longue période du mouvement lunaire," in *CT* (1813/July 1811), pp. 223–27; *OC*, **13**, pp. 79–84. Continued in [1812a]. See (76) and (94).

[1812a] "Sur l'inégalité à longue période du mouvement lunaire," in *CT* (1815/Nov. 1812), pp. 213–14; *OC*, **13**, pp. 85–87. A continuation of [1811c]. See (76) and (94). *RSC* 33.

[1813a] "Sur les comètes," in *CT* (1816/Nov. 1813), pp. 213–20; *OC*, **13**, pp. 88–97.

[P1814] Debate on a proposal to authorize the exportation of grain, Chambre des Pairs, 8 Nov. 1814, in *AP*, **13**, pp. 470–71. Laplace's opinion supported free trade. Not in *OC*.

[1815a] "Sur l'application du calcul des probabilités à la philosophie naturelle," in *CT* (1818/1815), pp. 361–77; *OC*, **13**, pp. 98–116. Read before the Institut on 18 Sept. 1815 (106). *RSC* 34.

[1815b] "Sur le calcul des probabilités appliqué à la philosophie naturelle," in *CT* (1818/1815), pp. 378–81; *OC*, **13**, pp. 117–20. Supplement to [1815a]. *RSC* 34.

[1815c] "Sur le flux et reflux de la mer," in *CT* (1818/1815), pp. 354–61; not in *OC*. Read on 10 July 1815 (105). *RSC* 35 mistakenly identifies this with [1820a], of which it was a preliminary abstract.

[1816a] "Sur la longueur du pendule à secondes," in *AC*, **3** (1816), pp. 92–94. An excerpt from [1817a]. Read on 28 Oct. 1816 (108).

[1816b] "Sur l'action réciproque des pendules et sur la vitesse du son dans les diverses substances," in *AC*, **3** (1816), pp. 162–69; *OC*, **14**, pp. 291–96. Read before the Academy on 25 Nov. 1816 (109). *RSC* 36.

[1816c] "Sur la transmission du son à travers les corps solides," in *BSPM* (1816), pp. 190–92; *OC*, **14**, p. 288. An abstract of [1816b]. See (110). *RSC* 40.

[1816d] "Sur la vitesse du son dans l'air et dans l'eau," in *AC*, **3** (1816), pp. 238–41; *OC*, **14**, pp. 297–300. Read at the Academy on 23 Dec. 1816 (110). *RSC* 37.

[P1816] "Sur une disposition du code d'instruction criminelle." Pamphlet, 15 Nov. 1816. Bibliothèque Nationale, Fp. 1187. On the mathematical equity of majorities required to find an accused guilty, in court of first instance and on appeal. Proposed modification of Article 351 of Code d'Instruction Criminelle as unfair to the accused. Not in *OC*.

[1817a] "Sur la longueur du pendule à secondes," in *CT* (1820/1817), pp. 265–80; *OC*, **13**, pp. 121–39. A fuller text than [1816a], with which *RSC* 38 mistakenly identifies it. Read on 28 Oct. 1816 (108).

[1817b] "Addition au mémoire précédent sur la longueur du pendule à secondes," in *CT* (1820/1817), pp. 441–42; *OC*, **13**, pp. 140–42.

[P1810] Recommendation on the cadastre, Chambre des Pairs, debate on the budget of 1817, 21 Mar. 1817, in *AP*, 2nd series, **19**, pp. 506–7; *OC*, **14**, pp. 372–74. Urged completion of cadastre tied to geodetic survey as the basis for a fair assessment of the land tax.

[1818a] "Application du calcul des probabilités aux opérations géodésiques," in *CT* (1820/1818), pp. 422–40. Excerpted in *AC*, *BSPM*, and *JP*. *RSC* 41. Read before the Academy on 4 Aug. 1817 (111). Incorporated, with additions (112), in *Théorie analytique des probabilités*, 3rd edition (1820), as Supplement 2; *OC*, **7**, pp. 531–80.

[1818b] "Sur la figure de la terre, et la loi de la pesanteur à sa surface," in *AC*, **8** (1818), pp. 313–18; *CT* (1821/1819), pp. 326–31. Abstracted from [1819b]. Read on 3 Aug. 1818 (114). *RSC* 44.

[1819a] "Sur la rotation de la terre," in *CT* (1821/1819), pp. 242–59; *OC*, **13**, pp. 144–64. Read before the Academy on 18 May 1818 (113). *RSC* 43.

[1819b] "Mémoire sur la figure de la terre," in *MASIF*, **2** (1817/1819), pp. 137–84; *OC*, **12**, pp. 415–55. Read on 3 Aug. 1818 (*OC* mistakenly has 4 Aug.). See (114). *RSC* 42.

[1819c] "Sur l'influence de la grande inégalité de Jupiter et de Saturne, dans le mouvement des corps du système solaire," in *CT* (1821/1819), pp. 266–71; *OC*, **13**, pp. 175–80.

[1819d] "Sur la loi de la pesanteur, en supposant le sphéroïde terrestre homogène et de même densité que la mer," in *CT* (1821/1819), pp. 284–90; *OC*, **13**, pp. 165–72.

[1819e] "Addition au mémoire précédent," in *CT* (1821/1819), p. 353; *OC*, **13**, pp. 173–74.

[1819f] "Mémoire sur l'application du calcul des probabilités aux observations et spécialement aux opérations du nivellement," in *AC*, **12** (1819), pp. 337–41; *OC*, **14**, pp. 301–4. An abstract of the 3rd Supplement of *Théorie analytique des probabilités*, 3rd ed. (1820); *OC*, **7**, pp. 581–616. Read on 20 Dec. 1819 (118).

[1819g] "Application du calcul des probabilités aux opérations géodésiques de la méridienne de France," in *BSPM* (1819), pp. 137–39; reprinted in *CT* (1822/1820), pp. 346–48. Incorporated in the 3rd Supplement of *Théorie analytique des probabilités*, 3rd ed. (1820); *OC*, **7**, pp. 581–85. Cf. *OC*, **13**, p. 188, for correction of an error of calculation. Read on 20 Dec. 1819 (118). *RSC* 47.

[1819h] "Considérations sur la théorie des phénomènes capillaires," in *JP*, 89 (Oct. 1819), pp. 292–96; **14**, pp. 259–64. Read on 13 Sept. 1819 (116). *RSC* 45.

[P1819] Recommendation on the lottery, Chambre des Pairs debate on the budget of 1819, 16 July 1819, in *AP*, 2nd series, **25**, pp. 683–84; *OC*, **14**, pp. 375–78. Urged suppression of the lottery on the grounds that it was inappropriate to raise public funds by encouraging illusions among the citizens.

[1820a] "Sur les inégalités lunaires dues à l'aplatissement de la terre," in *CT* (1823/1820), pp. 219–25; *OC*, **13**, pp. 189–97. Read on 19 Jan. 1820 (119).

[1820b] "Sur le perfectionnement de la théorie et des tables lunaires," in *CT* (1823/1820), pp. 226–31; *OC*, **13**, pp. 198–204. Read on 29 Mar. 1820 (121). *RSC* 49.

[1820c] "Sur l'inégalité lunaire à longue période, dépendante de la différence des deux hémisphères terrestres," in *CT* (1823/1820), pp. 232–39; *OC*, **13**, pp. 205–12. Read on 12 Apr. 1820 (122).

[1820d] "Mémoire sur le flux et le reflux de la mer," in *MASIF*, **3** (1818/1820), pp. 1–90; *OC*, **12**, pp. 473–546. An expansion of [1815b].

[1820e] "Addition au mémoire sur la figure de la terre, inséré dans le volume précédent," in *MASIF*, **3** (1818/1820), pp. 489–502; *OC*, **12**, pp. 459–69. The preceding memoir was [1819b]. Also printed under the title "Sur la figure de la terre," in *CT* (1822/1820), pp. 284–93. Read on 26 May 1819 (115). Cf. *OC*, **13**, p. 187.

[1820f] "Mémoire sur la diminution de la durée du jour par le refroidissement de la terre," in *CT* (1823/1820), pp. 245–57. Incorporated in *Mécanique céleste*, Volume V, Book XI, Chapters 1 and 4. *OC*, **5**, pp. 24–28, 82–88, 91–96. Cf. *OC*, **13**, p. 213. *RSC* 48.

[1820g] "Addition au mémoire précédent sur la diminution de la durée du jour," in *CT* (1823/1820), pp. 324–27. Incorporated in *Mécanique céleste*, Volume V, Book XI, Chapter 4, 88–91. Cf. *OC*, **13**, p. 214. *RSC* 48.

[1820h] "Sur la densité moyenne de la terre," in *CT* (1823/1820), pp. 328–31; *OC*, **13**, pp. 215–20. *RSC* 50.

[1820i] "Éclaircissements sur les mémoires précédents, relatifs aux inégalités lunaires dépendantes de la figure de la terre, et au perfectionnement de la théorie et des tables de la lune," in *CT* (1823/1820), pp. 332–37; *OC*, **13**, pp. 221–28.

[1821a] "Eclaircissements de la théorie des fluides élastiques," in *AC*, **18** (1821), pp. 273–80; *OC*, **14**, pp. 305–11.

[1821b] "Sur les variations des éléments du mouvement elliptique, et sur les inégalités lunaires à longues périodes," in *CT* (1824/1821), pp. 274–307; *OC*, **13**, pp. 229–64.

[1821c] "Sur la détermination des orbites des comètes," in *CT* (1824/1821), pp. 314–20; *OC*, **13**, pp. 265–72. Includes an example of the method applied to the comet of 1805 by Bouvard.

[1821d] "Sur l'attraction des sphères et sur la répulsion des fluides élastiques," in *CT* (1824/1821), pp. 328–43; *OC*, **13**, pp. 273–90. Read before the Academy on 10 Sept. 1821 (123). *RSC* 51.

[1821e] Précis de l'histoire de l'astronomie (1821). A separate publication of Book V of *Exposition du système du monde,* from the 4th edition (1813).

[P1821] Debate on Article 351 of Code d'Instruction Criminelle, Chambre des Pairs, 30 Mar. 1821, in *AP,* 2nd series, **30,** pp. 531–32; *OC,* **14,** pp. 379–81. Recommendations on the most equitable modes of reaching a decision in juries.

[1822a] "Développement de la théorie des fluides élastiques et application de cette théorie à la vitesse du son," in *CT* (1825/1822), pp. 219–27; *OC,* **13,** pp. 291–301. Submitted to the Bureau des Longitudes on 12 Dec. 1821. *RSC* 55.

[1822b] "Addition au mémoire précédent sur le développement de la théorie des fluides élastiques," in *CT* (1825/1822), pp. 302–23; *OC,* **13,** p. 302.

[1822c] "Sur la vitesse du son," in *CT* (1825/1822), pp. 371–72; *OC,* **13,** pp. 303–4. *RSC* 57.

[1822d] "Addition au mémoire sur la théorie des fluides élastiques," in *CT* (1825/1822), pp. 386–87; *OC,* **13,** p. 305.

[1823a] "De l'action de la lune sur l'atmosphère," in *CT* (1826/1823), pp. 308–17. Incorporated in *Mécanique céleste.* Book XIII; *OC,* **5,** pp. 184–88, 262–68. *RSC* 58.

[1824a] "Sur les variations de l'obliquité de l'écliptique et de la précession des équinoxes," in *CT* (1827/1824), pp. 234–37; *OC,* **13,** pp. 307–11.

[P1824] Debate on conversion of the public debt. Chambre des Pairs, 1 June 1824, in *AP,* 2nd series, **41,** p. 125; *OC,* **14,** pp. 382–83. Calculation of the effects of various schemes of amortization on the cost of servicing the debt.

[1825a] "Sur le développement en série du radical qui exprime la distance mutuelle de deux planètes, et sur le développement du rayon vecteur elliptique," in *CT* (1828/1825), pp. 311–21; *OC,* **13,** p. 312.

[1825b] "Sur la réduction de la longueur du pendule, au niveau de la mer," in *AC,* **30** (1825), pp. 381–87; *OC,* **14,** pp. 3l2–17.

[P1825] Debate on the conversion of the government bond issue, Chambre des Pairs, 26 Apr. 1825, in *AP,* 2nd series, **45,** pp. 144–45; *OC,* **14,** pp. 385–87. Favored reduction of the interest rate and an increase of capital.

[1826a] "Mémoire sur les deux grandes inégalités de Jupiter et de Saturne," in *CT* (1829/1826), pp. 236–44; *OC,* **13,** pp. 3l3–22.

[1826b] "Mémoire sur divers points de mécanique celeste: I. Sur les mouvements de l'orbite du dernier satellite de Saturne; II. Sur l'inégalité de Mercure à longue période, dont l'argument est le moyen mouvement de Mercure, moins celui de la terre; III. De l'action des étoiles sur le systéme planétaire," in *CT* (1829/1826), pp. 245–51; *OC,* **13,** pp. 323–30.

[1826c] "Mémoire sur un moyen de détruire les effets de la capilarité dans les baromètres," in *CT* (1829/1826), pp. 301–2; *OC,* **13,** pp. 331–33.

[1827a] "Mémoire sur le développement de l'anomalie vraie et du rayon vecteur elliptique en séries ordonnées suivant les puissances de l'excentricité," in *MASIF,* **6** (1823/1827), pp. 61–80; *OC,* **12,** pp. 549–66.

[1827b] "Mémoire sur le flux et reflux lunaire atmosphérique," in *CT* (1830/1827), pp. 3–18; *OC,* **13,** pp. 343–58. Published in Supplement to *Mécanique celeste,* Volume V; *OC,* **5,** pp. 489–505.

Sections J through N
Secondary Literature by Subject

J. Biographical and General. The only attempt at a comprehensive account is
Andoyer (1922), a modernized précis that is useful at this level. It is unfair to
this unpretentious little (162-page) work to quote here the comment with
which the editor of *L'action nationale* prefaced an excerpt from its methodological section (n.s. **18** [Jan.–June 1922], pp. 14–21), but the remark deserves rescue from the oblivion of that journal:

> At a moment when the spirit of snobbery and adventure, which seem to be the
> trademark of our times, believes it has found some undefined point of new departure
> for thought, some unsuspected route for scientific research, in the flashy [*tapageuses*],
> but not very comprehensible, theories of the German mathematician, Einstein, it is
> good to recall the effort, as innovative as it was intelligible, of our Laplace, whose
> great strength was accompanied by perfect modesty.

The most important éloge is Fourier (1831). J.-B. Biot and S.-D. Poisson
delivered eulogies at the funeral, and both texts are prefixed to the 1827
printing of *Exposition du système du monde*. Much later, Biot published a
reminiscence of Laplace as scientific father figure that he had delivered
before the Académie Française in Biot (1850). François Arago also composed
a biographical notice (1859). In this century innumerable anniversary pieces
have appeared, the only ones worthy of record being Danjon (1957); Taton
(1949); and Whittaker (1949).

It is still useful to consult the articles in certain well-known nineteenth-
century biographical encyclopedias; but before doing so, readers should bear
in mind that just as entries in these works concerning figures in previous
centuries derive from tradition and legend rather than from historical research, so the accounts of near-contemporary persons depend on gossip and
reminiscence rather than on scholarship. Hence, the decline of Laplace's
personal reputation is the main motif in Rabbe, Boisjolin, and Sainte-Preuve
(1834); Parisot (1811–1862); and Merlieux (1859). On Laplace's failure to
support the declaration of the Académie Française in favor of the freedom of
the press, see also Grassier (1906).

Notarial records and other official notices are reproduced in Boncompagni
(1883). Puisieux (1847) may still be consulted. Karl Pearson (1929) published
extracts from lectures on the life and work of Laplace and appended an
antiquarian article written at his request by the abbé G.-A. Simon (1929).
Bigourdan (1931) takes Laplace up to his election to the Academy. Lacroix
(1828) is useful for the context of Laplace's teaching in the 1790s. Laplace
joined with Legendre in a report read by the latter (11 nivôse an V [31 Dec.
1976], *PVIF*, **1**, pp. 154–57) on Lacroix, *Traité de calcul différentiel et intégral*,
2 vols. (1797), and printed in full at the end of vol. I of that work, pp. 520 ff.
There is much biographical information in Crosland (1967). Of more incidental interest are Sarton (1941); Hahn (1955); and Duveen and Hahn (1957).

There is information on Laplace's relations with Fourier, and further
bibliography, in Grattan-Guinness with Ravetz (1972). Paul Marmottan edited

the correspondence of Madame de Laplace with Elisa Bonaparte (1897). Their letters cover mainly the years 1807 and 1808.

K. Celestial Mechanics. It is worthy of remark that in the nineteenth century interest in Laplace centered on his celestial mechanics; that in the twentieth century it shifted to probability; that very recently attention has begun to be paid to his physics, although mainly for its institutional and political implications; and that surprisingly little scholarship has been addressed expressly to his mathematical work, which has been discussed mostly in sections of large-scale works on the evolution of overall aspects of mathematics.

Of near-contemporary astronomical and physical texts, the most helpful by far is Somerville (1831). Her exposition is somewhat more elementary than Bowditch's commentary [F], to which it makes a valuable supplement. Less satisfactory, although still worth mentioning, is Young (1821). More valuable historically are two Victorian monuments: Grant (1852) and Todhunter (1873). Indispensable for the background of the problem of the shape of the earth is Greenberg (1995), and equally so for the entire subject of planetary astronomy is Wilson (1980) and (1985).

Among other recent works, questions of the origin and development of the solar system bulk much larger than they did in the writings of Laplace himself. So it is in Bianco (1913), Vuillemin (1958), and Merleau-Ponty (1976). Jaki (1976) is more concrete on the development of Laplace's cosmogonical views, with special reference to the status of the nebular hypothesis in successive editions of the *Exposition du Système du monde*. Aiton (1953) is an authoritative study of tidal theory from Newton to Laplace.

L. Probability. The starting place is Todhunter (1865). The *Habilitationschrift* by Schneider (1972) is a comprehensive treatment of the entire subject. Still of value are the unsigned articles by de Morgan (1836–1837) and Czuber (1899).

Well known among modern books of general interest is Hacking (1975), a philosophically motivated, largely nontechnical, and very readable work that, however, gives Laplace rather less than his due. Daston (1988) is an elegant work of intellectual history. Porter (1986), a nontechnical and charming history of statistics, is closely relevant to probability. Gigerenzer et al. (1989) is a collective work giving a qualitative account of the role of probability in social and political affairs. Brian (1994) treats the interactions of Laplace with Condorcet in applications of probability to civic concerns, mainly demography. Very valuable for early attempts at application are the documents and commentary in Bru and Crépel (1994). Dale (1991) is a history of inverse probability.

Among writings on particular aspects are Maistrov (1974); Molina (1930), largely in defense of generating functions; Plackett (1972); Yamakazi (1971); Schneider (1975), which deals with Laplace's role in the emergence of the idea of treating kinetic theory by methods of the calculus of probabilities; Seal (1949); an important group of papers by Sheynin: (1973), (1976), which is mainly a summary critique of *Théorie analytique des probabilités*, and (1977); Gillispie (1963) and (1972); and Gillies (1987).

For statistical aspects the authoritative work is Stigler (1986a). Stigler has also published an invaluable series of papers on particular topics: (1973), (1974a), (1974b), (1975), (1978), (1986b). Other writings concerned with Laplace's importance for statistics are van Dantzig (1955); Eisenhart (1964); Lancaster (1966); and Lorey (1934). Also interesting for Lapalce on error theory, and for his differences with Fourier on that topic and statistical inference in general, is Callens (1997).

M. **Mathematical Physics.** The indispensable sources for Laplace's work in physical science are Bikerman (1975); Finn (1964); Fox (1974) and (1990); Frankel (1974); Frisingert (1974); Guerlac (1976); and Lodwig and Smeaton (1974).

N. **Mathematics.** Grattan-Guinness (1990a) is the standard work on French mathematics in Laplace's generation. In addition, help is available from works primarily concerned with topics in the history of mathematics and rational mechanics, notably Burckhardt (1908), an 1,800-page monograph of which section VI, pp. 398–408, deals especially with Laplace's method for integrating differential equations by means of definite integrals; Cajori (1928–1929), vol. II; Kornerz (1904), esp. pp. 52–54; Koppelmann (1971); Molina (1932); Muir (1906–1923); Petrova (1974) and (1975); Sheynin (1974); and Todhunter (1875). Chapter 3 in Panza (1991) is on Lagrange and Laplace.

Specially noteworthy is Deakin (1981–1982) on the history of the Laplace transform.

Section O
Secondary Works by Author and Date

Aiton, Eric J. 1953. "The Development of the Theory of the Tides in the Seventeenth and Eighteenth Centuries." Unpublished M.Sc. thesis, University of London.

Andoyer, H. 1922. *L'oeuvre scientifique de Laplace*. Paris.

Arago, François. 1859. "Notice," Oeuvres, **3**, pp. 459–515.

Arnauld, Antoine, and Pierre Nicole. 1664. *La logique, ou l'art de penser*, ed. Louis Martin (1970). Paris.

Bateman, Harry. 1911. *Report on the History and Present State of the Theory of Integral Equations*. London.

Bianco, B. O. 1913. "Le idee di Lagrange, Laplace, Gauss, e Schiaparelli sull'origine delle comete," *Memorie della Reale Accademia delle scienze di Torino*, 2nd series, **63**, pp. 59–110.

Bigourdan, G. 1931. "La jeunesse de Laplace," *Science moderne, no.* 8, pp. 377–84.

Bikerman, J. J. 1975. "Theories of capillary attraction," *Centaurus*, **19**, pp. 182–206.

Biot, J.-B. 1827. Éloge prefixed to the 1827 printing of *Exposition du système du monde* (5th ed.). Paris.

———. 1850. "Une anecdote relative à M. Laplace," *Journal des savants*, February, pp. 65–71.

Birembaut, Arthur. 1957. "L'Académie Royale des Sciences, vue par l'astronome suédois Lexell," *Revue d'histoire des sciences et de leurs applications*, **10**, pp. 148–66.

Boncompagni, B. 1883. "Intorno agli alti di nascità e di morte di Pietro Simone Laplace," *Bollettino di bibliografia e storia delle scienze matematiche e fisiche*, **15**, pp. 447–65.

Brian, Eric. 1994. *La mesure de l'État* (Paris, 1994).

Bromwich, T. J. I'A. 1916. "Normal co-ordinates in dynamical systems," *Proceedings of the London Mathematical Society*, 2nd series, **15** (1916), pp. 401–48.

Bru, Bernard. 1988. "Laplace et la critique probabiliste des mesures géodésiques," in Henri Lacombe and Pierre Costabel, eds., *La figure de la terre du XVIIIe siècle à l'ère spatiale*, pp. 223–44. Paris.

Bru, Bernard, and Pierre Crépel, eds. 1994. *Condorcet: Arithmétique politique, textes rares ou inédits, 1767–1789*. Paris.

Brush, Stephen. 1986. *The Kind of Motion We Call Heat*. 2 vols. New York.

Bucciarelli, Louis L., and Nancy Dworsky. 1980. *Sophie Germain: An Essay in the History of the Theory of Elasticity*. Boston.

Buchwald, Jed Z. 1989. *The Rise of the Wave Theory of Light: Optical Theory and Experiment in the Early Nineteenth Century*. Chicago.

Burckhardt, H. 1908. "Entwicklungen nach oscillirenden Funktionen und Integration der Differentialgleichungen der mathematischen Physik," *Jahresbericht der deutschen Mathematikervereinigung*, **10**, part 2: 3 vols., pp. 1–1803.

Cajori, Florian. 1928–1929. *A History of Mathematical Notations*. 2 vols. Chicago.

Callens, Stéphane. 1997. *Les maïtres de l'erreur: Mesure et probabilité au XIXe siècle*. Paris.

Caneva, Kenneth L. 1980. "Ampère, the etherians, and the Oersted connexion," *British Journal for the History of Science*, **13** pp. 121–38.

Carslaw, H. S., and J. C. Jaeger. 1941. *Operational Methods in Applied Mathematics*. Oxford.

Carson, J. R. 1926. *Electrical Circuit Theory and the Operational Calculus*. Reprint, New York, 1953.

———. 1922. "The Heaviside operational calculus," *Bell System Technical Journal*, **1**, part 2, pp. 43–55.

———. 1930. "Notes on the Heaviside operational calculus," *Bell System Technical Journal*, **9**, pp. 150–62.

Champagne, Ruth I. 1979. "The Role of Five Eighteenth-Century French Mathematicians in the Development of the Metric System." Facsimile edition of 1979 doctoral dissertation, Columbia University, available from University Microfilms, Ann Arbor, Mich.

Champion, F. C., and N. Davy. 1952. *Properties of Matter*. 2nd edition. London.

Chappert, André. 1977. *Étienne-Louis Malus (1775–1812) et la théorie corpusculaire de la lumière*. Paris.

Cohen, I. Bernard. 1971. *Introduction to Newton's "Principia."* Cambridge, Mass.

Coquard, Olivier. 1993. *Marat.* Paris.

Crosland, Maurice. 1967. *The Society of Arcueil.* London.

Czuber, E. 1899. "Die Entwicklung der Wahrscheinlichkeitstheorie und ihrer Anwendungen," *Jahresbericht der Deutschen Mathematikervereinigung,* **7**, no. 2, pp. 1–279.

Dale, A. I. 1982. "Bayes or Laplace? an examination of the origin and early application of Bayes' theorem," *Archive for History of Exact Sciences,* **27**, pp. 23–47.

———. 1991. *A History of Inverse Probability from Thomas Bayes to Karl Pearson.* New york.

Danjon, André, Alexandre Bigot, and Jean Chazy. 1957. "Pierre-Simon marquis de Laplace," Institut de France, Académie des Sciences, *Notices et discours,* **3** (1949–1956), pp. 19–38.

Daston, Lorraine. 1988. *Classical Probability in the Enlightenment.* Princeton.

David, F. N. 1965. "Some notes on Laplace," in J. Neyman and L. M. LeCam, eds. *Bernoulli, Bayes and Laplace,* pp. 30–44. Berlin.

Deakin, M. A. B. 1981–1982. "The development of the Laplace transform," *Archive for the History of Exact Sciences,* **25** (1981), pp. 343–90; **26** (1982), pp. 351–81.

Delambre, J.-B.-J. 1806–1810. *Base du système métrique décimal, ou mesure de l'arc du méridien compris entre les parallèles de Dunkerque et Barcelone.* 3 vols. Paris.

De Morgan, Augustus. 1836–1837. "Review of the *Théorie analytique,*" *Dublin Review,* **2**, 338–54; **3**, 237–48.

Dessi, Paola. 1982. "Laplace e la probabilità," *Rivista di filosofia,* **73**, October, pp. 313–32.

———. 1989. *L'ordine et el caso: discussioni epistemologiche et logiche sulla probabilità da Laplace a Peirce.* Bologna.

Dhombres, Jean. 1989. "La théorie de la capillarité selon Laplace: mathématisation superficielle ou étendue," *Revue d'Histoire des sciences et de leurs applications,* **62**, pp. 43–77.

———, ed. 1992. *L'École Normale de l'an III: leçons de mathématiques.* Paris.

Dhombres, Nicole and Jean Dhombres. 1989. *Naissance d'un pouvoir: sciences et savants en France (1793–1824).* Paris.

Doetsch, G. 1937. *Theorie und Anwendung der Laplacesche Transformation.* Berlin; reprint, New York, 1943.

Duveen, Denis, and Roger Hahn. 1957. "Laplace's succession to Bezout's post of Examinateur des élèves de l'artillerie," *Isis,* **48**, pp. 416–27.

Eisenhart, Churchill. 1964. "The meaning of 'least' in least squares," *Journal of the Washington Academy of Sciences,* **54**, pp. 24–33.

Eisenstaedt, Jean. 1991. "De l'influence de la gravitation sur la propagation de la lumiére en théorie newtonienne: l'archaélogie des trous noirs," *Archive for History of Exact Sciences,* **42**, no 4, pp. 356–71.

Everitt, Francis. 1974. "Maxwell," in *Dictionary of Scientific Biography*, ed. C. C. Gillispie, **9**, pp. 198–230.

Favre, Adrien. 1931. *Les origines du système métrique*. Paris.

Ferriani, Maurizio. 1979. "Dopo Laplace: Determinismo, probabilità, induzione," in Antonio Santucci, ed., *Eredità dell'Illuminismo*, pp. 201–304. Bologna.

Finn, Bernard S. 1964. "Laplace and the speed of sound," *Isis*, **55**, pp. 7–19.

Fischer, Joachim. 1988. *Napoleon und die Naturwissenschaften*. Stuttgart.

Fourier, Joseph. 1831. "Éloge historique de M. le marquis de Laplace, prononcé . . . le 15 juin 1829," *MASIF*, **10**, pp. lxxxi–cii.

———. 1835. "Histoire," pp. lxxxi–cii, prefixed to the 1835 printing of the 6th ed. of *Exposition du systéme du monde*; English trans. in *Philosophical Magazine*, 2nd series, **6** (1829), 370–81.

Fox, Robert. 1968–1969. "The background of the discovery of Dulong and Petit's Law," *British Journal for the History of Science*, **4**, pp. 1–22.

———. 1971. *The Caloric Theory of Gases from Lavoisier to Regnault*. Oxford.

———. 1974. "The rise and fall of Laplacian physics," *Historical Studies in the Physical Sciences*, **4**, pp. 89–136.

———. 1990. "Laplacian physics," in R. C. Olby, G. N. Cantor, J. J. R. Christie, and M. J. S. Hodge, eds., *A Companion to the History of Modern Science*, pp. 278–94. London.

Frankel, Eugene. 1974. "The search for a corpuscular theory of double refraction: Malus, Laplace and the prize competition of 1808," *Centaurus*, **18**, pp. 223–45.

Frisingert, H. H. 1974. "Mathematics in the history of meteorology: the pressure-height problem from Pascal to Laplace," *Historia mathematica*, **1**, pp. 263–86.

Gardner, M. F., and J. L. Barnes. 1942. *Transits in Linear Systems Studied by the Laplace Transformation*. New York.

Gigerenzer, Gerd, Zeno Swijtink, Theodore Porter, Lorraine Daston, John Beatty, and Lorenz Krüger. 1989. *The Empire of Chance: How Probability Has Changed Science and Everyday Life*. New York.

Gillies, Donald A. 1987. "Was Bayes a Bayesian?" *Historia Mathematica*, **14**, pp. 327–46.

Gillispie, Charles C. 1963. "Intellectual factors in the background of analysis by probabilities," in A. C. Crombie, ed., *Scientific Change*, pp. 433–53. London.

———. 1972. "Probability and politics: Laplace, Condorcet, and Turgot," *Proceedings of the American Philosophical Society*, **116**, no. 1, pp. 1–20.

———. 1979. "Mémoires inédits ou anonymes de Laplace," *Revue d'histoire des sciences et de leurs applications*, **32**, pp. 223–80.

———. 1980. *Science and Polity at the End of the Old Regime*. Princeton.

———. 1994. "Un enseignement hégémonique: les mathématiques," in B. Belhoste, A. Dahan Dalmedico, and A. Picon, eds., *La formation polytechnicienne, 1794–1994*, pp. 31–44. Paris.

———. 1996. "*L'Exposition du système du monde*, deux cents ans après sa publication," *La Recherche*, no. 292, November, pp. 74–79.

Grant, Robert. 1952. *History of Physical Astronomy from the Earliest Ages to the Middle of the 19th Century.* London; reprint, New York, 1966.

Grassier, E. 1906. *Les cinq cents immortels: Histoire de l'Académie Française.* Paris.

Grattan-Guinness, Ivor with J. R. Ravetz. 1972. *Joseph Fourier, 1768–1830.* Cambridge, Mass.

Grattan-Guinness, Ivor. 1990a. *Convolutions in French Mathematics, 1800–1840.* 3 vols. Basel.

————. 1990b. "Work for hairdressers: the production of de Prony's logarithmic and trigonometric tables," *Annals of the History of Computing*, **12**, pp. 177–85.

————. 1990c. "Thus it mysteriously appears: Impressions of Laplace's use of series," in Detlef D. Spalt, ed., *Rechnen mit dem Unendlichen: Beiträge zur Entwicklung eines kontroversen Gegenstandes*, pp. 96–102. Basel.

————. 1991. "Lines of methematical thought in the electrodynamics of Ampère," *Physics*, **28**, pp. 115–29.

————. 1995. "Why did George Green write his essay of 1818 on electricity and magnetism?" *American Mathematical Monthly*, **102**, pp. 387–96.

Greenberg, John L. 1995. *The Problem of the Earth's Shape from Newton to Clairaut.* Cambridge.

Guerlac, Henry. 1976. "Chemistry as a Branch of Physics: Laplace's Collaboration with Lavoisier," *Historical Studies in the Physical Sciences*, **7**, pp. 193–276.

Hacking, Ian. 1975. *The Emergence of Probability.* Cambridge.

Hahn, Hans. 1911. "Bericht über die Theorie der linearen Integralgleichungen," *Jahresbericht der deutschen Mathematiker-Vereinigung*, **20**, pp. 69–117.

Hahn, Roger. 1955. "Laplace's religious views," *Archives internationales d'histoire des sciences*, **8**, pp. 38–40.

————. 1994a. *The New Calendar of the Correspondence of Pierre-Simon Laplace.* Berkeley Papers in History of Science 16. Office for History of Science and Technology, University of California at Berkeley.

————, 1994b. "Le rôle de Laplace à l'École Polytechnique," in B. Belhoste, A. Dahan-Dalmedico, and A. Picon, eds., *La formation polytechnicienne, 1794–1994*, pp. 45–58. Paris.

Harman, P. M. 1992. "Maxwell and Saturn's rings: problems of stability and calculability," in P. M. Harman and Alan Shapiro, eds., *The Investigation of Difficult Things: Essays on Newton and the History of the Exact Sciences in Honour of D. T. Whiteside*, pp. 477–502. Cambridge.

Hawking, S. W., and G. F. R. Ellis. 1973. *The Large-Scale Structure of Space-Time.* Cambridge.

Heilbron, John. 1990. "The measure of Enlightenment," in Tore Frängsmyr, John L. Heilbron, and Robin E. Rider, eds., *The Quantifying Spirit in the Eighteenth Century*, pp. 207–42. Berkeley.

————. 1993. "Weighing imponderables and other quantitative science around 1800," *Historical Studies in the Physical Sciences*, **24**, pp. 150–65.

Henry, Charles. 1886. "Lettres inédites de Laplace," *Bollettino di bibliografia e di storia delle scienze matematiche e fisiche*, **19**, pp. 149–78. Reprinted in *OC*, **14**, pp. 340–71.

Herivel, John 1975. *Joseph Fourier, the Man and the Physicist*. Oxford.

Hines, A. M. 1994. "Laplace in Calvados," *Mathematics Intelligencer*, **16**, pp. 256–58.

Hofmann, James R. 1995. *André-Marie Ampère*. Cambridge.

Jaki, Stanley L. 1976. "The five forms of Laplace's cosmogony," *American Journal of Physics*, **44**, pp. 4–11.

Koppelmann, Elaine. 1971. "The calculus of operations and the rise of abstract algebra," *Archive for History of Exact Sciences*, **8**, pp. 155–242.

Kornerz, Theodor. 1904. "Der Begriff des materiellen Punktes in der Mechanik des achtzehnten Jahrhunderts," *Bibliotheca mathematica*, 3rd series, **5**, pp. 15–62.

Kula, Witold. 1970. *Miary i liudzie* (Warsaw). Trans. Joanna Ritt, *Les mesures et les hommes* (Paris, 1984); trans. R. Zreter, *Measures and Men* (Princeton, 1986).

Lacroix, S.-F. 1828. *Essais sur l'enseignement en général et sur celui de mathématiques en particulier*. 3rd edition. Paris.

Laissus, Yves. 1961. "Deux lettres de Laplace," *Revue d'histoire des sciences*, **14**, pp. 285–96.

Lancaster, H. O. 1966. "Forerunners of the Pearson χ^2," *Australian Journal of Statistics*, **8**, pp. 117–26.

Laskar, Jacques, 1995. "The stability of the solar system from Laplace to the present," in René Taton and Curtis Wilson, eds., *Planetary Astronomy from the Renaissance to the Rise of Astro-physics*, vol. **2**, part B, pp. 240–48. of *The General History of Astronomy*, ed. Michael Hoskin, Cambridge.

Lodwig, T. H., and W. A. Smeaton. 1974. "The ice calorimeter of Lavoisier and Laplace and some of its critics," *Annals of Science*, **31**, pp. 1–18.

Lorey, Wilhelm. 1934. "Die Bedeutung von Laplace für die Statistik," *Allgemeines statistisches Archiv*, **23**, pp. 398–410.

Lützen, J. 1979. "Heaviside's operational calculus...." *Archive for the History of Exact Sciences*, **21**, pp. 161–200.

MacDonald, H. M. 1902–1903. "Some applications of Fourier's Theorem," *Proceedings of the London Mathematical Society*, 1st series, **35**, pp. 428–43.

Maistrov, L. E. 1972. *Probability Theory: A Historical Sketch*. Trans. and ed. Samuel Kotz. New York.

Marmottan, Paul. 1897. *Lettres de Madame de Laplace à Elisa Napoléon, princesse de Lucques et de Piombino*. Paris.

Mellin, H. 1918. "Abriss einer einheitlichen Theorie," *Mathematische Annalen*, **68**, pp. 305–7.

Merleau-Ponty, Jacques. 1976. "Situation et rôle de l'hypothèse cosmogonique dans la pensée cosmologique de Laplace," *Revue d'histoire des sciences et de leurs applications*, **29**, pp. 21–49.

————. 1977. "Laplace as a cosmologist," in W. Yourgrau and A. D. Breck, eds., *Cosmology, History, and Theology*, pp. 283–91. New York.

Merlieux, E. 1859. "Laplace, Pierre-Simon, Marquis de," in F. Hoefer, ed., *Nouvelle biographie générale* (Didot), **29**, cols. 531–48.

Molina, E. C. 1930."The theory of probability: some comments on Laplace's *Théorie analytique*," *Bulletin of the American Mathematical Society*, **36**, pp. 369–92.

————. 1932. "An expansion for Laplacian integrals," *Bell System Technical Journal*, **11**, pp. 563–75.

Morando, Bruno. 1995. "Laplace," in René Taton and Curtis Wilson, eds., *The General History of Astronomy*, **2**, part B, pp. 131–50. Cambridge.

Muir, Thomas. 1906–1923. *The Theory of Determinants in the Historical Order of Development*, 4 vols. London.

Murnaghan, Francis L. 1962. *The Laplace Transformation*. Washington, D.C.

Murphy, R. 1833. "On the inverse method of definite integrals," *Transactions of the Cambridge Philosophical Society*, **4**, pp. 353–408; see esp. p. 362.

Panza, Marco. 1992. *La forma dell quantità*. 2 vols. Paris.

Parisot, A.-C.-M. 1811–1862. "Laplace, Pierre-Simon, marquis de," in L. G. Michaud, ed., *Biographie universelle, ancienne et moderne*, **70**, 237–60. Paris.

Pearson, Egon S., and M. G. Kendall, eds. 1970. *Studies in the History of Statistics and Probability*. London.

Pearson, Karl. 1923. "Laplace, being extracts from lectures delivered by Karl Pearson," *Biometrika*, **21**, pp. 202–30.

Petrova, S. S. 1974. "K istorii metoda kaskadov Laplasa" ("Toward the history of Laplace's method of cascades"), *Istoriko-matematicheskie issledovania*, **19**, pp. 125–31.

————. 1975. "Rannyaya istoria preobrazovania Laplasa" ("The early history of the Laplace transform"), *Istoriko-matematicheskie issledovania*, **20**, pp. 246–56.

Plackett, R. L. 1972. "The discovery of the method of least squares," *Biometrika*, **59**, 239–51.

Poirier, Jean-Pierre, 1993. *Antoine-Laurent de Lavoisier, 1743–1794*. Paris. English trans., *Lavoisier: Chemist, Biologist, Economist*. Philadelphia, 1996.

Poisson, S.-D. 1827. Éloge prefixed to the 1827 printing of *Exposition du système du monde* (5th edition). Paris.

Porter, Theodore M. 1981. "A statistical survey of gases: Maxwell's social physics," *Historical Studies in the Physical Sciences*, **12**, pp. 71–116.

————. 1986. *The Rise of Statistical Thinking, 1820–1940*. Princeton.

Puisieux, L. 1847. *Notice sur Laplace*. Caen.

Rabbe, A., J. Vieilh de Boisjolin, and Sainte-Preuve. 1834. "Laplace, Pierre-Simon, marquis de," in *Biographie universelle et portative des contemporains*, **3**, 151–53. Paris.

Sarton, G. 1941. "Laplace's religion," *Isis*, **33**, pp. 309–12.

Schneider, Ivo. 1972. *Die Entwicklung des Wahrscheinlichkeitsbegriffs in der Mathematik von Pascal bis Laplace*. Munich.

Schneider, Ivo. 1975. "Rudolph Clausius' Beitrag zur Einführung wahrschein-lichkeitstheoretischer Methoden in die Physik der Gase nach 1856," *Archive for History of Exact Sciences*, **14**, pp. 237–61.

Seal, H. L. 1949. "The historical development of the use of generating functions in probability theory," *Mitteilungen der Vereinigung schweizerischer Ver-sicherungs-Mathematiker*, **49**, pp. 209–28.

Sheynin, O. B. 1973. "Finite random sums," *Archive for History of Exact Sciences*, **9**, pp. 275–305.

———. 1974. "O poyavlenii delta-funktsii Diraka v trudakh P. S. Laplasa" (On the appearance of Dirac's delta-function in the works of P. S. Laplace), *Istoriko-matematicheskie issledovania*, **19**, pp. 303–8.

———. 1976. "P. S. Laplace's work on probability," *Archive for History of Exact Sciences*, **16**, 137–87.

———. 1977. "Laplace's theory of errors," *Archive for History of Exact Sciences*, **17**, pp. 1–61.

Simon, G.-A. 1929. "Les origines de Laplace: sa généalogie, ses études," *Biometrika*, **21**, pp. 217–30.

Somerville, Mary. 1831. *Mechanism of the Heavens*. London.

Stigler, Stephen. 1973. "Laplace, Fisher, and the discovery of the concept of sufficiency," *Biometrika*, **60**, pp. 439–55.

———. 1974a. "Cauchy and the Witch of Agnesi," *Biometrika*, **61**, pp. 375–80.

———. 1974b. "Gergonne's 1815 paper on the design and analysis of polyno-mial regression experiments," *Historia mathematica*, **1**, pp. 431–47.

———. 1975. "Napoleonic statistics: the work of Laplace," *Biometrika*, **62**, pp. 503–17.

———. 1978. "Laplace's early work, chronology and citations," *Isis*, **69**, pp. 234–54.

———. 1981. "Gauss and the invention of least squares," *Annals of Statistics*, **5**, pp. 465–74.

———. 1982. "Thomas Bayes's Bayesian inference," *Journal of the Royal Statisti-cal Society* (A), **145**, pp. 250–58.

———. 1986a. *The History of Statistics: The Measurement of Uncertainty before 1900*. Cambridge, Mass.

———. 1986b. "Laplace's 1774 memoir on inverse probability" (translation and commentary), *Statistical Science*, **1**, pp. 359–78.

Taton, René. 1949. "Laplace," *La Nature*, **77**, 221–23.

———. 1953. "Laplace et Sylvestre-François Lacroix," *Revue d'histoire des sciences*, **6**, pp. 350–60.

Todhunter, Isaac. 1865. *A History of the Mathematical Theory of Probability from the Time of Pascal to That of Laplace*. London. Facsimile reprint, New York, 1949, 1965.

———. 1873. *A History of the Mathematical Theories of Attraction and the Figure of the Earth from the Time of Newton to That of Laplace*, 2 vols. Facsimile reprint, New York, 1962.

———. 1875. *Elementary Treatise on Laplace's Function, Lamé's Function, and Bessel's Function*. London.

Truesdell, Clifford. 1964. "Whence the law of moment of momentum?" in *Mélanges Alexandre Koyré*, **1**, pp. 588–612. 2 vols. Paris.

Van Dantzig, D. 1955. "Laplace probabiliste et statisticien et ses précurseurs," *Archives internationales d'histoire des sciences*, **8**, pp. 27–37.

Vuillemin, Jules. 1958. "Sur la généralisation de l'estimation de la force chez Laplace," *Thalès*, **9**, pp. 61–76.

Whittaker, E. T. 1949. "Laplace," *Mathematical Gazette*, **33**, pp. 1–12.

Widder, D. V. 1946. *The Laplace Transform*. Princeton.

Wilson, Curtis A. 1980. "Perturbation and solar tables from Lacaille to Delambre," *Archive for the History of Exact Sciences*, **22**, pp. 53–304.

———. 1985. "The great inequality of Jupiter and Saturn from Kepler to Laplace," *Archive for the History of Exact Sciences*, **33**, nos. 1–3, pp. 15–290.

Yamakazi, E. 1971. "D'Alembert et Condorcet: quelques aspects de l'histoire du calcul des probabilités," *Japanese Studies in the History of Science*, **10**, pp. 59–93.

Young, Thomas. 1821. *Elementary Illustrations of the Celestial Mechanics of Laplace*. London.

Abeille, Louis-Paul, 150
Abel, Niels Henrik, 268
Académie Française, 177–78, 243, 277
Academy of Science, 4, 5, 9, 67, 68, 79, 93,
   97, 101, 143, 149–55, 190, 243, 249, 274;
   Institut de France, 166–67, 176, 178, 190,
   209–14 passim, 219n, 243, 263, 277
Adams, John Couch, 144
Airy, George B., 144
Alembert, Jean-le-Rond d', 3–9 passim, 31,
   34, 39n, 48, 52, 53, 55, 67, 111, 182, 184;
   opposition of, to probability, 14, 23, 25
Ampère, André-Marie, 167, 171, 244, 276
Arago, Dominique-François-Jean, 167, 179,
   210, 243, 249
Arcueil, Society of, 170, 179, 194, 210, 211,
   214, 216, 243, 249
Arcy, Patrick d', 36, 63, 97, 161, 172. _See
   also_ mechanics: areas, conservation of
astronomy, 4, 7, 46–50, 141–42, 156–65,
   179–83, 254–55; celestial mechanics, 7,
   27, 162; determinism of, 26–27, 145, 174,
   271; perturbation theory, 46, 172, 179,
   185–86, 193; physics, relation to, 32–35,
   51–53, 109–23, 193–96; precession of
   equinoxes (and nutation), 31, 55–56, 63,
   160, 172, 179–81; probability, relation to,
   14, 26–27, 39–43, 174–75. _See also_
   gravity; Jupiter; lunar theory; Saturn;
   solar system; tides
atmospheric refraction. _See_ physics

Bailly, Jean-Sylvain, 68, 138
Bayes, Thomas (Bayes's Rule), 16–17, 72,
   78, 221, 223, 228, 239
Bérard, Jacques-Étienne, 179, 202, 214, 248
Bernoulli, Daniel, 13, 18, 26, 33, 39n, 53,
   55, 62, 220, 237, 274
Bernoulli, Jakob, 13
Berthollet, Claude-Louis, 108, 168, 176–79,
   204, 210–11, 244
Bessel functions, 265
Bezout, Étienne, 7, 68, 97, 111
Biot, Jean-Baptiste, 154, 167, 176–79,
   199–200, 210, 243–44, 249, 263, 276–77

Bochart de Saron, Jean-Baptiste-Gaspard,
   109
Bonaparte, Napoleon, 166, 168, 170, 176,
   178, 243, 277
Boole, George, 16
Borda, Jean-Charles, 149–50, 152, 154–55
Bošković, Rudjer, 96–97, 189, 241
Bossut, Charles, 7, 35, 97, 149
Bouvard, Alexis, 181–82, 190, 192, 196, 241,
   254–56, 277–78
Bowditch, Nathaniel, 187, 192
Bradley, James, 55, 160, 196
Brisson, Mathurin-Jacques, 150, 155
Brissot, Jacques-Pierre, 67, 68
Buffon, Georges-Louis Leclerc, comte de,
   174, 204–5, 235
Burckhardt, Jacob, 277
Bureau des Longitudes, 155, 196, 205, 210
Burg, Johann Tobias, 191–92

Caen, University of, 3
calendar, revolutionary, 154, 170
calorimeter. _See_ physics
capillary action. _See_ physics
Carnot, Lazare, 176
Carnot, Sadi, 167
Carson, J. R., 269
Cassini, Jacques, 162–63
Cassini map of France, 93, 153
Cauchy, Augustin-Louis, 244, 268, 276
causality, 23–26. _See also_ probability: of
   causes
central limit theorem, 216–17, 219, 232,
   254
chance, games of, 7, 11–13, 18, 22–24, 26,
   73–78, 222, 228, 230, 236; lottery, 12,
   88–89, 228; theory of, 10–11, 13, 23, 26,
   39, 72–78, 89, 227–28. _See also_
   probability
chemistry, 105–8, 178–79, 204, 211, 244.
   See also physics
Clairaut, Alexis-Claude, 55, 181, 204–5,
   206n
Clément, Nicolas, 201
Collège de France, 5, 178
Collet-Descotils, Hippolyte-Victor, 179

cometary theory, 38–43, 51, 96–100, 186, 193; probability of cometary orbits, 38–43, 174, 217–19, 230–31
Condorcet, Marie-Jean-Antoine-Nicolas-Caritat, comte de, 5, 9, 14–16, 46, 48, 67, 78, 93, 149–51, 172, 275
Coulomb, Charles-Augustin, 155, 209, 244
Cousin, Jacques-Antoine-Joseph, 5
Crosland, Maurice, 179
Cuvier, Georges, 173, 251, 277

Dalton, John, 214, 246
Delambre, Jean-Baptiste, 142–43, 153–56, 158, 172, 181, 187, 192, 194, 220–21, 242
Delaroche, François, 202, 214, 248
Delaunay, Charles, 145
De Moivre, Abraham, 11–15, 18, 86n, 274
De Morgan, Augustus, 16, 272
Desormes, Charles-Bernard, 201
Dini, Ulisse, 269
Dionis du Séjour, Achille-Pierre, 39, 43, 94, 97, 150
Dulong, Pierre-Louis, 244

earth: cooling of, 251–52; shape of, 31–32, 51–53, 110, 113, 117–21, 160–62, 171–72, 180–81, 187–89, 192, 250. See also astronomy; geodesy; tides
École Militaire, 4, 12, 89, 176
École Normale, 95, 152–53, 166, 168–69, 217, 225, 237, 239
École Polytechnique, 68, 166–69, 177, 210, 276
Einstein, Albert, 170–71, 271, 276
electricity. See physics
error, theory of, 18–22, 76–77, 82–85, 94–95, 99, 161, 187–89, 216–23, 233–34, 241–42, 254–55. See also central limit theorem; least squares
Euler, Leonhard, 4, 9, 35, 37, 44–47, 55, 87, 170, 180, 182, 220–21, 227, 259–60, 273

Fisher, R. A., 242
Flamsteed, John, 140
forces. See physics
Fourcroy, Antoine-François de, 168
Fourier, Jean-Baptiste-Joseph, 169, 171, 209, 214, 219n, 241n, 244, 249, 251, 262–66, 272–73, 276–78; Fourier transform, 264–69
French Revolution, 145, 149–55, 166

Fresnel, Augustin-Jean, 167, 170, 243–44, 276–77

Gadbled, Christophe, 3
Galileo, 121
Galton, Francis, 216
Gauss, Carl Friedrich, 99, 216, 220–21, 223, 234
Gay-Lussac, Joseph-Louis, 179, 194–95, 200–202, 207, 210, 246–47
geodesy, 31–32, 117–21, 150–52, 161–62, 180, 187–89, 192, 241–42. See also metric system; pendulum, seconds
Germain, Sophie, 215, 243, 276
gravity, 29–37, 172, 179, 185, 195, 203, 211, 245, 252–53
Green, George, 110
Guerlac, Henry, 101, 107
Guyton de Morveau, baron Louis Brenard, 168

Halley, Edmond, 36–37, 126, 128, 138–41, 143
Hauksbee, Francis, 206
Haüy, René-Just, 168, 207, 210–11
heat. See physics
Heaviside, Oliver, 269
Héricy, marquis d', 3
Hermite polynomials, 266
Herschel, John, 275
Herschel, William, 100, 110, 141, 193
Huygens, Christian, 121, 211–13, 277

Jaki, S. J., 173
Jupiter: moons of, 128–33, 156–60, 172, 181, 186, 190, 192–93; and Saturn, inequality of, 29, 35–37, 38, 47, 50, 124–28, 137–42, 172, 181, 186, 191, 235

Lacaille, Nicolas-Louis de, 32, 153
La Condamine, Charles-Marie de, 32, 150
Lacroix, Sylvestre-François, 176
Lagrange, Joseph-Louis, 4, 8, 9, 15, 54, 70–72, 97, 100, 108, 110, 150n, 151, 153n, 154, 168, 177, 214, 226, 260, 267, 272–74; Lagrange variables, 47; mean value, 18, 82–83; methods of integration, 8, 87–88; secular inequalities, 29, 34–35, 37, 46–47, 126–27, 129, 133–34; theory of error, 76, 82–83
Lalande, Joseph-Jérôme Lefrançais de, 97, 129, 142, 154, 158–59, 162

Lambert, Johann Heinrich, 128, 200
La Michodière, François de, 94–95
Laplace, Pierre-Simon: early work, 4–5; education, 3; as educator, 4, 68, 95, 167–69; election to Academy, 5–6; *Essai philosophique sur les probabilités*, 11, 16, 26–27, 91, 95, 169–70, 216–17, 224–25, 237–41, 274; *Exposition du système du monde*, 36, 109, 153, 156, 166–75, 179, 183–84, 186, 204, 238–39, 249, 273, 278; *Exposition du système du monde* in relation to *Mécanique céleste*, 184, 186, 239, 250; family, 67, 155, 176–78, 242, 278; *Mécanique céleste*, 7, 29, 36, 95, 109, 113, 115, 117, 122, 124, 156, 161–62, 169–70, 176–96 passim, 204, 206, 208, 211, 216, 224, 227, 235, 239, 245–47, 250–52, 256, 271, 273, 277; *Mémoire sur la chaleur*, 102–8, 203, 214; scientific reputation and influence, 67–69, 177–78, 209–15, 243–45, 248–49, 271–78; social and political honors, 176, 243, 277; *Théorie analytique des probabilités*, 26, 86, 177, 216, 221, 224–42, 252–53, 255, 266, 277; *Théorie du mouvement et de la figure elliptique des planètes*, 7, 108–13, 203
Lavoisier, Antoine-Laurent de, 101–5, 107–8, 149–51, 155, 178, 199, 204, 214
least squares, 216, 220–21, 233–34, 241, 252, 254
Le Canu, Pierre, 3
Legendre, Adrien-Marie, 29, 88, 110–11, 113, 150n, 216, 220–21, 223, 234, 252, 274
Leibniz, Gottfried Wilhelm, 226
LeVerrier, Urbain-Jean-Joseph, 271
light, 160, 194–95, 205, 210, 243–45, 249; atmospheric refraction, 193–95, 205–6, 210–11; double refraction, 211–14; general theory of refraction, 204–6, 210–11
lunar theory, 34, 113, 124–25, 143–45, 172, 180–82, 191–92, 250; libration, 144, 186, 192

Maclaurin, Colin, 53, 112, 274
Magendie, François, 278
magnetism. *See* physics
Malus, Étienne-Louis, 167, 179, 210–14, 276–77
Marat, Jean-Paul, 67, 154–55
Maskelyne, Nevil, 192

mathematics, 168–69; calculus of operations, 70, 87, 269; cascade method, 44, 268; constants, variation of, 48, 51, 70, 162; difference equations, 4, 7–9, 15, 27, 87, 221–22, 225; differential equations, 5, 8–9, 44–50, 70–71, 90, 117, 260; differential equations, partial, 44, 87, 113–14, 226, 261; extreme value problems, 4–5; harmonic theory, 56; Laplace expansion, 50; Laplace's angles, 115; Laplace's functions, 110–11, 117; Laplace's theorem, 112; Laplace transform, 259–69; series, recurro-recurrent, 4, 5, 9–12, 14, 18, 27, 87; series, theory of, 9, 86–88, 217–18; spheroidal attraction theory (potential function), 31, 51–53, 187; successive approximations, method of, 46
Maupertuis, Pierre-Louis Moreau de, 32
Maxwell, James Clerk, 122n, 275
Mayer, Tobias, 34–35, 100
mean value, 18–22, 38–43, 76–78, 82–85, 216–23, 230–31, 233–34, 237, 254–56. *See also* error, theory of; least squares
Méchain, Pierre-François, 99–100, 153, 187, 242
mechanics, 109, 171–72, 184–85; areas, conservation of (d'Arcy's principle), 36, 63, 97, 161, 172; least action, principle of, 213; *vis viva*, 103, 162
metric system, 102, 117–18, 150–55, 159, 171, 187
Monge, Gaspard, 68, 150n, 151, 153n, 167–68, 176–77, 276
Morand, Jean, 95n
Murphy, Robert, 268

Navier, Claude-Louis-Marie-Henri, 276
nebular hypothesis. *See* solar system
Newton, Isaac, 32, 47–48, 51, 53, 55, 67, 96–97, 103, 111, 164, 173, 185, 199, 201–7 passim, 211, 226, 238, 245, 273, 276

observations, theory of. *See* probability
Oersted, Hans Christian, 244
Olbers, Heinrich Wilhelm Matthias, 99

Pascal, Blaise, 14, 86n, 240
pendulum, seconds, 32, 117, 119–20, 150–52, 161, 180, 189
Petit, Alexis Thérèse, 244

physics, 107–8, 170, 178–79, 209–15;
atmosphere, 195, 252–54; calorimeter,
102, 104–7; capillary action, 102, 196,
203–8, 210, 249; elastic surfaces, 214–15,
243; electricity, 102, 199, 209, 235,
244–45; forces, short-range, 31, 108,
203–8, 210, 213, 249; gases, 194, 199–202,
210, 214, 245–48; heat, 101–8, 199, 209,
219n, 245–48; magnetism, 235, 244–45;
sound, 87, 199–202, 210, 246–47; specific
heat, 103–7, 199, 214, 244. *See also* light;
mechanics
Pingré, Alexandre-Gui, 99, 156
Poincaré, Jules Henri, 269
Poisson, Simon-Denis, 16, 110, 115, 167,
179, 200, 208–10, 215, 243, 248–49, 251,
263, 265, 267, 272, 276–77
population, 78–81, 92–95, 218, 233, 235–36
potential function. *See* mathematics
Pouillet, Claude-Servais-Mathias, 243
precession of equinoxes. *See* astronomy
Price, Richard, 16, 78
probability, 7, 13–28, 72–85, 130, 222,
274–75; astronomy, relation to, 14,
26–27, 38–43, 222–23; of causes, 5,
15–18, 23–27, 38, 45, 72–73, 78, 82, 90,
228, 234–36, 256; definition of, 12–14,
16, 25–27, 225; early history, 13–14;
epistemology of, 23–27, 91–92, 233–34,
274; generating functions, 11, 71, 86–88,
114, 193, 219, 221–22, 224–25, 228, 266,
268; social applications, 15, 23, 68, 72,
78–81, 231–32, 235–37, 239–41, 275;
statistical inference, 72, 82, 235–36. *See
also* cometary theory: probability; error,
theory of; mean value; population
Prony, Gaspard Riche de, 153

Quetelet, Adolphe, 189, 275

Ramond, Louis, 254–55
Riemann, Georg Friedrich Bernhard, 268
Römer, Ole Christensen, 129
Romme, Gilbert, 154

Saturn: rings of, 53, 111, 118, 121–23, 172,
187. *See also* Jupiter: and Saturn
series. *See* mathematics
Simpson, Thomas, 39n
solar system, stability of, 36–37, 46–47,
124–45, 157, 172, 195, 271, 273; nebular
hypothesis, 172–75
sound. *See* physics
spheroid, attraction of. *See* mathematics
Stigler, Stephen, 21, 76, 86n, 140–41, 241,
252, 254
Stirling, James, 89–90, 217

Talleyrand, Charles-Maurice de, 151
Taylor, Brook, 264, 274
Thenard, Louis-Jacques, 179
tides, 53–64, 110, 118–21, 156, 162–65, 172,
189, 235, 250, 252
Tillet, Mathieu, 149–50, 152
Trémery, Jean-Louis, 207, 210
Turgot, Anne-Robert-Jacques, 79, 150
Turin, Royal Society of, 7

Uranus, discovery of. *See* astronomy

Vandermonde, Alexandre, 5, 97
Volta, Alessandro, 102

Wallis, John, 226
Wargentin, Pehr, 129, 131, 139, 158
Welter, Jean-Joseph, 201–2, 247
Wollaston, William Hyde, 211–12

Young, Thomas, 207, 213

About the Authors

Charles Coulston Gillispie is Dayton-Stockton Professor of History Emeritus at Princeton University, where he founded the Program in History of Science in 1960. He is the author of numerous books, including *The Edge of Objectivity* (Princeton). He is the editor of the sixteen-volume *Dictionary of Scientific Biography*. Robert Fox is Professor of History of Science at the University of Oxford. Ivor Grattan-Guinness is Professor of Mathematics at Middlesex Polytechnic University in London.